Remembrance of Things Present

Remembrance of Things Present

The Invention of the Time Capsule

NICK YABLON

The University of Chicago Press Chicago and London

The University of Chicago Press, Chicago 60637
The University of Chicago Press, Ltd., London

For more information, contact the University of Chicago Press,
1427 E. 60th St., Chicago, IL 60637.
Published 2019
Printed in the United States of America

28 27 26 25 24 23 22 21 20 19   1 2 3 4 5

ISBN-13: 978-0-226-57413-4 (cloth)
ISBN-13: 978-0-226-57427-1 (e-book)
DOI: https://doi.org/10.7208/chicago/9780226574271.001.0001

Any additional permissions/subsidy info needed—to come from
acquisitions/contracts/sub rights

Library of Congress Cataloging-in-Publication Data

Names: Yablon, Nick, author.
Title: Remembrance of things present : the invention of the time
capsule /Nick Yablon.
Description: Chicago ; London : The University of Chicago Press,
2019. | Includes bibliographical references and index.
Identifiers: LCCN 2018043831 | ISBN 9780226574134 (cloth : alk.
paper) | ISBN 9780226574271 (ebook)
Subjects: LCSH: Time capsules—United States—History. | Time
capsules—Social aspects—United States. | History—Sources—
Social aspects—United States. | United States—Civilization—
1865–1918. | United States—Civilization—1918–1945.
Classification: LCC E169.1 .Y24 2019 | DDC 973—dc23
LC record available at https://lccn.loc.gov/2018043831

♾ This paper meets the requirements of ANSI/NISO Z39.48-1992
(Permanence of Paper).

Contents

Introduction: Memory, History, Posterity

On September 23, 1938, the day of the autumnal equinox, the organizers of New York's upcoming world's fair joined executives from Westinghouse Electric at the edge of a hole, fifty feet deep, in Flushing Meadows Park, Queens. After striking the noon hour with a gong borrowed from a Chinese restaurant, they observed the lowering of an eight-hundred-pound cylinder into that "Immortal Well" (**fig. 0.1**). The hole was left open to allow fairgoers to peer down through a periscope and add their blessings (**fig. 0.2**). It was finally sealed in October 1940, and the site was subsequently marked with a black-granite monument dedicated by Robert Moses and Mayor Fiorello La Guardia. A plaque detailed some of the entombed contents: "22,000 pages of microfilm, 15 minutes of newsreel, and 124 commonly used articles and materials . . . [including] an alarm clock, tooth powder, bifocals, an asbestos shingle, . . . an issue [of] Harper's Magazine, a zippered tobacco pouch, beetleware [plastic dishes], carrots, & a Miami fashion show"—all bequeathed "to the people of Earth 6939 AD."[1]

While the value and intelligibility of such offerings to the distant future remains to be seen, Westinghouse bequeathed something undeniably significant to the immediate future: the term *time capsule*. Coined by public relations consultant George E. Pendray, who had originally suggested time bomb, it proved as popular as another of

FIG. 0.1 Anonymous untitled photograph, September 23, 1938. A. W. Robertson (*left*), Westinghouse Electric Company's chairman of the board, oversees the lowering of the Westinghouse Time Capsule into the "Immortal Well." President of the New York World Fair Corporation Grover Whalen looks on. MssCol 2233, New York World's Fair 1939–1940 records, 1935–1945, Manuscripts and Archives Division, New York Public Library.

his neologisms for Westinghouse, *laundromat*. So successful was his linguistic fabrication, not to mention the exhibit itself, it has obscured the longer history of transmitting messages and objects to the future via some kind of sealed container. Many assume that the tradition began with the Westinghouse Time Capsule along with the Crypt of

Civilization, sealed that same year in Atlanta by the president of Oglethorpe University, Thornwell Jacobs. Jacobs himself fostered this myth by declaring, "we . . . are the first generation equipped to perform our archaeological duty to the future" in advance.[2] Even those who acknowledge precursors have identified the years 1938–1940 as marking "the completion of the first major time capsule, the first major collection of an era's information to be deliberately preserved and sealed for long-term retrieval at a specified faraway date," thereby ushering in a "Golden Age" of time capsules.[3]

Yet Oglethorpe and Westinghouse were co-opting a civic practice that had thrived for more than six decades and had yielded more than thirty capsules. Although not as broadly publicized, these earlier capsules were comparable in scale and ambition. They systematically compiled collections that were microcosms of some larger political or social whole, albeit not necessarily of "American" or "modern civilization" and without the exhaustive, totalizing zeal of their successors.

FIG. 0.2 Anonymous untitled, undated photograph (ca. 1939). Fairgoers crowd around the periscope to view the time capsule in its unsealed tomb; Westinghouse pavilion in background. MssCol 2233, New York World's Fair 1939–1940 records, 1935–1945, Manuscripts and Archives Division, New York Public Library.

And while most had more modest time spans—typically one hundred years—by the 1910s they were considering more distant generations. Moreover, in the absence of a standard procedure, these time capsules avant la lettre were more varied than their successors. They bore diverse names such as Memorial Safe, Antiquarian Box, or Century Chest, which reflected the array of receptacles used, from bank safes to lead boxes to bronze chests. And they were stored in a variety of locations: libraries, colleges, banks, city halls, and even the Capitol. To avoid a teleology that would reduce earlier efforts to precursors or prototypes, we will call these aboveground deposits time vessels. Rather than anticipate the Westinghouse model, they represent an alternative and, in some respects, more democratic practice.

While tracing this longer genealogy, we must resist the conclusion that time capsules can be found throughout history and in all cultures. Certainly, the impulse to stash things away is ancient, perhaps universal. After their revolution, Americans searching for nation-building rituals embraced the cornerstone ceremony: the depositing of symbolic artifacts and documents in the first stone of a building or monument, usually in the presence of Masonic officials. This practice itself recalls the church consecration rituals of medieval Europe and the foundation and tomb deposits of Mesopotamia. In ancient Greece and Rome, coins and other objects were left in sacred places, or even placed in the mast steps of boats, as votive offerings embodying the depositor's "hopes, dreams, and anxieties." Similar rites have been found in Chinese, African, Native American, and other cultures.[4] And various future-oriented texts, from diaries to wills, have been secreted in private or public spaces since time immemorial. Even letters are, in a sense, time capsules sealed for at least a day, as are the postmarked envelopes inventors have used since the mid-nineteenth century as proof of priority, sometimes depositing them in libraries.[5]

Yet none of these forms appear to have specified when they should be opened. The inventor hoped his or her envelope never needed opening, while a cornerstone or foundation deposit could not be opened without damaging the building or a burial deposit without desecrating the grave. Those deposited artifacts were either offerings to gods, resources for the afterlife, or a kind of message in a bottle for some indefinite future time. The time capsule—defined as an intentional deposit with a preconceived target date—appears to be a relatively recent American invention, dating perhaps only to 1876.[6] During the first five years of the nation's second century, a spate of such timed devices, programmed to be opened a hundred years later, were launched in Washington, DC, Chicago, San Francisco, Boston, Amherst, and Ramapo, New York. The phenomenon cropped up in other towns and cities, in short stories and novels, and (by 1907) overseas. Related practices subsequently emerged, such as the procedure (established in 1952) of sealing US census data for a period of seventy-two years.

The time capsule's origins in nineteenth-century America's culture of democratic individualism might seem surprising. In an aristocratic country, Alexis de Tocqueville wrote, where "families maintain the same station for centuries and often live in the same place," one becomes "in a sense . . . contemporaneous" with past and future generations and willingly carries out duties and sacrifices out of a kind of "love" for one's remote descendants. By contrast, in the social flux and geographic restlessness of American democracy, "the woof of time is ever being broken and the track of generations lost. Those who have gone before are easily forgotten, and no one gives a thought to those who will follow." Even "ambitious men in democracies," he added, exhibit little concern for "the interests and the judgment of posterity. The [present] moment completely occupies and absorbs them." A disregard for posterity was also evident to Tocqueville in Americans' indifference toward preserving "traces" of the present, other than newspapers. The brief terms of elected officials, their tendency not to keep extensive written records, and the lack of public archives were a recipe for historical memory loss. "In fifty years' time it will be harder to collect authentic documents about the details of social life in modern America than about French medieval administration; and if some barbarian invasion caught the United States by surprise, in order to find out anything about the people who lived there one would have to turn to the history of other nations."[7]

Yet, the time capsule did emerge in nineteenth-century America and was the product of multiple concerns and desires. Superficially, time vessels marked, or ritually commemorated, some temporal passage—the centennial of a city, state, or nation; the anniversary of an event; or the dawn of a century—thus asserting historical continuity into the future. In that respect, they appear to illustrate Americans' overwhelming confidence in progress. Yet they betrayed an undercurrent of anxieties about potential threats to civilization, in particular from the mounting struggle between labor and capital. Various efforts to insulate time vessels from unruly mobs—through concealment, concrete, or remote mountain locations—reveal the gradual growth of secular apocalypticism, which the Westinghouse capsule ultimately disavowed.

Time vessels also indicate doubts about traditional sites and institutions of memory. Public monuments appeared to invite destruction by mobs or to preserve insufficient information about the present, while libraries and archives had to contend with paper's vulnerability to fire and decay. To ensure the survival of primary sources, time vessels drew on more durable materials such as vellum and on new methods of fireproofing and waterproofing. They were thus closely allied to a growing movement of historians, archivists, and historic preservationists struggling against impermanence, disposability, and forgetting.[8] Time vessels served as laboratories of media conservation, advancing techniques and materials not yet used in libraries, such as sealing documents in charcoal, vacuum bottles, or inert gases, or transcribing them onto terracotta, microfilm, or rag paper.

If archival entropy was one concern, another was its opposite: information overload. Before the 1930s, time vessels avoided the conceit that their contents would be the only surviving records of the present. Rather, they were responses to the tendency of modern civilization, with its burgeoning print culture and memorial infrastructure, to generate a superabundance of records that would overwhelm its future historian. Locking away a message was a means to cut through that noise. It represented a gamble that a century of concealment could paradoxically make it better remembered. The time capsule may thus be linked to a late-nineteenth-century fascination with the productive effects of forgetting: the potency of individual or cultural memories when they resurface.[9]

These vessels arose out of a host of other impulses and hopes, too. In attempting to freeze or capture "the present," they were reactions to the radical flux of modernity, with its accelerated rate of change. In seeking to thwart natural processes of decomposition, they echoed embalming, taxidermy, and canning, and the domestic arts of scrapbooking, hairwork, and flower pressing. In curating a representative sample of messages and objects that would microcosmically stand in for a larger whole, they exhibited the encyclopedic impulse that animated late-Victorian projects, from museums and world's fairs to private collections and biographical dictionaries, not to mention encyclopedias themselves. And in addressing a distant recipient, they embodied a fascination with the communicative possibilities opened up by new technologies such as the telephone. Their inclusion of photographs, phonograph records, and eventually film even hints at a pervasive fantasy of conscripting media to exert a presence from beyond the grave.

Ultimately, however, the time vessel served political ends. One such goal emphasized by its earliest advocates, was to foster national identity.

Although the nation is most often understood as a constructed space, a clearly bounded territory, scholars have begun to explore its temporal underpinnings. Studies of the creation of monuments, the preservation of historic sites, or the enshrinement of anniversaries have highlighted the fabrication of a shared, national past.[10] Nations have also been shown to cohere through the sense of a shared present—the experience of participating simultaneously in such daily rituals as reading the morning newspaper or (by the late nineteenth century) adjusting one's watch to the "standard time" conveyed by government-sponsored public clocks.[11] What requires further exploration, however, is how shared *futures* worked to create national subjects.[12] Through the ceremonial sealing of relics in a time vessel, citizens expressed solidarity with their successors and affirmed faith in the nation's endurance.

The time vessel's political uses were not limited to the inculcation of national identity; some sought to foster local identities, and at least one, conversely, advanced internationalism. Indeed, they proliferated because they appealed to a broad political spectrum. Through them, reactionaries could defend capitalist interests, deny social injustices, denounce labor unions and political unrest, or even advocate eugenics. Those hoping to shore up the status quo by addressing poverty and inequality linked them to social and political reform, charities, and temperance. At the same time, they attracted oppositional groups, from anti-imperialists, socialists, and agrarian populists to members of the labor, women's suffrage, and black activist movements. For each group, a sealed vessel was a totem of a future age in which their vision—conservative, reformist, or radical—would be realized. Unlike Westinghouse-style time capsules, which are tightly controlled corporate and industrial statements and which present "American civilization" as an organic whole, these earlier vessels could hold competing futures. In revealing how these multiple futures permeated their present, the vessels evoke a heterotemporality that calls into question the standard thesis of time's homogenization in modernity.[13]

Memories of the Present

Despite their enduring and widespread appeal, time capsules—especially those predating Westinghouse's—have been oddly understudied. Even with the memory studies boom, which has brought monuments, museums, and archives to the fore, no historian has investigated the practice of sealing artifacts for the future.[14] Consigned to archives after their opening, time vessels' contents were quickly forgotten and, in many cases, remained unconsulted. The lack of a standardized name before 1938 has further obscured the phenomenon's origins and extent, as has its reputation for containing apparently banal or meaningless artifacts.

The time capsule's neglect also stems from our deeper wariness toward such deliberate testimony. Since the mid-nineteenth century, historians have consistently emphasized the limits and biases of intentional evidence such as memoirs. We instead prioritize involuntary sources—those nontendentious traces (such as letters to contemporaries) that have inadvertently survived—on the assumption that their lack of intent to influence us allows us to penetrate deeper into the past.[15] Archaeologists exhibit a similar bias. Time capsules, complained one, cannot attain the status of an archaeological site. They are premeditated rather than accidental; they remove their objects from their everyday contexts and interrelations; and they tend to privilege new objects, thus presenting a temporally flattened world.[16]

Our privileging of unintentional sources has blinded us, however, to the evidentiary richness of intentional sources, which do not have to be read as transparent windows onto the past. Nor should we assume the time capsule is one thing. Read against the grain and as a changing form, time capsules can reveal the politics of memory: changing attitudes about who or what should be memorialized, in what form, for whom, and under whose direction. Time vessels, in particular, evinced dissatisfaction with the official sites and frameworks of memory. The founding of a monument, museum, historical society, or other mnemonic project was a long, complex process typically necessitating a voluntary association with an elected president, the appointment of an architect or artist (usually through a design competition), extensive fundraising, and government approval. By contrast, a time vessel could be compiled and launched quickly and easily. A portrait photographer, a Jewish merchant, an eccentric former dentist, a Hungarian immigrant, a retired seamstress, and an agrarian populist were able unilaterally to appoint themselves creator of a time vessel.

At the same time, those individuals realized the need to engage a larger community if only to avoid the impression of commemorating themselves. They enlisted not just elite white men but also women (who were particularly enthusiastic contributors, perhaps because they already presided over the domestic arts of memory), children, ethnic minorities, and workers. Time vessels thus became more collaborative than traditional memory projects—a quality that Westinghouse subsequently abandoned. Despite our tendency to distinguish between official and vernacular, or top-down and bottom-up, forms of memory, time vessels were a hybrid of both. As public assemblages of private messages, they also remind us that individual memory cannot be disentangled from nor conflated with collective memory.[17]

Time vessels also broached the issue of who gets to be memorialized. In depositing a message, photograph, or just a signature, ordinary members of the middle class, including women and immigrants, were asserting a right to be remembered alongside political and financial leaders. Indeed, time vessels elicited criticism for cheapening fame, not only by democratizing it but also by granting it in the present. Yet they could shore up that institution, too, by upholding traditional criteria of accomplishment in politics, science, and the arts, disbarring sportsmen, movie stars, and other popular celebrities, especially those who were female or ethnic. Even these exclusions were challenged, as time vessels prompted diverse audiences to contest the meaning of fame. They thus allow us to recover broader, popular attitudes toward the changing mechanisms of veneration.[18]

By the turn of the twentieth century, time vessels marked a further shift in attitudes to memory. In addition to memorializing individuals, they sought to preserve information (both textual and visual) about larger social wholes, whether an institution or city. Not content to allow a memory of their period to emerge organically over time, they attempted to supply it—or, at least, the raw materials for it—in advance. While we know that groups sought to construct (through monuments, historical pageants, or movies) various usable pasts, we have neglected how they sought to influence how future generations would remember the present—what we might call *prospective memory*.[19] As collaborative projects, moreover, time vessels represented a hidden battleground of multiple, competing perspectives on their own period, which they, unlike monuments or history books, did not integrate into a unified account. They thus indicated a growing awareness of how political and social power depended on mnemonic control over the past *and* present.

The recourse to time vessels further implied that libraries, historical societies, museums, and archives were failing as repositories of the present. Those institutions tended (until relatively recently) to exclude contemporary or recent materials.[20] They also tended (and still tend) to accrue their collections haphazardly, through the intermittent donation or purchase of artworks, books, artifacts, or the depositing of expired administrative records or personal papers.[21] And they have privileged the writings of the eminent over those of the anonymous, or noteworthy relics over everyday, manufactured artifacts.[22] In contrast, time vessels sought, especially by 1900, to encapsulate a broader social milieu, from its institutions to its quotidian aspects, through purpose-built records that pictured it at a single moment. Locals were commissioned to draft essays or letters on political, economic, cultural, and social conditions, including the built environment; to write journals of their everyday lives; or to compile glossaries of current slang. They also submitted artifacts without prior historical or aesthetic significance, such as shoes, stockings, or a telephone, thus transforming the time vessel into a pioneer of the public collecting of ordinary, contemporaneous material culture. Such offerings further attested to a growing sensitivity to how material artifacts, even mass-manufactured ones, could transmit collective memories.

Time vessels also gradually expanded the definition of document to include those generated by new media technologies.[23] Long before libraries and other collecting institutions in the United States fully embraced them, they enlisted cameras to document local urban and domestic spaces, phonographs to preserve popular songs and political speeches, and movie cameras to capture "historic" events of the day, thereby advancing the notion of multimedia archives.[24] Time vessels functioned also as proving grounds for new techniques and devices, such as composite photography (the overlaying of portraits to produce an "average" face) or the kinetophone (Edison's abortive attempt to marry sound and film). By the 1910s, these media began to displace more personal, written messages and to foment the fallacy that records produced by the camera and phonograph were necessarily objective and more vivid, and thus more useful "data" for future historians. Yet until then, vessels employed those media both to produce documents *and* to connect affectively with their recipients. They also commingled modern *and* ancient media so that the strengths of each compensated for the weaknesses of the other. In our own age of archival digitization, they exemplify how we might incorporate new media without fetishizing them.[25] And they further remind us how ordinary objects such as safes or chests can themselves function as media, ones that constrained but did not necessarily determine their content.

The sealing away of those collections in metal or wooden boxes—and the reinforcement of their inviolability through rituals of sanctification—implied a further critique of traditional institutions of memory. As open, ongoing collections, libraries, museums, and archives can jettison or "deaccession" materials, and are thus (like monuments) subject to changing academic, aesthetic, or even political beliefs. Time vessels, on the contrary, bypassed those intermediaries who collate, edit, or otherwise filter historical records so as to speak directly to a future generation. The locking, welding, or bolting of these boxes—and, by 1900, the additional provision of confidential, sealed envelopes for individual submissions—also encouraged contributors to furnish more candid and critical assessments of their present.

As sealed containers, time vessels may appear to reinforce the emerging notion that memory can be stockpiled, that it no longer needs to be actively and continually circulated, performed, spoken, or viewed to be transmitted across generations. Philosopher Bernard Stiegler, while acknowledging that humans have been displacing their memories onto things since the invention of tools (and increasingly so with the invention of writing), perceives externalized memory as becoming dangerously "dissociated" from living memory during the nineteenth century.[26] Yet early time-vessel advocates also sought to retain a connection between the two by linking their archives to ongoing public rituals, practices, and institutions, thus forging hybrid memory sites. Only with the Westinghouse and Oglethorpe projects do we see the ascendance of that conceit of *autonomously* transmitting memory, which presupposed that their contents—and indeed the very notion of a time capsule—could be semantically self-sufficient.

Half-Cooked History

Time vessels also violated the emerging tenets of academic history. Historians founded their profession in the late nineteenth century not just on a new faith in objective or "scientific" study of the past but also on a constricted focus on the lofty realm of political history—especially that of the nation state and its institutional evolution—as revealed in written sources, above all, government documents.[27] They stigmatized other kinds of history (social, cultural, intellectual) and other sources (visual images, folk traditions, or material artifacts) as low, feminine, and trivial and thus unscientific and ahistorical by dint of their association with Romantic, amateur, antiquarian, and female historians.[28] Time vessels' unorthodox embrace of domestic objects, clothing fashions, and audiovisual media was thus an implicit appeal for a radical expansion of what could count as history. Vessels presumed that by the year 2000 historians would have embraced the study of everyday life, the built environment, material culture, and by extension women's history and local history. There was indeed such a historiographical revolution in the latter half of the twentieth century, but its groundwork was arguably laid around 1900 by community and amateur historians, marginalized female historians, local historical societies, private collections, and time vessels themselves.[29]

The time vessel further challenged professional history in its presumption to know what sources future historians would want. A founding myth of the profession was that of the intrepid male researcher traveling to remote archives and heroically penetrating through the dust and disorder to unearth the crucial, "virgin" source. His powers of scrutiny and verification, knowledge of the past, and ability to sift out the insignificant were what retroactively transformed documents, after a sufficient lapse of time, into "sources."[30] Yet time-vessel contributors presumed to identify and even produce sources in advance. They thus cast the future historian as the passive recipient of preselected documents, all conveniently packaged, labeled, and inventoried. In his letter declining to contribute to a 1900 time vessel at Harvard, William James scoffed that it was no longer "enough," apparently, to leave behind the "raw" ingredients of history; "now we must half cook it for the future historian."[31] Certain contributors, including some of James's colleagues, typed up their sources, thus obviating even the challenge of deciphering obscure handwriting.

Some vessels neglected professional historians altogether, addressing themselves instead to future communities as a whole. The essays written for several time vessels could be construed not just as raw (or parboiled) testimony for future historians to cook up but as preliminary drafts of the history of the present. Such "contemporary histories," moreover, were imagined as collaborative ventures: the product of a community for that community's future members. A loose, multi-authored, heterogeneous collage of texts, images, and other artifacts, this vernacular (or public) history implicitly challenged the emerging positivist distinction between the historian and the archivist.[32] Still today, historians tend to reinscribe hierarchical and institutional distinctions between history and memory, and nonacademic practices of collecting and displaying the past are typically relegated to the latter realm.[33]

Time vessels' indifference to such distinctions sometimes constituted a rejection of historiography altogether. Some evoked a broader notion that photographs (and subsequently films and phonograph records) were not merely sources for future historians but themselves a superior form of historical writing, inscribed by light (or sound) without intermediating words. Yet a number of time-vessel organizers also stressed the need for contributors to include historical perspectives in their letters. In combining personal messages with public history, they further repudiated historical orthodoxy, which was (and still is) predicated on the muteness of the past, its irretrievability except through its unintended traces. Sources that speak directly in the ear of the historian were an affront to the latter's premise of objective distance.[34]

Open and Embodied Futures

For all its contributions, the burgeoning field of memory studies has inevitably reinforced a broader fallacy that societies and communities are primarily organized around constructions of their past.[35] By focusing on time vessels and situating them in relation to utopian and science fiction along with indirect acts of anticipation such as tree planting, this book seeks to correct that overemphasis on the retrospectivity (or, for that matter, the present-mindedness) of American culture. The French phenomenologist Eugène Minkowski, like Martin Heidegger, suggested that humans in general are fundamentally oriented toward the future. Memory of the past should "occupy only a secondary place in an analysis of lived time."[36]

Those scholars who have explored how societies (including our own) conceive the future tend to emphasize the limits of the temporal imagination. Historians, philosophers, and theorists have shown how the modern, Western conception of historical time as linear and progressive has stifled the ability to imagine the future as anything other than an extrapolation of current trends: a closed future.[37] In particular, they have charged "scientific" forecasters, from H. G. Wells to futurologists, with attempting to "colonize" the future, to "contain" its otherness by reducing it to the framework of their present or by focusing on the immediate rather than distant future.[38] This colonization has been accompanied, some say, by a process of abstraction. Traditionally embedded in religious worldviews and lived contexts, the future has become disembodied and decontextualized in modernity. Thus emptied of content, it can even be exchanged as a commodity, most obviously in the futures market.[39]

In some respects, time vessels have been complicit in this foreclosure of the future. Many contributors used them to celebrate contemporaneous technological, economic, and cultural progress and to predict what further, uninterrupted progress would achieve. They also reflected and reinforced the abstraction of the future in their specification of a numerical time span (typically, a century) rather than a generational one. Indeed, the sealing of a time vessel implied a will to colonize the future by prescheduling its opening. In addition to containing predictions, the time vessel is itself a prediction, the prediction being that it will be opened. This is of course a self-fulfilling prophecy, as the future will never know about the forgotten time capsules that failed to fulfill their mission.

These aspects of time vessels, however, did not bar them from critiquing the social order and expressing utopian hopes for a very different future. Several contributors—women, labor leaders, a black activist, and those sympathizing with them—referred to the persistence of factors that perpetuated gender, class, or racial prejudices and inequalities in their present and envisioned their eradication by the time their message was opened. Even some time-vessel architects broke with class and political loyalties to criticize aspects of the status quo—such as capitalist greed, working-class poverty, or imperialist expansion—and to imagine alternative futures. The task of writing to a distant time appears to have momentarily "freed" them, in Minkowski's words, "from the embrace of the immediate future"—from the grip, that is, of self-interest—and to have opened up a future that is "further, more ample, full of promises."[40]

Unlike ordinary predictions, time vessels could instill a more open and embodied understanding of the future by encouraging contributors to address themselves not to posterity in general but to specific putative individuals. Rather than imagining the future as a blank screen or an empty container, they could conceive it, more concretely and corporeally, in terms of actual people. The addressees might be their own descendants, evincing genealogy's growing popularity, especially among Anglo-Americans wishing to reassert status and bloodlines in the face of immigrant incursions.[41] But they increasingly designated nonrelatives such as the president or a mayor, or, in later vessels, the representatives of a church, university, profession, or ethnic society. Indeed, the time vessel attracted individuals who did not have children, offering them a kind of surrogate posterity. The future was thus personified, rendered tangible through the bodies of those who would open the vessel and its envelopes.

An open future could also be grasped in terms of the bodily absence or otherness of their recipients. Several contributors wondered not only whether their message would reach its addressee but also whether the latter would ever exist, as the specified organization might not have survived. They thus realized they were, to borrow the communications theorist John Peters's phrase, "speaking into the air."[42] Even if the message were successfully transmitted, they doubted whether it would be of any interest. Some struggled to imagine their recipients, their unfathomable conditions of life, and their attitudes toward their ancestors. As literary historian Gillian Beer writes, posterity is always "ghostly, unimaginable," even when personified as specific individuals.[43] Lack of knowledge about their recipients led several contributors to procrastinate or even consider abandoning the task. These uncertainties and contradictions—inherent in any effort to communicate across time—were dramatized in the fiction of the period; several authors, including Mark Twain and Jack London, and science fiction writers, such as Alvarado Fuller and George Allan England, described time vessels that failed, in various ways, to transmit their contents as intended. Such narratives of failure only increased the allure and fascination of speaking to the future.

Even when addressing the more generic audience of "posterity," time vessels resisted abstracting the future. This expanded, collective sense of posterity as all future members of a city, nation, or even the human race retained the echoes of its older, now archaic meaning, namely, the direct descendants of a single individual. In some respects, this etymological derivation renders the term problematic; like its synonym *future generations*, posterity is a temporal concept founded on dreams of patriarchal continuity and reproductive sexuality.[44] Yet *posterity* does at least imply an embodied relationship to future ages and perhaps even a notion of passing on an inheritance, one that was political, cultural, and environmental rather than merely financial and biological.

The direct address of recipients, moreover, prevented expressions of hope for a better future from lapsing into empty or (to use Marxist philosopher Ernst Bloch's term) *abstract* utopian gestures. By orienting those hopes to a fixed point in the future and by articulating those hopes to the bodies (or bodily absence) of their intended recipients, time vessels rendered them concrete. Female, black, and working-class contributors could address their future counterparts, or even figuratively embrace them, thereby forging an embodied bond of solidarity across time and a sense of continuing struggle for freedom. Several time vessels were tied to actual schemes to bring about that utopia, such as wealth redistribution or international cooperation, thus rooting it in what Bloch called the "real-possible."[45] In specifying a target date, furthermore, they eschewed the abstract eternity to which monuments—and indeed, nations—aspired. Time vessels only asked to be remembered for a finite number of years.

In Praise of Posteritism

With their frequent references to posterity, time vessels enable us to trace some critical shifts not just in conceptions of the future but also in conceptions of the present's obligations to it. The rhetoric of posterity has become ubiquitous, especially in debates about our environmental or fiscal responsibility to future generations. Yet even as philosophers, environmentalists, politicians, and economists articulate an ethics of posterity, there has been little consideration of how such a notion emerged and spread.[46] While this book cannot trace its full history, I will adumbrate some crucial moments in its development in the United States.

The few, fragmentary writings on this topic posit a gradual growth of this sentiment in the West. In classical antiquity, historian Carl Becker wrote, *posterity*—although derived from the Latin *posteritas*, meaning those who come after—lacked the force of an ethical stimulus. Medieval religious beliefs that "the fate . . . of all mankind" was predetermined further impeded the idea of "labor[ing] . . . for posterity," as did the Renaissance humanists' assumption of the world's ongoing decline since the fall of Rome.[47] Becker claimed that it was not until the Enlightenment, with its theories of human progress and perfectibility and its quest for an authority that transcended that of the monarchy and the church, that posterity emerged as a motivating force, one that proved central to the revolutions in France and America and then to Romanticism.[48] This quasi-religious cult of posterity, according to historical geographer David Lowenthal, reached its zenith in Victorian and Edwardian England, with its "railroads, aqueducts, sewer systems, libraries, parks, and gardens," all "intended to endure for centuries to come."[49] The philosopher Hans Jonas offered a somewhat different account yet still presumed a steady expansion of the idea. Insofar as parents sought to prepare their children for life or statesmen to safeguard their state, they have always acted for posterity. What has changed is "the reach of those duties," which "used to be short"—"hardly exceeding the lifetime of the newborn"—but, with the growth of "modern technology" and the concomitant growth of our perceived capacity to influence the future, has lengthened.[50]

In this book I posit the time vessel as a sign not so much of a command over posterity than of a crisis of posterity. The will to communicate directly with future generations—evident also in Walt Whitman's hailing of his posthumous readers—was paradoxically driven by a mounting sense of disconnection from the future. Just as the past was increasingly perceived as a "foreign country," cut off by ruptures such as the Civil War, so, too, did the future appear estranged.[51] With the increasing impermanence of modern media, such as wood-pulp paper, photography, phonography, and film—media better suited, in Harold Innis's influential formulation, to disseminating knowledge over space than over time—one could no longer take the cultural transmission of the present for granted.[52] Their ephemerality was exacerbated by the great fires that ravaged American cities, by growing industrial pollution, and by the lack of state and national archives. Even without fires, capitalist speculation precipitated the demolition of relatively recent office buildings. Meanwhile, those structures that would endure, some feared, might not be intelligible to future archaeologists. These and other problems were attributed above all to Americans' seeming dereliction of duty to posterity.

In some respects, time vessels reinforced this temporal myopia. They stimulated paeans to posterity that often lapsed into what might be called temporal chauvinism, a presumption that later generations would look back with gratitude and admiration. One can also detect in them a growing tendency to reduce the duty to posterity to a merely archival duty, as if it were enough to preserve a smattering of documents, photographs, and artifacts; often inexpensive or redundant, these materials represented no great sacrifice. Moreover, by the 1910s, as vessels increasingly relied on technological media, they included fewer personal letters to posterity. Similarly, as their time spans expanded from a century to millennia by the 1930s, their participants struggled to cathect with their recipients. Although prominent critics perceived the 1970s as a watershed decade in which Americans renounced earlier "commitments to posterity" (Henry Steele Commager) or became unable to "identify with posterity" (Christopher Lasch), time vessels suggest this trend developed in earlier decades.[53]

The rhetoric of posterity that some groups adopted at time-vessel sealings also indicated presentist ends. While ostensibly addressed to future recipients, time vessels targeted contemporary audiences through media publicity, disclosure of sample messages, and lavish dedication ceremonies. This emphasis on publicity and display may have been motivated by a desire to prevent the vessels from being forgotten and thus overshooting their intended future. But several time-vessel pioneers used their devices to advertise their own name and reputation, their publications, their political or professional affiliations, or their city's commercial prospects. This promotionalism became the very raison d'être of Westinghouse's time capsule, which was created to disseminate the ideology of technocracy to fairgoers and newsreel audiences.

Yet the earlier time vessels also worked in surprisingly strong ways to instill a duty to posterity. A time-vessel sealing could inspire, among both participants and witnesses and arguably even among recipients, a meditation on their individual or collective legacy, a sense of the present's relative backwardness or insignificance, or a yearning to serve the future. Even the most meager offerings could betoken that commitment, while the time vessel itself could be imagined (in Innis's terms) as performing a "time-binding" function, its heaviness compensating for the "spatial bias" and cult of the "ephemeral" engendered by modern society's reliance on lighter media capable of being transported or broadcast.[54] Concerns about the degradation of the environment, for instance, led one citizen to conceive a time vessel for Colorado Springs as a tool for promoting what he called *posteritism*. His "century chest" was to unite his community in an enduring, collective pledge to solve problems for the sake of those to come. This effect of time vessels was not limited to the immediate community. News coverage of the century chest prompted labor advocates back east to appropriate the term *posteritism* in their campaign for laws to protect industrial workers and their families.

Such pledges of duty to posterity were often articulated in terms of a physical or even affective bond. Rather than address posterity in a vague, rhetorical sense, time-vessel contributors imagined themselves to be speaking in person to the not yet born. One common trope was that of a voice from the grave or of a body (or hand) magically extending beyond its death so as to touch, embrace, or shake hands with posterity. Another was that of the vessel as a gift, affirming the bond between generations; indeed, several included monetary or other offerings. Some even exhibited emotional closeness, confessing personal matters to or affection for their recipients. This was especially the case with contributors who felt alienated from their own generation; just as nostalgia and retrospection have offered refuge from the present, so, too, has the idea of posterity. Their expressions of love or friendship, moreover, evoke the notion of "intimacy at a distance." Developed over the past half century by communications theorists exploring the affective bond between media personalities and their dispersed audiences, this concept might be applied to communication across time.[55] We can trace a weakening of this capacity to identify sympathetically with posterity in the early twentieth century as time vessels became conduits for conveying impersonal, mass-mediated "data" across ever-larger time spans. In our own time, when climate change and nuclear weapons imperil the very existence of future generations, it is more pressing than ever to recover that affective capacity. Statistical, legalistic, or even philosophical arguments for our duty to posterity, however well reasoned, cannot suffice.

A commitment to posterity, however, does not necessarily imply political progressivism. Time vessels attracted two reactionary movements in particular: the moral crusade to prohibit alcohol, arguably to ensure the productivity and docility of workers, and the racial program of selective breeding known as eugenics. Temperance and eugenicist pamphlets served not only as votive tokens of hope for the purified utopia their movements would secure by the time their vessels were opened but also as hints to their recipients to trace that utopia back to these seemingly minor publications. Time vessels also served to relegate various social "others" to historical oblivion, either by excluding them or, conversely, by including them as specimens of a group presumed to be vanishing. Those dedicated to "salvaging" American Indian customs for the edification of future white audiences, including the photographer Edward Curtis himself, thus embraced the device. Lowenthal's vision of the late nineteenth and early twentieth century as the high point of posteritism, or what he calls "stewardship of the future," thus obscures how this ideal was mobilized toward conservative or even racist ends.[56]

And yet the time vessel also animated and was animated by efforts to redefine the present's duty to posterity in less repressive ways, sometimes even by those very same eugenicists and temperance reformers. We will encounter several progressive programs, such as schemes to redistribute capitalist fortunes through philanthropy or profit sharing. Vessels also inspired workingmen to affirm the labor union as the agent that would eradicate class inequalities, and they induced socialists, Christian socialists, and advocates of cooperative communities to foresee capitalism's collapse as the precondition for a better society. Time vessels sustained efforts to solve for posterity the problem of race, too, not through eugenicist purification but through legal fights against segregation, repudiations of "scientific" racism, or dreams of a postracial society. So, too, with gender: while some contributors called for women to return to domestic and maternal roles, others rededicated themselves to universal suffrage and the equality of the sexes. We even find in time vessels challenges to imperialism and militarism; to the destruction of forests and animals; and to the growing dominance of technology over everyday life. These diverse advocates for posterity did not necessarily presume they themselves could overcome these problems. More often, they envisaged a protracted and ongoing struggle by current and future generations.

In this book I trace the hesitant beginnings and subsequent twists in the time-vessel tradition across six decades, a journey that will take us from Philadelphia and Chicago (chap. 1) and San Francisco (chap. 2) to Kansas City, Colorado Springs, Detroit, and Harvard University (chaps. 3 and 4) to New York (chap. 5) and eventually to Atlanta, Denver, and the Ozarks (chap. 6). Along the way, we will explore both the motives that inspired individuals to communicate with the future through a sealed box and the decisions about what those boxes were to contain. We will consider who was permitted to contribute and how their (often last-minute) contributions transformed the projects. And we will also hear from noncontributors, such as those who criticized the very idea of a time vessel. Among those critics will be authors—from Edgar Allan Poe, Mark Twain, and Jack London to lesser-known science fiction writers—who, in predicting the failure of fictional time capsules, exposed the limits of transtemporal communication. While each chapter examines the ceremonial sealing of vessels, we will have to wait until chapter 7 to learn whether they reached their target date and, if so, how those late-modern contexts informed their reception. Finally, in the epilogue I reflect on the time capsule's diminishing potential since 1940 but also on recent efforts to reinvigorate it as a means to inspire long-term thinking and intergenerational responsibility.

ONE

Safeguarding the Nation: Photographic Offerings to the Bicentennial, 1876–1889

The bells that rang out across a rainy, windswept Philadelphia to signal the Centennial Exposition's opening on May 10, 1876, also heralded the dawn of a cultural innovation. Most of the 186,272 visitors who inundated the fair on opening day elbowed their way toward the chief sights, above all President Ulysses Grant and Brazilian Emperor Dom Pedro II pulling the levers of the mammoth Corliss steam engine to breathe life into thousands of electrical exhibits throughout Machinery Hall. But elsewhere at the exhibition, and with less fanfare, a smaller, quieter invention was revealed to the public. In a corner of the Art Gallery, the New York magazine publisher Anna Deihm displayed her Century Safe, a five-foot, four-inch iron safe containing a collection of photographs, autographs, and other mementos for the year 1976 (**fig. 1.1**), while next door in Photographic Hall, the Chicago portrait photographer Charles Mosher unveiled his own embryonic collection of photographs, which he later deposited in his Memorial Safe, also for the bicentennial (**fig. 1.2**). Although Deihm's and Mosher's exhibits would barely have stood out from the more than thirty thousand other businesses flaunting their wares to nearly ten million visitors to Fairmount Park that summer, they can retrospectively be identified as pioneers of a distinct practice.[1]

What was it about the political, social, cultural, or technological circumstances of the American Gilded Age—and, more specifically, the year 1876—that generated this idiosyncratic, invented tradition? How did the nation's cultural legacy, conventionally entrusted openly to posterity via public monuments, museums, and libraries, come to be condensed into representative samples locked up in small metal boxes? We will first examine the immediate catalysts of the centennial and its attendant world's fair, which stimulated not only compendia of the present but also anticipations of the future, anticipations that—given the social, economic, and political problems of that year—were not entirely optimistic despite assumptions about Victorian Americans' confidence in progress. We will then consider the time ves-

FIG. 1.1 Anna Deihm's Century Safe, showing autograph and photograph albums, framed photographs, and other items. From C. F. Deihm, ed., *President James A. Garfield's Memorial Journal* (New York: C. F. Deihm 1882), 196.

FIG. 1.2 Unknown designer, "Mosher's Memorial Offering to Chicago." Detail from backmark of one of Charles D. Mosher's "memorial photographs," Chicago, n.d. Chicago History Museum.

sel as an outgrowth of the communications revolution; a response to a crisis of posterity stemming from concerns about the inadequacy of paper-based records, archives, monuments, and the built environment; and a symptom of a larger democratization of fame. In each instance, the time vessel appears part of a broadening interest in communicating directly and self-consciously with posterity.

Particularly crucial to the emergence of the time vessel was photography. The centennial vessels were, first and foremost, troves of photographic portraits. Deihm originally conceived her Century Safe as a collection of autographs, but she subsequently assembled a vast album containing photographs of the president and his cabinet, Supreme Court justices, and every member of Congress while also soliciting photographs from the nation's leading figures in "Science, Literature, and the Fine and Mechanical Arts."[2] Mosher, a photographer by profession, was even more ambitious. Growing out of his exhibit at the Philadelphia exposition of 590 portraits of leading Chicagoans, the Memorial Safe expanded to include politicians, generals, entrepreneurs, inventors, literary figures, and other "notables" (and their wives) from across the nation and ultimately any middle-class American willing to

FIG. 1.3 *"Mosher's Centennial Historical Album 1876,"* one of five albums deposited in Mosher's Memorial Safe. Chicago History Museum.

pay. He eventually deposited almost ten thousand portraits, some inserted in mammoth photographic albums (**fig. 1.3**).[3]

As photographic collections intended for future viewers, Mosher's and Deihm's vessels pose a challenge to the emphasis in recent decades on the circulation of images. The social-historical approach to this medium has foregrounded the networks through which photographs, as material artifacts, were distributed, viewed, and interpreted. We are now familiar with how they were exhibited in galleries, studios, and eventually museums; distributed from vending machines and cigarette packs; exchanged by collectors and salesmen; projected as slides for lectures and travelogues; reproduced as engravings and halftones in magazines and newspapers; disseminated across and beyond the nation as postcards; and, in our own era, emailed, texted, shared, and (re)tweeted. This unfettered, promiscuous circulation—what Allan Sekula called the traffic in photographs—has been posited as a defining feature of the medium, central to its reputation both as a tool of political and commercial persuasion and as an instrument of ideology and power.[4]

The emphasis on contemporaneous circulation, however, occludes photographs' orientation to the future, which has often entailed their temporary withdrawal from circulation. From its very conception, writers have imagined a special relationship between photography and posterity. Photography, wrote the *Athenaeum* in 1845, has "enabled us to hand down to future ages a picture of the sunshine of yesterday, or a memorial of the haze of to-day."[5] This focus on future generations was especially central to projects that sought to record cultures or structures that seemed about to disappear, such as the Smithsonian's mid-nineteenth-century collection of portraits of American Indians or Britain's early twentieth-century surveys of old buildings. We thus need to augment the social-historical approach by attending to how photographs are earmarked for various futures.

In addition to exploring the social and cultural forces that prompted Deihm and Mosher to take several thousand photographs out of circulation, we will consider their projects' political significance. Despite the rhetoric of altruism and impartiality that often accompanies photographic preservation (and archival practices in general), Mosher's and Deihm's collections were closely allied to specific political movements. That allegiance appears to have shifted over time. Initially, their vessels hinted at a conservative faith in the continued dominance of the capitalist elite, the Republican Party, and republican institutions more generally. Mosher and Deihm intended them to serve as ritual objects inculcating a sense of belonging to a nation, understood as an entity extending not just across vast spaces but also into a remote future. Yet by the time they were sealed in the aftermath of the Great Railroad Strike of 1877, they assumed a more ambivalent stance and even expressed social democratic hopes for a future characterized by gender, racial, and class equality.

Exhibiting the Present

While the Century and Memorial safes were remarkably similar in format and content, their creators appear to have invented the time vessel independently, arriving at it from different motives and career paths. Anna Mary Deihm was a forty-seven-year-old Civil War widow and magazine publisher from New York City (**fig. 1.4**).[6] Five years younger, Charles Delevan Mosher was one of Chicago's leading photographers, with a gallery on State Street and the honor of having taken Abraham Lincoln's portrait (**fig. 1.5**).[7]

It would be tempting to characterize these early vessels—as some contemporaries did—as mere publicity stunts. The Century Safe certainly advertised Deihm's publishing business by showcasing her magazine, *Centennial Welcome*, and her newspaper, *Our Second Century*. Mosher similarly promoted his gallery and studio by offering the bait of a complimentary photograph for his memorial safe, thus luring customers to purchase duplicates for themselves at five dollars per dozen—a scheme he appears to have devised after the exposition.[8] Even amid the profusion of exhibitors, both won praise at the Centennial. Deihm reputedly "attracted . . . much attention," and Mosher

FIG. 1.4 Unknown photographer, *Mrs. C. F. Deihm*. Silver-framed photograph, deposited in Anna Deihm's Century Safe. Records of the Architect of the Capitol.

FIG. 1.5 Charles D. Mosher, *C. D. Mosher Photographer 125 State St.* Albumen print on cabinet card, n.d. Chicago History Museum.

won first prize for "excellence in art photography."[9] Mosher proclaimed the exposition a "golden opportunity" for photographers to attract customers, the "best cash investment" they could make.[10] His idea of photographing customers for posterity, moreover, promised to elevate portraiture, traditionally subordinated to other genres of art, to the prestige of a civic deed.

To dismiss them as self-promotional schemes, however, would be to overlook the deeper cultural forces that engendered the time vessel. It is no coincidence that both were introduced at the Centennial Exposition. Just as a world's fair was, in the words of contemporaries, a "world in miniature" or a three-dimensional "illustrated encyclopedia of civilization," so did a time vessel represent a microcosm of a larger whole.[11] Mosher's and Deihm's drive to assemble a "complete" set of photographs and autographs of a city's or nation's "representative men" mirrored the world's fair organizers' efforts to gather a sample of artifacts representing the technological, aesthetic, and material accomplishments of all nations. Informed by a broader Victorian obsession with collecting, the exposition also imposed a complex taxonomy on its exhibits.[12] Mosher similarly classified his photographs by profession, with sixteen separate categories for politicians, doctors, clergymen, lawyers, journalists, businessmen, teachers, and so forth. These categories were inscribed on the safe's inner door and on a printed catalog deposited within.[13] In imposing this internal order on its contents, the time vessel departed from the older, more indiscriminate depositing of token objects in building cornerstones.

The two projects also exemplified the exposition's role as an "exhibitionary complex," a laboratory for new strategies of display. Like subsequent fairs, Philadelphia's elaborated an aesthetic of the commodity by drawing on (and, in turn, refining) department stores' and museums' display techniques.[14] Deihm's safe—whose iron doors were to remain open while its inner, plate-glass door remained sealed so that its contents would always remain visible—particularly resembled display cabinets or vitrines in the fair's main building that fetishistically enshrined commodities.[15] By the time it was sealed, the safe showcased a silver inkstand and two gold pens specially made and engraved by Tiffany and Company (**fig. 1.6**) along with a patented photograph album bound in rosewood, ebony, and cut-glass by the luxury stationers W. W. Harding.[16] The safe was itself an object of display, richly decorated with presidential portraits and gilt lettering, lined with royal purple velvet, and topped with an ornate pediment. It thus served as an advertising tie-in for the Marvin Safe Company, which Deihm commended for the "perfection" of its products.[17]

Although conceived as centennial commemorations, Deihm's and Mosher's exhibits preserved no relics from the nation's birth but instead confined themselves to contemporary materials. This narrowing of temporal focus—a feature that distinguishes time capsules from older collections such as cabinets of curiosities—again echoed the exposition. Despite evidence of growing nostalgia in Victorian culture, there were surprisingly few historical relics on display there. Restricted to a two-dimensional spatial grid and indexed by "type of object" or "country of origin," architectural historian Bruno Giberti writes, the Centennial had "literally no space for the past. . . . [It] was a statement

FIG. 1.6 Silver inkstand specially made and engraved by Tiffany and Company for Deihm's Century Safe, to accompany two gold pens. Future officials were to use the pens and inkstand to sign Deihm's autograph album. Records of the Architect of the Capitol.

of *now*; history was an insubstantial sideshow in comparison to the great spectacle of the present condition."[18]

This emphasis on the contemporaneous, however, did not imply an indifference to the future, as some have suggested.[19] The vessels were clearly galvanized by thoughts of the bicentennial. "A century hence," Deihm's *Centennial Welcome* declared, copies of the magazine will have become a "relic memorial" of our "civilization."[20] Placing these and other items behind glass, she enabled fairgoers to view them as relics in the making. Her and Mosher's safes were not the only centennial-inspired efforts to communicate with those living a hundred years hence. During the year 1876, a lead box containing stamps, coins, and local products and deposited in a granite vault under a sapling in the town of Ramapo, New York; a copper box containing family records in a Presbyterian Church in Rahway, New Jersey; and a tin box containing a centennial album in the bank vault of Ohio's state treasury in Columbus were each addressed to celebrants of the bicentennial.[21] A year earlier, a group of North Carolinians, believing their Mecklenburg Resolves of May 20, 1775, constituted the first declaration of American independence, planned to deposit a copper box containing similar items, which would make them the true originators of the time vessel were there evidence the plan was actually carried out.[22] Such efforts were not limited to material deposits. Throughout the spring and summer of 1876, there was an extensive, nationwide campaign to collect "[historical] facts for future historians of the country," prompted by a joint resolution of Congress requesting that every town and county "assemble" its history and deliver it to their county clerk and to the Librarian of Congress, thus ensuring "a complete record . . . of the progress of our institutions during the first centennial of their existence."[23] Some even envisaged the permanent buildings of the Philadelphia exposition as a kind of cumulative time vessel to be bequeathed to their successors. One centennial commissioner proposed (unsuccessfully) that the buildings be topped up annually with subsequent inventions and discoveries so that "when our posterity comes to do honor to the close of the second century of American civilization they will find in them trophies of the [last] hundred years."[24] The centennial may not have generated much retrospection, but it certainly generated prospective gestures.

Many of these anticipations of the bicentennial involved optimistic, even utopian predictions regarding the nation's progress by that time, particularly in consolidating the allegiance of southerners. The editor of the *Washington Star-News* recorded his "day dream" about America in 1976: "Regionalism is now gone. All Americans are now one in their patriotic devotion to the Nation."[25] In *Democratic Vistas*, first published in 1871 but revamped and republished five years later to commemorate the centennial, Walt Whitman was even more specific in his prophecies:

> Long ere the second centennial arrives, there will be some forty to fifty great States, among them Canada and Cuba. . . . Much that is now un-dream'd of, we might then perhaps see establish'd, luxuriantly cropping forth, richness, vigor of letters and of artistic expression. . . . Intense and loving comradeship, the personal and passionate attachment of man to man . . . will then be fully express'd.

Democratic Vistas was a time vessel itself, as Whitman intended it "perhaps for future designers" rather than contemporary readers—an accurate prediction, as it was virtually ignored in its time and only rediscovered in the 1950s.[26]

The anticipatory acts and texts prompted by the centennial were not, however, consistently or unequivocally optimistic about the nation's future. Although the exposition has been characterized in terms of a postbellum "mood of self-congratulation" and an unshakeable confidence in American progress, there were expressions of lingering doubt.[27] On December 31, the *New York Times* speculated that future historians would describe 1876 as a "year of depression and trial."[28] During that year, US cavalry were annihilated at Little Bighorn; the presidential election failed to produce an undisputed winner; and the Grant administration, already rocked by the Whiskey Ring graft trials and Credit Mobilier scandal, confronted revelations of profiteering by Secretary of War William Belknap. Meanwhile, Southern "Redeemers" were resorting to electoral fraud, intimidation, and murder to terminate Reconstruction. The "Long Depression" also reached its nadir in the centennial year, with unemployment peaking at 14 percent. And the formation of the antimonopolist Greenback Party, the outbreak of industrial strikes, and the execution of twenty "Molly Maguires" in Pennsylvania's coal fields augured conflict between capital and labor.[29] Even the normally ebullient Whitman had voiced anxieties in certain passages of *Democratic Vistas* about what lay ahead. Apprehensive about how the new experiment in universal male suffrage would turn out, and alarmed by the corrupting and vulgarizing effects of wealth, the growing power of capitalists, and the corresponding impoverishment of laborers, Whitman concluded that "the problem of the future of America is in certain respects as dark as it is vast" and that "athwart and over the roads of our progress loom huge uncertainty, and dreadful, threatening gloom."[30]

If pundits sensed some kind of crisis—constitutional, sectional, spiritual, or social—looming in the near future, they largely remained hopeful it would be resolved by the distant future. The bicentennial thus represented a horizon of reconciliation, a utopian state in which all conflicts and contradictions will somehow have been dissolved. Whitman hinted at this deferred resolution throughout *Democratic Vistas*. In those hazy "vistas" of the future, America will have finally realized its democratic potential. In his view, democracy—in the fullest sense of the word—has never existed, even in ancient Greece, and now exists only in embryonic form, as a "crude and latent" germ in the "hearts" of ordinary, native-born Americans. The task of the critic was to extrapolate that glorious future from "this present vagueness," to apprehend that eventual state from those "certain limnings, . . . more or less fading and watery, [that] have appeared."[31]

Technologies of Temporal Communication

Although a simple device, the time vessel was in many respects an extension of some of the more advanced technologies of communication on display at the Centennial Exposition. The *Times* lamented that "no momentous discoveries, no highly-promising inventions, have been made during the year," yet several machines launched there would have revolutionary consequences.[32] Visitors to Machinery Hall witnessed for the first time the light bulb, a commercially viable typewriter, a monorail, and a player piano. Edison introduced his automatic telegraph, which transmitted messages faster than hand-operated ones, and an obscure teacher of deaf children named Alexander Graham Bell demonstrated a prototype of the "acoustic telegraph" (or telephone)—although it reportedly "attracted less notice than the packages of magic tricks on sale nearby."[33]

If the telegraph and telephone made it technically possible to communicate over ever greater distances, thereby "annihilating" space, they also made it possible to imagine new ways of bridging the chasms of time.[34] John Peters and others have explored how the telegraph, and later the telephone, inspired spiritualists (including Bell's own assistant, Thomas Watson) to summon the ghosts of the past; indeed, séances peaked in popularity during the 1870s.[35] Yet those media also aroused a desire to contact unborn generations. One contributor to a time vessel sealed in Boston in 1881 remarked that it "is a manifest improvement on the telephone, when you wish to commend a glowing sentiment to remote ages."[36]

The communication promised by the telegraph and the telephone, moreover, was one-to-one, bringing two people into a kind of remote presence. So, too, did the time vessel promise a direct, albeit one-way, communication between depositors and recipients, supposedly free from any interference or slippage of meaning. The idea of speaking to someone not physically present could breed anxiety; early users of the telephone expressed misgivings about being unable to determine their interlocutor's facial expressions or even their identity.[37] But such technologies also bred new fantasies of intimacy, of projecting one's disembodied (or "acousmatic") voice into another's domestic sphere and metaphorically touching them; indeed, a new genre of electromechanical romance immediately blossomed, including one involving a visitor to the Centennial Exposition.[38] Mosher hinted at this idea of a quasi-physical intimacy in the depiction of two generations reaching out to shake one another's hands (**fig. 1.2**). The formality of that handshake also evokes the new codes of etiquette introduced to address the risks of social impropriety provoked by such communication devices.[39] Similarly, the physical objects contained in a time vessel countered the dematerialization of sentiments into electrical impulses, while the slowness of a hundred-year transmission countered the instantaneity of wired communication.

The communications revolution, however, was not driven entirely by technological invention. The introduction of the prepurchased US postage stamp in 1847, the manufactured envelope in 1849, and the public mailbox in 1858 along with innovations in the collecting, sorting, and delivering of mail enabled Americans to imagine letters, too, as private, one-to-one communications.[40] A genealogical link can be traced from the mailbox to the centennial vessel; both were nontechnological or low-tech devices (constructed out of heavy cast-iron to deter vandals or thieves) that expanded the horizons of communication. Indeed, the creator of the Rahway time vessel adopted the language of the mails, inviting citizens to submit anything they wished to send to the future at the postal rate of "a dollar for each half ounce." Three years later, a vessel in San Francisco was advertised as a "great Antiquarian Postoffice."[41]

Posing for Posterity, or "Shadows of the Living"

The technology that most inspired the time vessel, however, was one of inscription rather than transmission. Photography, still relatively new, startled contemporaries not only with its accuracy, detail, and apparent objectivity but also with its capacity to preserve its subjects by bearing their physical imprint—the light reflecting off their body—on the surface of the image, rendering that image an indexical sign, to use the contemporaneous philosopher Charles Sanders Peirce's term.[42] Photography's indexicality appeared to grant the sitter a kind of afterlife, thus triggering associations with resurrection or time travel. Through the photographs in his safe, Mosher proclaimed, the "living of to-day" will cast "shadows" over the celebrations of 1976, thus transcending their bodies' temporal constraints.[43]

Photography was not the first medium to perpetuate a direct trace of the human body. It grew out of earlier efforts, including the physionotrace, a machine invented in 1783–1784 to trace an individual's silhouette; taxidermy, which Charles Willson Peale imagined as a means to "hand down to succeeding generations, the relicks of . . . great men"; wax or plaster casts of the faces (or bodies) of the dead (death masks) or living (life masks), a procedure applied to politicians and generals such as Lincoln and Grant and to literary celebrities such as Whitman; and embalming, an ancient art revived during the Civil War to allow the dead to be transported to their home towns for burial and improved by the invention of formaldehyde in 1867.[44] Victorian America was in many respects a culture obsessed with preservation, continually seeking new means—mechanical, chemical, or manual—to arrest the decay of the human body, or indeed that of animals and food (Martha Maxwell's modern taxidermy techniques and an automated canning machine were also exhibited at the Philadelphia exposition).[45] Bodily traces could be preserved even without advanced techniques. Objects like the clothing or material possessions of historic figures were believed to bear their "impress" (albeit an invisible one) and were collected as relics precisely for that sense of material persistence from, and thus connection to, the past.[46] Even historic houses (such as those of the founding fathers) were by midcentury perceived as portals that could transport the visitors through time, and thus worthy of preservation.[47]

So, too, was something as uncomplicated and trivial as a signature considered a kind of imprint left behind by the body—perhaps the reason Mosher and Deihm included them in their vessels. Deihm collected the signature of every member of all three branches of government and of "eminent men" from every state, along with "protographs" (whole documents handwritten by their author, now termed *holographs*). Mosher compiled an equally impressive collection of signatures, mounted beneath each portrait in the albums.[48] More than simply captions, they were to contribute to the extension of the individuals' physical presence beyond the present. The indexicality of the autograph called attention to—and supplemented—that of the photograph and vice versa.

But autographs and photographs did more than embalm the body. Despite (or perhaps because of) the standardization and regimentation of handwriting instruction by the 1870s, autographs were deemed distinctive, valuable, and collectible substantiations of notable individuals' moral and intellectual qualities. Those signed knowingly for collections such as Deihm's and Mosher's were considered somewhat less revelatory and valuable than those scribbled spontaneously, but they still enabled the beholder to enter into an "almost mystical encounter" with the illustrious.[49] Photographs, too, were believed to extract and preserve the essence of an individual's character and to provide direct, unmediated access to his (or, less often, her) innermost soul. "By the early 1850s," writes the historian of photography Barbara McCandless, "the standard for a truly accurate likeness had become not merely to reproduce the subject's physical characteristics but to express the inner character as well."[50] The introduction of cabinet cards in 1866—the format both Mosher and Deihm employed—permitted even "greater attention to detail and expression of character," as they were significantly larger (at 4 × 5½") than the older cartes de visite.[51] And if graphology (popularized by Edgar Allan Poe a generation earlier) claimed to deduce character and class from the idiosyncrasies of an autograph, phrenology and physiognomy looked to the shape of the head and the facial features (or posture), often as represented in photographs.[52]

Drawing on such beliefs about photographs' (and autographs') expression of character, Mosher and Deihm merely extrapolated them into the future. Photographs of eminent individuals could not only inspire contemporary viewers across the globe but also "transmi[t]," Mosher wrote, a "manly face and noble character . . . to posterity."[53] He devised plans for his photographs to be removed from the safe in 1976 and displayed in glass cabinets in a vast "memorial gallery" or "temple" he proposed for Chicago's downtown lakefront just south of where the Art Institute now stands. The structure would "outdo the grand Memorial Hall at the Centennial, not only in its architecture and elegance, but in the heaven honored purposes for which it will be dedicated and sustained."[54] Visitors during (and after) the bicentennial would be able to perceive those men's "very thoughts" from these "speaking likenesses" and learn about their lives from the accompanying biographies, thereby imbibing their virtues.[55] Mosher went so far as to claim that pregnant mothers, by "studying intently" his portraits, would be able to "photograph" their subjects' physical, moral, and intellectual features onto their fetus's "face and brains," foreshadowing recent ideas about prenatal imprinting or in-womb education.[56] His vision of a future museum of photography was ahead of its time. Through the end of the nineteenth century, photographs—although displayed at world's fairs and in private clubs—were used by public museums merely to document aesthetic, biological, and anthropological specimens or to provide contextual backdrops to exhibits. They did not yet appear there as exhibition objects or art works in their own right.[57]

At the same time, Mosher's prospective "memorial gallery" grew out of the commercial photographic galleries of his own period, the most famous of which was Mathew Brady's. Opening in 1844, Brady's Daguerrean Miniature Gallery quickly became a leading attraction of

New York. His collection of portraits of the nation's leading figures, begun the following year and numbering about ten thousand by 1861, anticipated—and possibly inspired—Mosher and Deihm in several ways.[58] It comprised one of the first efforts to photograph every member of Congress and the administration, an undertaking that led him to open a branch in Washington; indeed, Brady later supplied the photographs for Deihm's vessel.[59] He evaded the charge of undemocratic hero-worship by presenting his gallery as a kind of neoclassical temple inculcating "the civic virtues required by the Republic," in the words of one contemporary.[60] Its depiction of politicians from both parties and both sides of the Mason-Dixon line simultaneously dodged the charge of political or sectional partisanship. The gallery grew, moreover, to embrace not just politicians and generals but also businessmen, intellectuals, artists, reformers, clergymen, and even theater stars, along with the occasional European celebrity.[61] While Brady was not averse to copying portraits taken by other photographers, he mainly accumulated his collection by inviting dignitaries and celebrities to his gallery to have their portraits taken, usually by an assistant—a model that Mosher copied.[62] Concealing those photographic processes in a back studio, Brady transformed the front rooms into a quasi-domestic space where visitors could scrutinize the portraits and socialize with one another unencumbered by the hazards and gender (im)proprieties of the street.[63] Mosher similarly transformed his studio on Chicago's State Street into an "elegantly appointed gallery" with multiple, salon-like rooms exhibiting life-size portraits of presidents and other leading figures. "Finished in the latest style of architecture and adorned with the copies of thousands of pictures of eminent men and women," enthused one visitor, "it is simply a paradise."[64]

Brady further paved the way for the time vessel by raising ideas about the future utility of photographic portraits. While evidently driven by short-term profit motives, he presented his collection ultimately as a gift to posterity. This was particularly apparent in the *Gallery of Illustrious Americans* (1850), Brady's attempt to render his gallery as a book, issued to subscribers in semimonthly parts, containing lithographic reproductions, autographs, and biographical sketches. "We wish before those great men . . . are gone," the introduction announced, "to catch their departing forms, that through this monument of their genius and patriotism, they may become familiar to those who they will never see."[65] However self-serving, Brady's pose as a noble patriot "endeavor[ing] to anticipate the awards of posterity" was enthusiastically endorsed by the press, with one newspaper stating that the *Gallery* "will furnish a monument of art and patriotism for coming times."[66]

If Brady's collection was imagined as a gift to posterity, it was not necessarily for the benefit of historians. Photography was not considered a legitimate historical source until well into the twentieth century.[67] Certainly, commentators viewed his photographs as "materials for history," one that made them long for "such a portrait gallery of the revolutionary days."[68] And they celebrated photographs as more revealing than "all our books, all our newspapers, all our private letters,— though they are all to be weighed yearly by the ton."[69] Indeed, Walt Whitman claimed to have inspired Brady's project by suggesting in the 1840s how just "three or four" portraits of Roman emperors or philosophers would trump the mass of "contradictory records" left "by witnesses or historians": "that would be history—the best history—a history from which there could be no appeal."[70] Yet in all these instances, the photograph's memorial capacity is imagined as rendering written history redundant. Rather than a source for future historians, photography *becomes* history. As expressed by the Scottish optical scientist David Brewster in 1856, "the sun will thus become the historiographer of the future. . . . Truth itself will be embalmed and history cease to be fabulous."[71]

Rhetoric aside, Brady in fact did very little to ensure the transmission of his photographs. He made no plans to set aside or donate any of his photographs during his lifetime. High costs, poor sales, and a shifting political climate led him to abandon the final twelve issues of his *Gallery*. Facing mounting financial difficulties in the late 1860s stemming largely from the expense of covering the war, he tried unsuccessfully to sell his vast collection, first to the New-York Historical Society and then to Congress.[72] After he was declared bankrupt in 1873, his negatives were dispersed—some claimed by creditors, others acquired cheaply by the War Department, while many were damaged, destroyed, or lost. Subsequent transfers of ownership during the ensuing decades further compromised the integrity of Brady's corpus.[73] His dreams of bequeathing a great gift to the nation were thus deflated by the realities of the capitalist marketplace. The lesson for Deihm and Mosher would have been clear: only by taking images out of circulation could one be confident of preserving them for posterity.

The time vessel's genealogy includes not just commercial galleries like Brady's but also private modes of photographic display. Relishing the medium's apparent power to consecrate friendships, catch the fleeting moments of childhood, and embalm the deceased, Victorian Americans self-consciously gathered and curated their own collections. They viewed those photographs through a variety of frames: actual picture frames, jewelry, leather cases (some colored black with embossed funerary design for postmortem photographs), and albums. Mosher's and Deihm's photographic albums were merely enlarged versions of those kept by middle-class Americans—especially women, the appointed guardians of memory—as a symbol of bourgeois status and domestic harmony to be displayed on the parlor's center table alongside the family bible. Introduced in 1861, the photograph album evolved out of earlier kinds of blank books, from the commonplace book and friendship album of the early modern era to the autograph albums and keepsake or "memory" albums of the nineteenth century. Each of these promised to freeze the ephemeral, whether a saying, a lock of hair, or a pressed flower. Although cabinet cards were originally intended to be displayed on a drawing-room cabinet (hence the name) and did not fit the older albums designed for cartes de visite, special albums were soon introduced to accommodate them, with as many as twenty patented by 1873.[74]

Mosher himself explicitly advocated the collecting of cabinet cards in such albums.[75] He encouraged customers to fill them with both family pictures and celebrity portraits—dedicated family albums being a twentieth-century innovation.[76] He also sought to correct flaws in their design. To prevent the "great annoyance" of photographs being "frequently removed without authority," perhaps by domestic servants or children, he invented and patented his own photograph album, which "secur[ed] the pictures in place" while providing a slot for each subject's name or autograph.[77] With the further introduction of hasps and miniature padlocks (**fig. 1.3**), such albums anticipated the hermeticism of the time vessel. Both were arguably responses to the ongoing problem of photographs straying beyond their original context and thus losing meaning and value (a similar problem arose with signatures, which collectors excised from letters and documents). Unlike lockets and other forms of jewelry that allowed photographs to circulate in public, albums were designed to keep photographs in place. Mosher's and Deihm's projects thus grew out of, and in turn promoted, everyday practices of collecting and storing photographs.

Withdrawal from circulation may have been a response to another problem specific to photographs: their tendency to fade or stain when exposed to light, humidity, atmospheric pollutants, or oil from viewers' fingers. This instability had been a known issue from photography's inception, always threatening to negate its reputation for miraculous preservation. It is photography's "one fatal drawback," observed the *Illustrated London News* in 1859; "what the sun gives, the sun will take away."[78] One could narrate the medium's technical history as a quest not just for greater detail, quicker exposure times, and increased affordability, portability, and usability but also for greater permanence of the final, printed photograph. In 1855, the Photographic Society of London appointed a special committee to determine the causes of fading and thus restore public confidence in photographs' longevity. Salt prints of the 1840s and 1850s were especially notorious for rapid and almost total fading—one reason they failed to supplant the daguerreotype as a process for portraiture.[79] Albumen prints, popularized in the late 1850s and employed by Mosher and Deihm, were more durable yet still susceptible to fading, staining, and yellowing, ultimately contributing to their own obsolescence by the 1890s. In 1866, a leading US photographic scientist detected unavoidable and destructive traces of silver in albumen prints. Experts therefore recommended they be stored in a dry, secluded place insulated from air, light, and touch.[80] A time vessel offered such insulation, albeit for an extreme length of time.

The depositing of photographs in bank safes was, at the same time, a reaction to the implications of technological reproduction. The invention of photography did not, of course, immediately strip images of their aura. Examining the earliest daguerreotypes, Walter Benjamin perceived a lingering, auratic quality, which he defined not simply as uniqueness but as a "semblance of distance," a distance that is temporal as well as spatial. Although that generation had not been "obsessed with going down to posterity in photographs," instead withdrawing "rather shyly . . . into their private space in the face of such proceedings," their daguerreotype portraits conferred on them an "air of permanence" and a sense of "fullness and security" right down to the creases in their coats. It was as if the length of the exposure enabled the haut-bourgeois male subject to "focus his life in the moment" and "[grow] into the picture." This aura was soon displaced not just by the ability to produce multiple copies but also by the introduction of faster exposure times and smaller cameras, which could catch a fleeting incident or a casual pose, and by the "industrialization," or mass commodification, of photography.[81] (Autographs lost some of their aura, too, with the introduction in the 1860s of rubber-stamp signatures.)[82] But instead of embracing the implications of these new technological developments, Benjamin argues, studio photographers continued to "simulate" the lost aura—by retouching the negative or paper (as

Mosher did); by adopting printing processes and techniques that imparted a "penumbral tone" or "artificial highlights"; by dressing sitters in evocative clothes and encircling them with draperies, palm trees, or even classical columns; or by mounting the pictures in leather-bound, gilt-edged albums.[83] By the 1860s, art photographers, too, were summoning up the lost aura by using soft focus, combining multiple negatives, and creating other pictorialist effects.

This compensatory desire for faux aura was arguably what prompted photographers like Mosher to encourage their sitters to self-consciously pose for posterity. In addition to advertising how he manipulated lighting to produce a "delicate modeling of the features" and how his artists

FIG. 1.7 Charles D. Mosher, *Mrs. Elizabeth Cady Stanton Woman['']s Rights*. Albumen print on cabinet card, Chicago, n.d. Chicago History Museum.

retouched the negative to erase "unsightly freckles and blemishes of the skin," he advised them how to present themselves to posterity's gaze, suggesting certain clothing and poses.[84] His sitters appear to have internalized that future gaze. Suffragist Elizabeth Cady Stanton remarked in a letter to Mosher that she was less happy with his "side face" portraits of her because they revealed her "double chin" and were "too animal looking, to go down to future generations" (**fig. 1.7**).[85] Moreover, while his and Deihm's photographs were printed from glass plate negatives, sealing them away for a hundred years as though they were one-of-a-kind relics granted them a compensatory "cultic value." Locked up in a safe, the collection eluded what Benjamin described as the impulse of the "masses" to "possess the object in close-up in the form of a picture, or rather a copy."[86]

Rhetorics of Posterity and the Perishability of the Printed Word

The discourse of posterity was not confined to photographic galleries and time vessels. Mosher's self-appointment as "National Historical Photographer to Posterity" and Deihm's statement that her vessel's raison d'être was "principally" to relay the pantheon of distinguished Americans "to posterity" aligned them with other future-oriented enterprises.[87] In the 1870s, New York, Boston, and Philadelphia founded their first municipal art museums, justifying them partly on the grounds of social and cultural benefits to future generations. Such rhetoric was also mobilized by the founders of the first public library in Boston in 1852, the Chicago Public Library in 1872, and the American Library Association, launched in Philadelphia in 1876; by advocates of architectural and historic preservation, a movement that finally coalesced in the 1850s with a national campaign to rescue George Washington's house at Mount Vernon for the enjoyment of "future generations"; and by proponents of municipal and national parks such as Frederick Law Olmsted, who urged the California legislature in 1865 to respect the "rights of posterity" to inherit a pristine Yosemite.[88]

That environmental ethic of posterity extended to the planting of new trees. In *Man and Nature* (1864), George Perkins Marsh urged Americans to protect their forests *and* to replant those they had destroyed, not simply to conserve resources for themselves but as an act of "magnanimity," a "self-forgetting care for the moral and material interests of our . . . posterity."[89] Tree planting, which Marsh considered inherently altruistic, as trees (especially oaks) grow too slowly to profit their planters, now emerged as a ritual of devotion to posterity. Arbor Day, first celebrated in Nebraska in 1872, soon spread across the nation as a means to restore trees not just to deforested countryside but also to treeless cities. During the centennial, the planting of a tree became a popular ceremony in numerous towns. Like the depositing of a time vessel, it constituted—in the words of one proponent—a collective gift to our "offspring a century hence," and thus it was a "fitting tribute" to the selflessness of patriots a century earlier.[90]

A growing sense of duty to posterity was even evident in the great infrastructural projects of mid-nineteenth-century cities, which were conceived as civic monuments as well as fixed capital assets. In the "formative period" of the commercial city, there had been "relatively little investment in durable structures," according to Lewis Mumford. "Only in the nineteenth century" did British cities "produce the great succession of warehouses and docks."[91] David Lowenthal cites such undertakings to identify the Victorian era as the apogee of a concern for posterity—or what he terms "stewardship of the future"—that first emerged in the Enlightenment.[92] American cities, too, witnessed a dramatic investment in sewers, reservoirs, tunnels, bridges, public parks, and railroad stations rendered in the enduring forms of the classical, Gothic, or Egyptian—although in New York it was Boss Tweed's corrupt siphoning of money into infrastructure that led to the choice of expensive, durable materials, such as vitrified clay pipes rather than bricks for sewers.[93]

The concept of posterity as the collective descendants of a city, nation, or humanity certainly existed earlier, even in classical antiquity, but it had little significance as a motivating force. Not until the emergence of a progressive temporality during the Enlightenment, argued Carl Becker, could "this pregnant idea . . . play its part in the world." Alienated from their own present, philosophes such as Denis Diderot began to invoke posterity as the supreme "judge and justifier" of their convictions, as a transcendent source of inspiration ("the sole encouragement, the sole support, the sole consolation"), and as their ultimate duty and responsibility. Thus, despite the philosophes' repudiation of superstition, they erected posterity into a substitute religion, "personified, reverentially addressed as a divinity, and invoked in the accents of prayer." In Becker's view, only by co-opting and secularizing these Christian notions of God and the afterlife could they hope to appeal to the "common run of men."[94]

This cult of posterity proved crucial to the legitimation of subsequent political and cultural revolutions. American patriots invoked it to justify their opposition to and ultimate break with Britain; indeed, the Declaration of Independence was implicitly addressed not just to George III, to potential supporters at home and abroad, and to humanity in general ("the opinions of mankind"), but also to future generations, who would come to their own judgments. As well as a judge, posterity figured as the beneficiary of the revolutionaries' sacrifices and actions. The Virginia Declaration of Rights affirmed the inalienable rights not just of "all men" but also of "their posterity," while the drafting of a durable American constitution would, its preamble famously promised, "secure the Blessings of Liberty to ourselves and our Posterity."[95] The leaders of the revolution were arguably propelled by a desire for enduring fame generated by the revolution itself. They derived their notions of posterity from a classical tradition revived by Renaissance and republican political theorists according to which this desire, far from being a dangerous impulse, could spur virtuous acts and transmute self-interest into public service.[96] By the Romantic era, posterity had become the object of an aesthetic as well as a political cult. Poets such as Coleridge and Wordsworth saw themselves as writing for a posthumous audience and conceived literary genius as something that could only be recognized belatedly.[97]

But if time vessels grew out of this blossoming cult of posterity, they were also signs of its withering. In their very efforts to ensure posterity's freedoms, American revolutionaries had paradoxically weakened intergenerational ties. One target of revolutionary fervor was the power to transfer a financial legacy (or debt) across multiple generations. New state constitutions abolished "perpetuities," that is, efforts to control property for long periods beyond one's own lifetime through legal instruments such as wills, contracts, and entails. They also abolished the feudal law of primogeniture, which had enabled families to keep their estates whole through time.[98] The right to pass on political office was similarly dissolved, most notably by the US Constitution's Emoluments Clause, which (among other things) forbade the inheritance of titles or ranks. The outcome of these abolitions, Alexis de Tocqueville later argued, was a weakening of the family as an agent of historical continuity or self-perpetuation, and consequently a shrinking of temporal horizons: "[When] the family is felt to be a vague, indeterminate, uncertain conception, each man concentrates on his immediate convenience; he thinks about getting the next generation established in life, but nothing further."[99] Jefferson even extended this principle to laws and constitutions, arguing that they themselves should expire after nineteen years, because no generation has the right to bind its successors—"the earth belongs always to the living generation."[100] The revolutions of the late eighteenth century and the wars that followed them (including the American Civil War) disrupted assumptions of historical as well as familial and legal continuity. As unprecedented moments of rupture, they rendered time discontinuous. The present was thereafter perceived as caught between an unrecoverable past and an unknowable future. Cut adrift, moderns would need surrogate means to retether themselves to predecessors and successors.[101]

The Vicissitudes of Print

By the 1870s, doubts were also growing about modern civilization's ability to transmit its textual legacy. Paradoxically, the triumph of print appeared to pose an obstacle to the survival of its most treasured ideas, documents, and names. Although some prominent Americans, from Thomas Jefferson to Wendell Phillips, had viewed the printing press, with its capacity to democratize and diffuse knowledge across society, as a prophylactic against cultural oblivion, others expressed concerns about the medium on which it depended.[102] In 1858, the *National Magazine*, while acknowledging the "merits" of paper, emphasized its "defects" compared with earlier media. "It is not strong enough or durable enough for important legal documents; and its fragility renders it incapable of bearing the wear and tear of the school-room or lending library."[103] A decade later, its durability was further undermined by the introduction in the United States of a mechanized process for manufacturing paper out of wood pulp rather than plant-based textiles such as hemp, linen, and cotton.[104]

Mosher's and Deihm's schemes might be viewed, then, as responses to the ascendancy of what the pioneering communications scholar Harold Innis called space-binding media (in this case, paper based, but subsequently electronic) over time-binding media such as clay, metal, and parchment. The latter, heavier and more durable, favor the transmission of information across time and were thus employed by monasteries and priesthoods to perpetuate monopolies of knowledge. The former, with their lightness and circulability, excel instead at disseminating information across space and thus facilitate the expansion and administration of empires. Civilizations become unstable, Innis argued, when they begin to exhibit an imbalance. An excessive bias toward time gave rise to the hierarchical "rigidities" of ancient civilizations, and an equally excessive bias toward space had, by the nineteenth century, begun to induce "present-mindedness" in Anglo-American civilization.[105] Time vessels may have been unconsciously intended as a counterweight to that spatial bias.

Indeed, paper's fragility appears to have prompted an interest in alternative materials of inscription. Mosher drew on innovations such as indelible ink and parchment paper (the latter invented in 1857 by immersing paper in sulfuric acid) and modern preservatives such as powdered charcoal, which he used to encase his thirty-five packages of photographs. Thus, wrote the *Chicago Inter-Ocean*, would his documents "be permanently preserved for the historians of the future."[106] But the centennial vessels also reverted to older media (characterized by Innis as time-binding), such as the parchment scroll Deihm had inscribed by every member of Congress.[107] They thus anticipated the revival of premodern materials by the "private presses" (such as William Morris's Kelmscott Press and Elbert Hubbard's Roycroft Press) that grew out of the transatlantic Arts and Crafts movement of the 1890s, but without the fetishism of the archaic that characterized those antimodernists. This mingling of emergent and residual media would become a recurrent theme in the time vessels of the Progressive Era.

A civilization's capacity to transmit itself through time, however, is not determined entirely by its media but has much to do with the social practices and institutions governing them. The libraries that housed books and other paper-based documents were, it was clear by the 1870s, themselves vulnerable to the conflagrations that periodically ravaged American cities. The Great Fire of 1871 destroyed, among other things, the Chicago Historical Society, incinerating all one hundred thousand books, pamphlets, and manuscripts, including a copy of the Emancipation Proclamation handwritten by Lincoln. Its collection was partially restored through book donations from other cities and countries only to be devastated by another fire in 1874.[108] Mosher, whose first gallery and studio succumbed to the 1871 fire and whose new studio narrowly survived another a decade later, cited other hazards, such as the polluted air of industrial cities, which prematurely ages and discolors paper documents. Librarians were becoming increasingly alert to further perils, including fluctuations in temperature and humidity, insects, pests, mold, and flooding, not to mention theft and negligence.[109]

If documents entrusted to public institutions were at risk, those in private hands appeared even less likely to reach posterity, as the dispersal and near loss of Mathew Brady's collection underscored. Despite

Oliver Wendell Holmes Sr.'s 1859 prophesy of "vast" libraries of stereographic photographs, photographs continued to be neglected as a resource to be accumulated and preserved.[110] So were government public records. The lack of a national archive (or even a centralized inventory of privately held historical manuscripts) in the United States was a major concern by the Gilded Age, especially to the growing ranks of professional historians who embraced Leopold von Ranke's emphasis on primary sources.[111] While France's Archives nationales was founded as early as 1790 and Britain's Public Records Office by 1838, efforts to establish a similar repository in the United States—initiated in 1810 and revived by President Rutherford Hayes and others in the 1870s—consistently failed. Only in 1934 did Congress create the National Archives (state archives were almost as belated, emerging only by 1901).[112] Until then, the written records of all branches and agencies of federal government—records which proliferated exponentially with the expansion of government during and after the Civil War—were liable to be lost, discarded, or, given the two hundred and fifty-four fires that ravaged federal offices between 1833 and 1915, reduced to ash.[113]

Ironically, while little was done to preserve government records, there were major advances in the safeguarding of financial instruments and documents. Invented in New York in the 1830s for banks and businesses, the safe quickly superseded the simple oak strongboxes and chests that had been imported from Europe. To resist fire, its inventors nested an iron box inside a larger one and filled the intervening space with nonconductive materials (initially gypsum). To thwart burglars, they reinforced it with a harder metal and introduced complex locks.[114] By 1876, Mosher and Deihm could acquire a safe equipped with improved fire-retardant, rustproof, and damp-resistant materials, such as alum and dry plaster filling, and with the combination lock, invented the previous year by the appropriately named Joseph Loch.[115] Their safes' ornateness may also have been influenced by new models that were disguised as parlor furniture for domestic use. In fact, the Victorians were fascinated with all kinds of lockable receptacles—keepsake and relic boxes, bridal chests, writing boxes, cabinets or secretary desks—which they (especially women) used to sequester their diaries, letters, family heirlooms, and other memory objects, often in nested inner compartments. Not merely containers, both safes and private repositories must be considered media of communication in their own rights, their heaviness and durability serving to offset the spatial bias of a paper-based culture.

Deihm and Mosher may also have sealed their materials for a century so that they would ultimately prevail over all other contemporaneous traces. If personal souvenirs accrue mnemonic force through being locked in a bureau, so might public documents attract attention by virtue of having been unseen for a century. The expansion of print (and photographic) culture had generated concerns not only about the ephemerality of paper-based materials but also, conversely, their superabundance. "The materials of historical and literary research have accumulated so rapidly," warned *Harper's Weekly* in its inaugural issue, "they have fairly outgrown the reach of individual acquisition or even knowledge," giving rise to journals such as itself that "digested" the plethora of books for busy readers but also for the potentially overwhelmed "future historian."[116] The centennial vessels performed a similar duty in preselecting a limited number of documents from the deluge of contemporary materials. They offered themselves as a kind of filter against what has come to be called information overload.

Depositing documents in a container does not by itself guarantee their endurance. It could just as easily lead them to become irrecoverable, as happened to the cornerstone George Washington laid in 1793 for the US Capitol, which still remains unlocated despite having been adorned with an engraved silver plate.[117] The time vessel, however, emerged as a potential solution to the limitations of the cornerstone. Whereas the latter could be obscured by a subsequent extension to the building (as may have occurred with the Capitol), the former could remain fully visible in prominent interior spaces, thus persisting in public memory. Mosher planned to install his in the lobby of Chicago's new city hall and county courthouse, a plan the mayor eventually approved.[118] Even more boldly, Deihm chose Statuary Hall, the circular chamber in the US Capitol in which the House of Representatives had assembled for half a century and that had been repurposed as a hall of fame a decade earlier. She envisaged the safe functioning there as a permanent tourist attraction, with generations of visitors coming to gaze on the sacred relics behind the glass inner door.[119] The ability to open (or partially open) a safe was a further advantage over the cornerstone. Through Deihm's biennial topping-up ceremonies and Mosher's quarter-century ceremonies, future publics would be reminded of the safes' existence (**fig. 1.2**).[120] (While the term *time capsule* today connotes a container that is completely sealed and concealed until its target date, there was no such expectation in its early years.) The locking of the vessel also offered an opportunity to conscript future officials in the task of perpetuating its memory. Mosher entrusted the combination to the mayor of Chicago with instructions to transfer it to his successors, while Deihm intended to consign her spare key (along with a duplicate of the instructions for its opening) to the Smithsonian.[121] She also designated her heirs—or, "in case I should not have any," those of two other families—as key holders.[122]

These arrangements served the additional function of weaving the time vessel into the fabric of lived memory. As an artificial container for removing memorial materials from public access, the time vessel would appear to eschew the need for embodied, living acts of remembering such as social rituals of mourning. It would thus typify what Bernard Stiegler perceives as a growing dissociation between hypomnesis (memories displaced onto external things) and anamnesis (the direct process of recollection)—a trend that began in the nineteenth century with the appearance of photography and other "mnemotechnologies" (supraindividual industrialized technologies that produce and organize memories).[123] The time vessel is, in some respects, an agent and symptom of that reformulation of memory. Yet in devising periodic, commemorative rituals for their vessels, Mosher and Deihm retained a vital connection between inert, archival memory and active, ongoing practices of memory.

The City as Flawed Vessel for Transtemporal Communication

The Gilded Age crisis of posterity extended beyond written texts to buildings. Expanding the term *medium*, as Peters proposes, would allow us to consider how the built environment communicates (or fails to communicate) across time. Indeed, Innis viewed the pyramids not just as expressions of monarchical "control over time" but also as (disastrously expensive) bearers of hieroglyphs.[124] Mumford went further, describing the city itself as having the key function, since antiquity, of "enlarging the boundaries of the community in time" as well as space. The city, which emerged as a "self-contained unit" at the same time that writing emerged as a means of long-term, symbolic storage, has served as a stable "container" both of texts and of "durable buildings and institutional structures," and thus it binds "times past, times present, and times to come." Mumford evokes the time vessel in his conception of the city as "a special receptacle for storing and transmitting messages."[125] By the 1870s, however, confidence in American cities' capacity to perform this critical function had eroded. What kinds of architectural and archaeological remains, contemporaries began to ask, would their built environments bequeath to future generations? What would cities, conceived as media, communicate to posterity?

One perceived challenge to the city's mnemonic function was the phenomenon of demolition, in particular its increasing scale and frequency. The effects of "creative destruction" are often considered a concern that arose around 1900, when patricians in New York began to work with architects and urban planners to protect certain historic landmarks and elite neighborhoods.[126] But such concerns were already evident by the time of the centennial, evoked, for instance, by William Dean Howells's account in *Their Wedding Journey* (1872) of "the eternal building up and pulling down" in New York.[127] A preservationist impulse can be detected even in the antebellum period in the complaints of the philosopher Henry P. Tappan about merchant's "palaces" being "turned into boarding-houses, then pulled down and replaced by warehouses"; the retired cartman Isaac Lyon, for whom the "onward march of the spirit of gain" was annihilating "all the ancient and time-honored landmarks of the city"; the ex-mayor Philip Hone, who bemoaned the loss of both the old Trinity Church and his own "poor, dear house" at 235 Broadway; and even Walt Whitman himself, who, in one of his earliest articles, decried "the pull-down-and-build-over-again-spirit," that impulse to "level to the earth all the houses . . . not built within the last ten years," and predicted that future historians will note his generation's restless indifference to its architectural heritage.[128] Demolition was decried not just because it prevented contemporaries from enjoying the transtemporal experience that old buildings (even private ones) afforded but also because it denied that enjoyment to future generations. These complaints extended even to funerary architecture, as cemeteries were increasingly removed in New York and other cities to make way for commercial properties or to widen streets. Loss of confidence in the inviolability of urban burial sites prompted not only the rise of the rural cemetery movement (with more elaborate grave markers and improved upkeep) but also, perhaps, an interest in alternative forms of memorialization such as the time vessel.[129] Indeed, the autograph albums Deihm and Mosher adopted echo the memorial registers that cemeteries employed to record the names of the interred.

Obliteration of older structures was not the only issue. Cities were also failing to communicate with posterity through newer structures. The capitalist logic of real estate, with its cycles of demolition and reconstruction, was believed to be downgrading architecture to a short-term, provisional investment. Novelist and conservative critic James Fenimore Cooper came to this realization as early as 1837, while sitting on the tallest of Rome's seven hills. Whereas the city below him was one of "palaces, monuments, and churches, that have already resisted centuries," he mused, New York was one of "architectural expedients, that die off in their generations, like men." A mere two hundred years, especially given our "climate," would "probably" suffice to "obliterate every mark of our possession [of our land]. We have a few forts, and a sea wall or two, that might resist the wear of a few centuries; but New York would not leave a trace, beyond imperishable fragments of stone." Cooper attributed this failure to "buil[d] for posterity," as the ancients had, to a larger temporal condition: driven by their "estee[m]" for "money," Americans, especially New Yorkers, live in a state of flux, without any ties to tradition or place. They are "so eager for the present as to compress the past and the future into the day."[130]

By the Gilded Age, the ephemerality of the modern built environment found perhaps its ultimate expression in world's fairs. Pavilions were typically constructed on a temporary basis on shallow foundations and with cheap materials such as wooden frames or laths covered with a thin veneer of sculpted plaster or staff (fiber-reinforced plaster).[131] These faux-marble, neoclassical structures would be dismantled as soon as the fair closed, or worse—in the case of the Columbian Exposition—plundered by "relic hunting vandals," allowed to fall to ruin, and then consumed by fire.[132] Of the more than 180 buildings erected for the Philadelphia exposition, only two major structures were intended as permanent additions to Fairmount Park and therefore built out of brick, stone, and iron: Horticultural Hall and the Art Gallery (subsequently renamed Memorial Hall), in which Deihm's vessel was exhibited. Within five years, little else remained as the site reverted to parkland.[133]

If American cities were failing to accrue permanent structures, they were also found lacking in monuments—the most obvious medium for long-term messages, as historians of antiquity were already recognizing. Even after launching major public monuments in the second quarter of the nineteenth century, the United States lagged behind Britain and France as a result of its reliance on private capital rather than state funds. Tappan warned of the historical oblivion that would result from the prioritizing of individual profit over collective works: "Were New York now to experience the fate of Athens, or Rome, or Venice, she would leave to the world no memorials whatever."[134] Even more troubling were the numerous monuments launched in a fit of enthusiasm but then suspended as various obstacles—public indifference toward the past, republican ambivalence about deifying leaders, aesthetic controversies, financial depressions, and civil war—hampered fundraising. The Bunker Hill monument, the Washington monuments in Richmond and the capital, the Battle of New Orleans monument, the Forefathers monument in Plymouth, and the Confederate monument in Raleigh, North Carolina, all remained unfinished for years or even decades. Others were aborted altogether, such as the Washington monuments in Philadelphia and New York, the Mary Washington monument in Fredericksburg, and the Jefferson Davis monument in Richmond. The "ruins-in-reverse" of those half-completed structures were read as signs of a deep disregard not only for history but also for posterity. In the letter she deposited in her safe, Deihm herself denounced the "disgrace" of the Washington Monument, whose stump was embarrassingly visible to foreign visitors during the centennial.[135]

The failure to complete such monuments, moreover, would render the inscriptions and relics embedded in their cornerstones incomprehensible to future generations, or so Edgar Allan Poe implied in "Mellonta Tauta" (1850), a title he derived from the Greek for "things of the future." Presuming (correctly) that the Washington monument on Manhattan's upper east side would rise no higher than its cornerstone, Poe speculated about the fate of that granite block. His narrator, writing aboard an air balloon, describes its accidental rediscovery in AD 2848, which raised archaeologists' hopes that its contents—"a leaden box filled with various coins, a long scroll of names . . . [and] newspapers"—and especially its exterior inscription would finally shed light on the "manners" and "customs" of the "Knickerbocker tribe of savages" that had supposedly inhabited the island of Paradise, now the emperor's pleasure garden. Yet like so many encrypted texts (hieroglyphs, cryptograms) in Poe's works, the inscription resists deciphering. The narrator interprets the words, "surrender of Lord Cornwallis to General Washington," as a reference to the sacrifice of some "wealthy dealer in corn" to a tribe of cannibals, presumably to be turned into "sausage." Nonetheless, she unwittingly arrives at the virtual truth that "a thousand years ago actual monuments had fallen into disuse . . . the people contenting themselves . . . with a mere indication of the design to erect a monument at some future time; a corner-stone being cautiously laid by itself . . . as a guarantee of the magnanimous intention." Thus, the Washington Monument Association (also referred to in the inscription) must have been merely "a charitable institution for the depositing of corner-stones." The cornerstone's illegibility stems partly from the reliance on second- or thirdhand information; its inscription had been transcribed, published, translated, and then copied into this future letter, which (like the message in "M.S. Found in a Bottle") is corked up and jettisoned from a doomed vessel, then rediscovered by Poe and translated back into English by his friend. But the illegibility also stems from the absence of a completed monument that might have provided the necessary context and clues.[136]

Speculations about future archaeologists finding few completed monuments in the ruins of American cities vied with counterclaims that they would find abundant unintended ones in the form of commercial structures. The perception of America's contemporaneous, utilitarian structures as surrogate monuments to their age dates back to the 1830s, when connoisseurs of ruins such as Thomas Cole imagined the "Bridges Aqueducts Warehouses and huge mills" materializing along the Erie Canal as America's "castles" that "in future ages shall tell the tale of the enterprise and industry of the present generation."[137] Others, however, questioned their future archaeological and historical value. Tocqueville postulated that democracy stimulates its citizens to erect, on the one hand, a large number of "trivial things" (such as log cabins, which fail to endure as legible ruins) and, on the other, "a few very large monuments" (the US Capitol or civil engineering projects) whose "scattered remains" will reveal something about the power of the state to "concentrate" vast labor power on a "single undertaking" but "nothing about the social conditions and institutions of the people who put them up" or about how "happy" or "enlightened" they were. "Between these two extremes," the trivial and the colossal, "there is nothing."[138]

The Brooklyn Bridge's completion in 1883 prompted a particularly trenchant critique of the limits of America's colossal structures as unintended monuments by the prominent architectural critic Montgomery Schuyler. He did not doubt that the bridge—or at least its masonry towers—would "outlast every [other] structure" and still be standing when some postapocalyptic traveler rediscovers the city. He also did not object to the possibility that "our most durable monument" was likely to be a functional structure. Rather, the problem was that it did not sufficiently express its function. Because the towers have a flat rather than sloping roof, they would not reveal their original function as a support (or "saddle") for the cables, which might by then have "rusted into nothingness." That postapocalyptic traveler would therefore learn "nothing of its uses"—not unlike the Persian archaeologists in John Ames Mitchell's postapocalyptic story *The Last American* (1889), who remain "at a loss to divine its meaning." The bridge would testify to the civilization's "architectural barbarism": its architects' inability to fully articulate and express their buildings' function through their form, and thus their inferiority to their medieval predecessors. The traveler might even assume that European cathedrals had been built in a later and more advanced age. Schuyler thus foregrounded the "question" of the built environment's future legibility, "a question" that should be considered by "every man who builds a structure . . . meant to outlast him." Evoking the Gothic was not sufficient.[139]

The connection between the urban environment and the desire to reach out and touch future generations is a theme Walt Whitman explored in his poem "Crossing Brooklyn Ferry" (1857). The sight of New York and its harbor from a ferry turned his mind toward posterity. He imagined the multitudes of commuters and pleasure-seekers who, "fifty years hence . . . , a hundred years hence, or ever so many hundred years hence," will glimpse the same sunset, the same ebb and flow of the tide, the same islands and ships, as they too are ferried back across the East River to their Brooklyn homes. Initially citing them in the third person, he addresses them directly by the middle sections.

> I am with you, you men and women of a generation, or ever so many generations hence,
> I project myself—also I return—I am with you, and know how it is. . . .
> I project myself a moment to tell you—also I return. . . .
> What is it, then, between us?
> What is the count of the scores or hundreds of years between us?[140]

Whitman critics have characterized his yearning to transcend temporal distance and mystically merge with future generations either as an exuberant celebration of America's infinite progress and timelessness, a disavowal of his own mortality, or an effort to compensate for his poetry's limited popularity by anticipating a more receptive posthumous readership, readers "who look back on me, because I looked forward to them."[141] The poem might instead be read as an expression of a more widespread anxiety about the present's legacy to posterity. Like those in the emerging library and preservationist movements, Whitman recognized that the communion between the living and the yet to be born depended on the continued existence of the city's physical objects, the "dumb, beautiful ministers" he addresses in the poem's final section.[142] Words alone cannot bind the generations.

Enshrining the Business Class

This growing desire to repair a broken connection to the future, whether through literary or physical vessels, did not necessarily imply a sense of duty and responsibility to that future. Whereas the eighteenth century, posterity was imagined as the recipient of an essential and meaningful legacy—conceptions of human reason or founding documents—and as the inspiration for great political acts, by the mid-nineteenth century, it could be invoked to express a desire just to be remembered. Posterity, in this usage, serves merely to perpetuate its predecessors' memory.

This attenuation of the discourse of posterity was, in fact, a gradual development, beginning in the early republic. According to political scientist Michael Lienesch, revolutionary leaders had articulated a classical Greek conception of posterity whereby one acts in the interest of future citizens by providing examples for them to emulate in their own struggles against tyrants and corruption; as protégés, the latter were obliged to follow those examples even to the point of revolution. But in their subsequent efforts to bind the populace to the new status quo, the drafters of the constitution substituted a Roman conception of posterity in which ensuing generations were merely to follow the founders' rules, a theory of obedience rather than obligation. Thus reimagined as passive subjects rather than active agents and as a conglomeration of private individuals rather than a collective citizenry, future generations would have little role to play in history. With the crucial gift—the constitution—already given, there was little else to do for posterity.[143] The discourse of posterity was further eroded by concurrent developments of the idea of progress. If, as Becker asserted, the Enlightenment cult of posterity had been predicated on the possibility of progress, the emergent nineteenth-century belief in the scientific inevitability of progress (material or cultural) obviated any necessity—or indeed capacity—to intercede for the sake of future generations.

Nevertheless, these early vessels did retain the Enlightenment conception of posterity as a historical judge evaluating the successes and failures of the present. Indeed, they must be understood as interventions in a larger political struggle over how their era should be remembered. The question of where they stood in that struggle, however, is more complex. Initially, they appeared to adhere to a conservative, even reactionary, agenda. Despite the social and political unrest already breaking out in 1876—workers striking in Pennsylvania and New England, "tramps" taking over Illinois towns, and farmers uniting across the South and Midwest under the antimonopolist Greenback Party—Mosher and Deihm affirmed the status quo simply by presuming that the nation would endure for another hundred years.[144] They thus disregarded critics, like the radical agrarian Henry George, who pointed to signs of imminent ruin. With the growing divide between the "colossal fortunes" of a few and the "hopeless poverty" of the masses, he declared in a speech that year, the "public soup house has become as fixed an institution in our great cities as the distribution of food in the declining days of Rome." Those considering voting for the Republican Rutherford Hayes in the presidential election should "turn to the history of other times and other peoples . . . hear the thud of the ax which severs the neck of Charles . . . hear the drum beat under the scaffold of Sir Henry Vane. See the uprising of the French people against an intolerable tyranny . . . pass into a wild orgy."[145] As one magazine editor observed, some were predicting "that we shall never, as a nation, celebrate another centennial year."[146]

In this context, a time vessel, for all its later associations with fears of civilizational collapse, could represent a refusal of such fatalism. By making advanced arrangements for the opening of their vessels during the bicentennial celebrations and the public reading of their contents by officials, Mosher and Deihm asserted a conservative faith in the ruling elite's ability to overcome the challenges of labor and populism.[147] They addressed their vessels to individual officers assumed to be still in existence: in Deihm's case, the chief magistrate of the United States, an anachronistic term for the president, and in Mosher's, the mayor of Chicago. The latter went one further, calling for his vessel to be topped up and resealed, like a chain letter, for the nation's tercentennial.[148] An article in Deihm's *Centennial Welcome*, deposited in her safe, echoed this optimism, declaring that "our children's children . . . may still avow themselves citizens of a storied Empire Republic" and presumably share the same values and assumptions. Not only would they celebrate a bicentennial, they would celebrate it with "still prouder jubilation than this," as would, the contributor hoped, "our descendants to the remotest generation."[149] The vessels also reveal this confidence in historical continuity through their omissions: they apparently felt no need to deposit enduring documents such as the constitution.

The compiling, sealing, and topping up of Mosher's and Deihm's vessels, moreover, were to ensure this continuity by functioning as nationalist rituals. Through them, the nation could be imagined as a community extending not only across space and backward in time but also indefinitely into the future, binding current and succeeding citizens as participants. Deihm, for instance, reserved space in her autograph and photograph albums for subsequent elected or appointed officials, and she placed those albums outside the glass door, in a separate vessel of rosewood, ebony, and plate glass so that they could add their portrait and signature every two or four years.[150] With their blank pages and their vacant inner recesses, the vessels projected the "homogeneous, empty time" of the nation, an abstract temporal void to be filled with the names and deeds of its coming "great men."[151] Even as the vessels invoked other, more embodied and embedded temporalities of the church, the family, or the locality—in their appropriation of genealogy, family albums, and other domestic practices of memorialization, and in their "bodily" preservation of their subjects—their hundred-year time spans helped to displace a generational notion of posterity with a purely numerical notion.[152] If the generational conception implied a respect for the future's alterity—its belonging to another generation—the numerical one clears the way for its colonization by the present, its reduction to immediate interests, especially financial ones. This abstraction of the future has arguably impeded efforts to imagine the rights of posterity ever since.[153]

Moreover, while subsequent time vessels were oriented around particular cities, these were national in scope, seeking to conscript all Americans as participants. Deihm planned to take her autograph album from Philadelphia to other "leading cities" before sealing it on the last day of the year.[154] She also sent loose blank sheets "to every State and Territory in the Union." Under each signature, a line was reserved for one's nearest descendant to inscribe their name in 1976—a reversion to the reproductive definition of posterity.[155] Similarly, Mosher, an extreme advocate of national centralization who believed that federal laws should supplant all state laws, identified himself not as a photographer of Chicagoans but as a "*National* Historical Photographer."[156] Both reframed a personal project, then, as a collective one—a "gift from the American people," in Deihm's words.[157] The contribution of one's portrait or signature thus constituted a pledge of national allegiance as well as an act of self-memorialization. Perhaps to confer a sense of tradition on this invented ritual, Mosher appointed Masonic grand masters to mark the "sealing with the highest honors and sacred rites of their order" and their successors to officiate at its opening.[158]

This conservative bias is also evident in the content contributed by others. Deihm's safe included a sermon on the centennial donated at the last minute by a local Methodist preacher. He warned there of the "antagonisms and inimical forces" that "sometimes, openly, and at other times secretly, . . . destroy [the] foundations [of republics] and lay their proud columns in the dust." The threat to the American republic came not from "foreign wars" but from internal threats such as the democratic excesses brought about by "our liberal enfranchisement of every class" and the flooding of "our shores" by "heterogeneous populations from the old world," with their "foreign customs and modes of thoughts," such as Catholicism, ultramontanism, and communism. "To have our valued citizenship . . . preserved from generation to generation," he urged the need for "true conservatism" and "eternal vigilance."[159]

Mosher appears to have shared these fears about the contamination of Anglo-Saxon Protestant stock. His belief in the possibility of imprinting great men's character onto fetuses stemmed from his larger belief in "stirpiculture," or what was subsequently called eugenics. In addition to portrait galleries disseminating the "science of stirpiculture," Mosher called for a "National Progenerate College" and local, compulsory "progenerate schools" to educate the young in motherhood and fatherhood; a federal law requiring marriage applicants to obtain a certificate proving they have "no hereditary incurable diseases of the body or brains, that could be transmitted to the unborn generations"; and another law banning dime museums from exhibiting and breeding genetic "monstrosities." By the close of the twentieth century, these measures will have produced a "beautiful world with a noble race of men and women . . . with perfect forms . . . [and] noble characters"—a world "without misery or crime."[160] Mosher thus gave physical form to his racial visions, rendering his eugenicist utopia concrete through the vessel. His portrait collection had in fact grown out of a photograph album of "very distinguished" Americans he had been asked to produce in 1873 as a gift for T. H. Huxley, a British evolutionary biologist who had divided humanity into four separate races and undertaken a photographic project for the British Colonial Office to distinguish those races visually.[161] Eugenicists were, indeed, keen advocates and practitioners of photography. The British founder of the movement, Sir Francis Galton, called on families to keep photographic as well as written records of their health. Albums containing periodic portraits of "yourselves and of your children" (both "standard profile and full-face") would provide indispensable medical histories for "[your] descendants in more remote generations."[162] Mosher, too, invited his customers to include a "certified family record" in the safe, ostensibly to "prove heirships to inheritance" but evidently also for a biological purpose. He was thus in the vanguard of the movement, which remained marginal in the United States before the 1910s. Certainly, he was one of the earliest to mobilize photography to promote scientific breeding of the human race.[163]

The conservative utopian vision of the republic's future expressed in these vessels was founded, above all, on a resolute faith in the business class's ability to maintain its hegemony. Successful businesspeople themselves, Deihm and Mosher celebrated the commercial elite and sought its endorsement. Deihm received assistance from corporations such as Southern Express Company and the Pennsylvania Railroad, and she invited a bank president to perform the locking of the safe. Her magazines and newspapers, included in the safe, were also touted by leading businessmen.[164] Mosher similarly appealed to the largesse of Chicago's leading capitalists, nominating Marshall Field (his landlord) president of his memorial project and Philip Armour, George Pullman, Cyrus McCormick, among others, as vice presidents; there is no evidence, however, that they accepted.[165] Although exploitative wages and poor working conditions at McCormick's and Armour's plants had already provoked workers to strike in the 1870s, Mosher made flattering references in the vessel to the "sagacity of [Chicago's] business men."[166]

The time vessels' commercial bias is borne out by the class of individuals enshrined within them. Despite fears that such projects cheapened fame, they maintained hidden biases. Insofar as one needed to belong to one of the sixteen professions (including businessmen) to be admitted into his albums, Mosher's safe was effectively a memorial to the propertied middle class. Merely to sign one's name, occupation, and address in his "memorial autograph register" cost one dollar (more than $20 today).[167] Equally telling are the absences from his photograph albums: labor leaders such as Terence Powderly of the Knights of Labor, leaders of the first socialist party in the United States (founded 1876) such as Philip Van Patten, agrarian radicals such as Ignatius Donnelly and James B. Weaver, and the single-tax advocate Henry George. Nor is there any discernible working-class presence. The Memorial Safe's exclusivity was reinforced by the expense of the cabinet cards customers were expected to purchase. This new format allowed wealthy patrons to distinguish themselves from those who could only afford the older, smaller cartes de visite.[168] Through well-placed advertisements, recommendations of luxury hotels for his customers, and denunciations of cheap photographers, Mosher further promoted his studio as the preserve of high society.[169] His vessel thus resembled another invention of the Gilded Age: "mug books." These claimed to encompass a city's most distinguished citizens, enshrining them with a portrait, biographical entry, and often an engraved autograph, yet they were commercial schemes immortalizing only those willing to pay a substantial subscription fee.[170] Mosher's and Deihm's vessels similarly obfuscated their criteria of inclusion. As a manifestation of expropriated, surplus capital, the bank safe was thus a fitting container for these shrines to the capitalist elite.

This allegiance to the business class was most evident in the tributes Mosher and Deihm paid to the political party that represented its interests, the Republicans. Despite the semblance of political impartiality they inherited from the daguerreotype galleries, neither concealed their loyalties. Mosher was subsequently appointed official photographer to the Republican National Convention in 1880 and issued an advertisement endorsing the Republican ticket in the 1884 presidential election.[171] Deihm's vessel was apparently funded by—and contained a Tiffany gold pen dedicated to—the industrialist and inventor Peter Cooper, a Republican since 1860 (although by 1876 he was running for president under the Greenback Party).[172] She further expressed her affiliations by announcing her attention to depict the current Republican president Ulysses Grant on the safe's inner doors in a kind of holy triptych with Lincoln and Washington. To maintain Grant in the national pantheon in 1876 was to disregard the mounting scandals. Deihm's *Centennial Welcome* merely conceded that "we have had some occurrences recently, that when viewed with a jaundiced eye, might strengthen . . . doubt . . . about the permanency of our institutions," ultimately concluding to the contrary.[173] She also accepted a contribution to the safe from the treasury secretary who had resigned over the Sanborn incident, one of the worst of the Grant cabinet scandals.[174] In this light, Deihm's photographic memorial to what was (until recently) the most corrupt administration—and congress—in US history represents a massive act of political denial. Like Mosher, she enlisted her vessel to advance a benign prospective memory of such contemporaneous conflicts and problems.

Envisioning a "Paradise" of Equality

Although conceived as centennial gifts, neither vessel was launched that year. Perhaps because of her partisan affiliations, her gender, or simply the novelty of her invention, Deihm struggled to generate interest. Several distinguished individuals she hoped to include were apparently "reluctant or indifferent."[175] She therefore continued to accumulate autographs and photographs—sending out solicitation letters, bringing her album to "all parts" of the country, and exhibiting her safe in the manufacturer's New York offices—for a further two and a half years, by which time two elections had rendered her collection of congressional portraits somewhat obsolete.[176] Mosher postponed his sealing even longer, most likely because the scheme was too lucrative. His collection, provisionally stored in the vaults of Chicago's First National Bank, continued to grow through the 1880s, while his safe remained on display in his photographic gallery as a "monument to his own enterprise and generosity."[177] Whereas recent time capsules tend to be hastily assembled, these were long-term efforts.

As a result of this temporal extension, the vessels came to register the intensified social unrest of the postcentennial years as well as Mosher's and Deihm's own growing ambivalence about the capitalist elite. While organized labor remained in the wings during the Philadelphia celebrations, it leapt to center stage during the Great Railroad Strike the following year. So, too, did the agrarian movement. Focused only on monetary policy, the Greenback Party had underperformed in 1876, with Cooper winning only eighty-two thousand votes. It now broadened and radicalized its platform to include a graduated income tax and government labor bureaus, and at the midterm elections of 1878, it secured fourteen congressional seats and over a million votes.[178] Even more troubling to capitalist interests was the imminent "fusion" of industrial labor and agrarian populism. Local labor or workingmen's parties with populist planks emerged across the country, while the Greenbacks in turn added the word *Labor* to their name, espoused labor causes such as a shorter work week and eight-hour day, and opposed the use of state troops or private agents to break strikes.[179] The industrial worker and the agricultural laborer, it appeared, were finally uniting to confront the forces of capital.

Just as their postponement forced the vessels to confront the eruption of class conflict, so were they shaped in unexpected ways at the final sealing ceremonies. Having assumed sole control over every aspect of her project, Deihm allowed others to contribute their own items at the ceremony, perhaps to reinforce the sense of a collective gift. The addition of several last-minute offerings inevitably undermined the ideological coherence of the vessel, opening it up to multiple, contradictory futures.

Thus, alongside more conservative warnings, Deihm's safe contained a book and pamphlet that fully embraced both the labor and agrarian movements. These materials appear to have been contributed at the last minute by Elizabeth Rowell Thompson, who played a key role at the ceremonial sealing, placing the gold Tiffany pen in the safe. Also deposited were several of Thompson's letters and visiting cards, her passport, and two framed photographic portraits of her.[180] After marrying into wealth, Thompson dedicated herself to causes such as medical and scientific research, women's suffrage, kindergartens, and the temperance movement, the latter represented in the safe by an advance copy of her book *The Figures of Hell; or, The Temple of Bacchus*. Like eugenics advocates, temperance reformers such as Thompson liberally employed a rhetoric of posterity. If the young are the "seed out of which will grow future society and future government," she wrote, then prohibiting alcohol will secure not just moral and physical but also social and political perfection. Failure to do so would be to allow the demon rum to destroy "not only the generations of the present, but of those who are to come."[181] Unlike many other bourgeois temperance reformers, however, Thompson addressed the social and economic relations that drove the poor to drink. Spurred by the utopian socialist ideas of Robert Owen and the anarchist individualism of Stephen Pearl Andrews and perhaps by her experience growing up poor, she used her wealth to challenge the capitalist status quo, supporting utopian colonies in the west and funding *The Worker*, a monthly journal promoting cooperative colonization.[182]

The clearest statement of Thompson's radical political philosophy was the prospectus for the American Worker's Alliance, which she coauthored in 1879 with the landscape architect and horticulturist William Saunders and deposited in Deihm's vessel. Saunders, himself memorialized with a framed photograph in that vessel, was cofounder and master of the Grange, a national organization to protect farming communities through mutual cooperation and to lobby against the corporate monopolization of railroads and grain elevators. In their prospectus, Thompson and Saunders championed industrial as well as agricultural labor. "All labor troubles would be vastly diminished, . . . [or] cease to exist" if workers cooperated with one another to organize mutual aid, spread "useful" knowledge, and secured "an equitable distribution of the profits of labor."[183] Thompson also signed another radical offering for Deihm's vessel, a copy of the book *The Crowned Republic*, published by "the Workers" in 1879, possibly at her own expense. Warning that the "just discontent" of destitute and unemployed laborers "threatens the very existence and stability of civil society" and that it was "folly" to hope "these things would adjust themselves," the book advocated a complete reorganization of society. In the new order, ushered in by workers' councils or "Bands of Workers," there would be a strong government prohibiting monopolies and securing the "right of labor . . . to employment" and to a proportional "share" of profits, common ownership of transportation and communications, a liberal arts education for all, equal rights for women and nonwhites, and a participatory democracy in which the people could directly elect and impeach officials and govern through referendums. Although neither the prospectus nor book prompted action—there is no other trace of an American Workers' Alliance—in preserving these rare publications, Deihm's vessel enshrined an ideal largely overlooked by subsequent labor historians who stress the divisions between workers: an alliance of farm- and factory workers, men and women, whites and blacks.[184]

The inflamed political climate compelled Mosher, too, to attach his vessel to a reformist agenda. His projected "Memorial home" on Chicago's lakeshore, housing not just the portrait gallery but also a library, concert hall, historical society, art supply stores, and lecture rooms, would uplift the poor—"Protestant, Catholic, Jew, or Infidel"—by rescuing them from the "misery and crime" of the streets; indeed, he envisaged a canteen in the basement to feed "apprentices, sewing girls, and others that subsist on small salaries."[185] It would redeem the rich, too, by distracting them from the narrow pursuit of wealth. To address that "serious question before the people," namely, the "centralization of millions of dollars into one person's possession," it would also contain a Memorial Security Legacy Bank that would distribute businessmen's mounting "surplus capital" to charities as well as their children.[186] Mosher's writings are replete with further, somewhat paternalistic proposals to mitigate the inequities of capitalist society throughout the nation along with scathing denunciations of the government's "shameful neglect" to do so. Arguing that government should protect the working classes as it does the "rich man's interests, the corporation and companies," and that "the daily toilers of the nation are the creators of all wealth," he envisioned National Savings Banks to teach them "the art of saving," a National Labor Bureau of Servitude to provide work (a Greenbacker idea), and a national, tax-funded public library and school system. "Such immortal legacies," he concluded, "would be the crowning glory of our Government."[187]

Mosher further democratized his project by opening up his albums to undistinguished Americans. Unlike Brady, who had denounced and resisted the trend toward cheaper, standardized photographs in the 1850s and 1860s, Mosher sought a larger, middle-class clientele.[188]

FIG. 1.8 Charles D. Mosher, *C. W. Wald[r]ick Salesman Garfield Lodge*. Albumen print on cabinet card, n.d. Chicago History Museum.

His vessel enshrined not just military leaders, political figures, and cultural celebrities (including Mark Twain, P. T. Barnum, and Edwin Booth), but also thousands of middle-class men and their wives. He even interpreted "profession" broadly, to include butchers and salesmen (**fig. 1.8**). Such a collection thus dissolves the distinction—made by Alan Trachtenberg, among others—between a public sphere of photographic galleries such as Brady's, where portraits were consumed for their didactic or "emulatory" function, and a private sphere of the parlor, where smaller, cheaper, pocket-size images of family and friends were consumed for "memorial" purposes.[189] In Mosher's memorial safe, and his customers' albums, the two were combined.

Deihm similarly democratized her vessel by allowing anyone to sign her "Citizens' Autograph Album."[190] While such latitude would become a standard feature of time vessels, it appeared at the time as a break with the assumption that only the illustrious deserve public memorialization and that illustriousness depended on prior accomplishments and, ultimately, posterity's judgment. One journalist who visited the Centennial Exposition viewed Deihm's exhibit as a symptom of a larger erosion of the institution of fame.

> For a ridiculously small sum, a mere pittance, anybody may have his autograph locked up in this iron box, kept from the tooth of time until the year 1976, and then held up to the gaze of posterity. Rarely is such a chance given to become illustrious on cheap terms. It is economical, universal, infallible. It brings immortality within the reach of the humblest of us, and the price is, we will say, five dollars a head. Think of it! Five dollars for a century of fame! Was ever anything so tempting heard of?

Even an illiterate boy, he sneered, can inscribe an X in the "sacred tome" and "without more ado [he] sails down the stream of glory." The journalist went on to compare Deihm's safe with other promises of "archiveal [*sic*] immortality," such as the Washington Monument Association's offer to engrave blocks of stone with the names of donors, or the naming of schoolhouses or charities after their founders. Venerating living individuals, he warned, was not merely unrepublican but also imprudent as they may prove to be "mixed up with . . . Crédit Mobiliers" or other scandals. But those other forms of celebrity set their "price . . . too high." Deihm's success demonstrated that "what our countrymen want is cheap fame."[191]

In addition to democratizing memorial culture, Mosher and Deihm also tied their projects to an assimilationist racial agenda. Even after the collapse of Reconstruction, Mosher continued to believe in the federal government's responsibility to "erec[t] school houses, suppor[t] schools, [and] educat[e] illiterate poor colored people's children in the South and North" and to establish "equal rights and equal privileges" for African Americans. This evidence of Mosher's belief in racial uplift and assimilation complicates our assumption that eugenicists believed in fixed racial differences and hierarchies, as does his inclusion of photographs of several black leaders, including Frederick Douglass and Mifflin Wistar Gibbs, a former abolitionist who had recently become the first black municipal judge elected in the United States (**fig. 1.9**).[192] Deihm's photographic album was also racially inclusive. Her decision to commemorate every member of Congress was an implicit celebra-

FIG. 1.9 Charles D. Mosher, *M. W.* [Mifflin Wistar] *Gibbs, Little Rock Ark.* Albumen print on cabinet card, n.d. Chicago History Museum.

tion of its unprecedented diversity. Despite the erosion of Republican control of the South, the forty-fourth Congress (1875–77) was the most diverse, a record not surpassed until 1969.[193] Among its eight black members were Joseph Rainey (SC), a former slave who in 1870 became the first African American elected to the US House of Representatives, and Robert Smalls (SC), whose heroic seizure of a Confederate ship in 1862 persuaded Lincoln to admit black Americans into the Union Army (**fig. 1.10**). Their inclusion quietly defied the Centennial Exposition's tacit policy of excluding African Americans from the celebrations.[194]

The utopian future projected by Mosher's and Deihm's vessels embodied not just class and racial equality (intermixed with capitalism, paternalism, and eugenics) but also gender equality. As a project conceived and managed by a successful businesswoman, Deihm's safe

FIG. 1.10 Mathew Brady, portraits of US congressmen, including Joseph Hayne Rainey (*top left*) and Robert Smalls (*lower right*). Albumen prints on cabinet card, ca. 1876. From Deihm's album, "Photographs of the Great American People of 1876." Records of the Architect of the Capitol.

stood alongside other exhibits at the Centennial Exposition—mostly housed in a separate building—that demonstrated women's capacity to perform work in public.[195] Installed in Statuary Hall, it would call into question that room's exclusive dedication to statues of men. (The first female statue, to Frances Willard, was not added until 1905.[196])

Deihm's appointment of Elizabeth Thompson to officiate at the sealing also hints at a feminist agenda. Thompson was a supporter of women's causes ranging from the plight of working women in the United States to child widows in India. She funded the multivolume *History of Woman Suffrage* conceived by Elizabeth Cady Stanton and Susan B. Anthony in 1876, and as president of the Women's Memorial Association, she led efforts to erect monuments to leading women.[197] She was also one of several women whose participation in the Grange movement pushed its male leaders (including Saunders) to embrace women's suffrage.[198] Mosher also commemorated Stanton, Anthony, and other suffragists in his safe, and he continued to correspond with them.[199] And if the *New York Times* criticized Deihm's safe for including "few" photographs of "ladies," Mosher's gender ratio was more balanced, if only because he extended his memorial photographic services to his subjects' wives.[200] Mosher's apparent sympathies are corroborated by his subsequent writings, which denounced the denial of suffrage to women as a "burning disgrace" and imagined a future "Paradise" in which "man and woman [are treated] equally before the law."[201] His vessel, together with Deihm's, thus embodied progressive visions for the bicentennial, with deposited items constituting tokens of hope, as if their withdrawal from circulation invested them with magical powers.

Turning the Lock

Given the publicity they generated at the exposition and the length of time they took to fill, the vessels' sealing proved anticlimactic. Arriving in the Capitol's Statuary Hall on February 22, 1879 (Washington's Birthday), Deihm discovered that several invitees, including President Rutherford Hayes, had declined to attend. Proceeding without them, Thomas Ferry, president pro tempore of the Senate, turned the key and "beat a hasty retreat to the Senate chamber." The spectators, "expect[ing] some more extended ceremony," took "some time" to realize it had already concluded.[202] Deihm's disappointment would have been compounded by the controversy that erupted three days later, when the *New York Tribune* accused her of mercenary motives, alleging she had charged each congressman a five-dollar fee to be enshrined in her safe and was now asking Congress to pay an additional $1,500 for the so-called gift. In an open letter she denied receiving "a dollar from anyone," claiming that the "enterprise cost me $15,000."[203]

If the newspapers suggested Deihm's sealing was a flop, they failed to report Mosher's at all. Exactly a decade later, Mosher also chose Washington's Birthday for the first phase of the process: the safe's installation in the lobby of Chicago's city hall, embedded in a marble framework.[204] On May 18, 1889, the second phase took place: the depositing of the first eight thousand photographs. At that event, Mosher raised expectations of a grand finale "a few months" later, when he would deliver a last batch of two thousand photographs and fasten the doors. "When I am ready to deposit [them]," he promised the assembled throng, "we will have some ceremonies, with an oration by some prominent man."[205] There is no record, however, of such an event.

The two vessels were thus cast on the sea of time without much fanfare. If their launching went almost unnoticed by contemporaries, what chances did they have of exciting future generations? Nevertheless, hopes remained. "It is presumed," wrote the *National Republican* after Deihm's ceremony, that the safe will be opened by a jubilant nation on the Fourth of July, 1976, and with greater interest than we can possible [*sic*] imagine now." For now, it "stands in the rotunda, surrounded by the statues of the fathers of the republic, who appear like guardian angels to keep watch and ward over it."[206] Perhaps one or two of the thousands of Chicagoans passing daily through city hall's marble lobby cast a similarly protective eye on Mosher's vessel as it began its own uncertain voyage toward the bicentennial.

TWO

"P.O. Box to the Future": Temperance, Insurgence, and Memory in San Francisco, 1879

In the middle of Washington Square Park—an oasis of calm in San Francisco's bustling North Beach neighborhood—stands an odd monument (**fig. 2.1**). Obscured behind a ring of poplars, it depicts not the first president but rather Benjamin Franklin. While that cast-zinc figure is instantly recognizable, the inscriptions on the granite pedestal remain obscure. The words *Vichy, Congress,* and *Cal. Seltzer* on three of the four sides suggest it once dispensed varieties of drinking water. But the inscription above the word Vichy hints at more: "P.O. Box with Mementos for the Historical Society. In 1979. From H.D.C." And a somewhat morbid epigraph adorns the fourth side: "Presented by H.D. Cogswell to Our Boys and Girls Who Will Soon Take Our Place and Pass On." The monument attracts little attention from the park's regulars—homeless people, Frisbee throwers, tai chi practitioners—or from the urban professionals who more hurriedly traverse the square on their daily commute between Telegraph Hill and the business district.

Like those urban professionals who have gentrified this once solidly Italian neighborhood since the 1980s, and like the Chinese Americans who have migrated from adjacent Chinatown, Franklin is a transplant. He originally

FIG. 2.1 Henry D. Cogswell's Benjamin Franklin monument and drinking fountain, Washington Square, San Francisco. Photograph by author, 2010.

stood downtown, in the heart of a gambling, prostitution, and gang-ridden district known as the Barbary Coast. On a triangular shaped lot forged by the intersection of Montgomery (now Columbus) Avenue with Kearny Street, almost five thousand people assembled on the afternoon of June 14, 1879—less than four months after Deihm's ceremony in the Capitol—to witness his unveiling. Spectators arrived long before the ceremony to ensure a good vantage point, while others watched from a specially constructed platform or from the windows of neighboring buildings, which were adorned with banners and flags. After prayers, benedictions, speeches, and patriotic music by a local band, the master of ceremonies gave the signal to remove the flag covering Franklin. In so doing, he revealed the pedestal in which a lead casket weighing a hundred pounds and measuring 16″ × 20″ × 20″ had been installed to transport "mementos" to the curators of the "historical society" a century hence.[1] To a chorus of loud cheers, the time vessel was dispatched for special delivery to the future.

The brainchild of an eccentric dentist and millionaire, Dr. Henry D. Cogswell (**fig. 2.2**), the Franklin vessel departed from Mosher's and Deihm's in significant ways. Cogswell did not enshrine his in a governmental edifice nor did he seek an official imprimatur from civic or national leaders. There is no record that he even invited the mayor or the San Francisco Board of Supervisors. There were no plans for further ceremonies to periodically open and top up the vessel; Cogswell wished his deposit to remain sealed until its target date, which he set not for the glorious bicentennial but for an arbitrary date one hundred years into the future. Cogswell, in other words, did not conceive his project as a memorial to the politicians, merchants, and literati of his day (in fact, he deposited messages denouncing the city's leaders). Instead, he

I am now 59 years of Age [illegible]
march 3/79 and have Founded
the "Cogswell Dental College"
of the University of State of Califo- in
Also the "Cogswell" Chair of Moral
and Intellectual Philosophy
of the State University
Also the Cogswell
Students Relief
Fund to help
needy Students
to persue their Studies
in said University of State of California
Artist I. W. Taber & Co.
My last resting place I am now
building at Mountain View
Cemetery on the City of Oakland
Side of the Bay on the top of the hill
Size of Lot 75 × 75 Monument
44 feet high & at present the finest
on this Pacific Coast to rest until
the Resurrection of the Great Day H.D.C.

FIG. 2.2 I. W. Taber & Company, portrait of Henry D. Cogswell, photograph, ca. 1879, with Cogswell's autobiographical statement, March 3, 1879, deposited in his time vessel. MSP 559, California Historical Society.

appears to have intended it largely as a monument to himself, entombing evidence of his philanthropic endeavors. At the same time, he recognized the potential unpopularity of such a project and thus the need to open it up to collective participation. He extended an invitation to San Franciscans of all ages to preserve some aspect of their lives: a written message, a photograph, or "such small articles as they desire" that could fit in his "Antiquarian Box."[2]

In its collaborative, democratic approach to civic commemoration and its acceptance of everyday objects, Cogswell's vessel departed not just from Mosher and Deihm's but also from other public memory projects. Monuments typically celebrated victorious generals, great statesmen, or founding fathers; libraries preserved the words of published authors; and historical societies collected the papers of notable figures or the finest specimens of material culture (and even then, American historians remained uninterested in the latter).[3] Cogswell's vessel, by contrast, embraced the messages and keepsakes of ordinary individuals and the commercial and printed ephemera of their city. It further departed from those and other institutions of retrospective memory, such as public archives, in its effort to actively construct a prospective memory of the present. Although Cogswell retained a desire to memorialize the pioneers of 1849 (of which he was one), he restricted himself almost entirely to current material such as newspapers, magazines, correspondence, and other written documents. His contributors went even further, envisioning contemporaneous ephemera such as trade cards or promotional calendars as historical documents in the making, or modern, mass-produced artifacts as future antiquities.

As a monument to the present, Cogswell's vessel ostensibly depicts a unified, harmonious civic community, ignoring the intense conflicts over ethnicity and class that were convulsing the city. On closer examination, however, those conflicts become visible. His decisions regarding the fountain's design, location, and specifications—as well as his implicit decisions *not* to include certain things—betray a deeper ethnic and class agenda. Moreover, although the messages deposited within rarely referred directly to the ongoing social unrest, the invocation of politically charged terms such as *intemperance* and *hoodlums* positioned it as a reaction to the specter of class insurrection embodied by the agitator Denis Kearney and his Workingmen's Party of California. This partisanship may even have contributed to the unpopularity of Cogswell's project, which, like other fountains he donated to the city and nation, aroused acts of verbal and physical abuse.

A Hybrid Archive

On a warm Sunday a fortnight before the dedication, Cogswell stayed home all day, according to his diary, assembling packages for his vessel. He also composed an eight-page cover letter to be read aloud to the "Historical Society" in 1979.[4] Rather than enlightening his projected audience about conditions in 1870s San Francisco or giving his predictions for the future, he wrote largely about himself: his personal motives for constructing the fountain and time vessel and, above all, his life story. He proudly traced his ancestry back to Sir John Cogswell, a London merchant, immigrant to America in 1635, and progenitor of many illustrious Cogswells over the subsequent two centuries, including "17 graduates in the different N. England colleges, 14 physicians & 3 clergymen." Despite this noble pedigree, Dr. Cogswell was born into poverty in Tolland, Connecticut, in 1820. He spent his "fatherless" childhood laboring first on a farm, "haying and milking by moonlight to pay for my food," and then in a textile factory, operating a spinning mule fourteen hours a day. (He omitted to mention that he also spent time in the poorhouse.) He managed, however, to educate himself at night, and eventually he qualified as a dentist in Providence, Rhode Island. Business must have been slow, because when news arrived of the discovery of gold at John Sutter's sawmill, Cogswell became an argonaut. In 1849, he boarded a clipper ship in Philadelphia, arriving 152 days later in San Francisco, which he recalled, with the typical exaggeration of a forty-niner, as having at that time "not more than ten building[s]." Unlike most gold rush migrants, Cogswell apparently had no intention of mining gold but rather sought to mine the miners by establishing a dentist office and supply store in Stockton, on the edge of the Sierra Nevada goldfields. When the gold rush proved short-lived, he returned to San Francisco, where he continued to practice dentistry, while also pioneering new techniques such as the use of chloroform as an anesthetic and the vacuum method of securing gold dental plates in the mouth. Just as miners flaunted their success by capping their teeth in gold, Cogswell did so by adorning the facade of his dental office with a giant gilded molar, like the one outside McTeague's in Frank Norris's eponymous novel of 1899; indeed, Cogswell has been cited as a possible source for that fictional San Francisco dentist. Unlike McTeague, Cogswell wisely reinvested his income from dentistry in real estate, mining stocks, and local enterprises, accumulating a fortune of over two million dollars (almost fifty million dollars today) by 1855. At that point, he retired from dentistry and dedicated himself to philanthropic endeavors, such as the provision of drinking fountains. Cogswell concluded his letter to future historians with a recommendation to visit Oakland's Mountain View Cemetery, where they would find the "resting place" of this "benefactor . . . to posterity." For good measure, he attached a drawing of his tomb, a grandiose, sixty-foot obelisk surrounded with fountains and marble statues, which he had designed more than two decades before his actual death in 1900—an act of self-regard that rivaled Franklin's formulation of his own epitaph at the age of twenty-two.[5]

In some respects, the Franklin fountain and time vessel was as much a monument to Cogswell himself as his own tomb. His diary (not included in the vessel) reveals that he played a similarly active role in its conception, closely and single-handedly supervising all aspects of the project, from the initial design to the foundation laying, stonecutting, plumbing, and even carving of the inscription.[6] He also retained absolute control over the selection and packaging of materials for the antiquarian box, which allowed him to deposit reports of his philanthropy, his dental journals, photographs of himself and his wife Caroline, forty-four inscriptions of his signature, and six copies of his business card.[7] He even assumed future antiquarians would want to pore over voided copies of his will; paid checks, money orders, and promissory notes; lapsed insurance policies, leases, and indentures; and old receipts, memos, and tax assessments—all presumably testifying to his affluence and fiscal rectitude.[8] In employing the time vessel as a shrine to himself, Cogswell thus exceeded even Mosher's and Deihm's self-promotional efforts while also eschewing the deliberative approach adopted by monument associations. As long as time vessels remained tools for personal aggrandizement, they would not gain widespread acceptance.

Cogswell's vessel might be linked to the Victorian practice of preserving personal effects or heirlooms (defined as inalienable possessions) in a chest to be handed down the generations except that, having no children, he bequeathed his box to the public. The sealing ceremony thus approximated the postfuneral rite of passage, conventionally administered by women, of transferring heirlooms to next of kin.[9] The idea may also have been inspired by his experience of ancient sites during a four-year world tour with his wife. In an article justifying his fountains, he cited an epigraph "of one of the ancient Egyptians (one of whose pyramids I visited, A.D. 1874)."

Behold! The puny child of man sits by Time's boundless sea;
Takes a few drops in his hand, and calls it history.[10]

Cogswell's interest in the pyramids was representative of a larger, popular fascination with ancient tombs and the deep temporal mysteries they evoked. Recent archaeological excavations, most notably Heinrich Schliemann's 1873 discovery of a wooden treasure chest that he erroneously attributed to King Priam of Troy, were widely reported in the American press and may have contributed to the time vessel's invention later that decade.[11]

Despite his apparent Pharaoh complex, Cogswell did seem to recognize the inadvisability of framing his fountain as a monument to himself or his time vessel as a personal trove of Cogswelliana. For all his wealth, he remained marginal within the San Francisco elite, with limited political or cultural influence. He appears not to have belonged to the leading clubs (not even the Society of California Pioneers, though he arrived before the end of 1849 and thus qualified), nor did he live in a mansion on Rincon Hill or in the new plutocratic enclave of Nob Hill. Cogswell's name was absent, too, from directories of the city's elite and from early histories of the gold rush era.[12] He hoped Alonzo Phelps would correct these omissions in his forthcoming illustrated biography of the pioneers, *Contemporary Biography of California's Representative Men*, one of those "mug books" that immortalized the wealthy for a fee, in this case $250. He proudly included the prospectus in his vessel and was the first to contribute.[13] Some disagreement, however, led to a scuffle on Market Street, a complaint against Phelps for assault, and a countersuit against Cogswell for defamation.[14] Cogswell waged similar feuds against other San Franciscans, including Mayor Andrew Bryant and even his own lawyer, establishing a reputation for cantankerousness. Had he devoted the time vessel to himself, its dedication would likely not have attracted thousands of spectators. Similarly, had he offered his papers to any historical society of his own day, they would likely not have been accepted.

Yet by adopting the device of the time vessel, which meant allowing others to contribute, Cogswell could convey his own name, image, and deeds to posterity. He therefore spent weeks soliciting items from residents of "this city and vicinity." He submitted a notice in the local newspapers, announcing "this great Antiquarian Postoffice" and encouraging the public to drop off their packages by June 14.[15] He also extended a special invitation to the "Teachers of the Public Schools + their pupils"—an early recognition of the pedagogical value of time capsules—while making additional visits to specific companies and individuals.[16] Photographs, autograph letters, and short autobiographical accounts were particularly welcome; the only condition was that the contribution be "small and compact, . . . thoroughly dry and without any adhesive gum."[17] Although presumably retaining some control over its contents rather than playing the neutral role of "mail sorter," he transformed the time vessel into a collective and public venture, more collaborative and heterogeneous than other memory projects of the period. And whereas Deihm's and Mosher's schemes, and indeed the mug books, were allegedly motivated by profit, Cogswell's P.O. Box demanded no postage fee for transmitting its subjects' names and exploits to future generations.

Memory of the Present

Cogswell further bolstered his time vessel by tying it to the commemoration of San Francisco's founding fathers, just as Mosher's claimed to memorialize Chicago's. He originally intended to include the names of all members of the "Societies [*sic*] of California Pioneers," and invited leading forty-niners to the dedication. The founder of San Francisco's First Presbyterian Church led the prayers, and the captain of an early gold rush ship unveiled the statue.[18] The vessel may have been launched in 1879 to commemorate the thirtieth anniversary of the arrival of those first steamships of gold seekers. The city's astronomic growth in the intervening years—from around twenty thousand residents to almost a quarter of a million (an increase that took Boston two and a half centuries) and from a landscape of wooden shanties, canvas tents, and adobe dwellings to a cityscape of hotels, mansions, factories, tenements, and offices—had bestowed an aura of antiquity on that recent past.[19] The forty-niners' almost mythical status would have lent Cogswell's project an air of authority.

This compression of time also conferred a sense of evanescence on the present, as if it, too, was prematurely receding into antiquity. Capitalist urbanization thus contributed to the time vessel's emergence as a device for rescuing that present from the stream of time. Cogswell himself described his antiquarian box as a "method of preserving a small fragment of our *present* history."[20] If the pace of change had made the past seem like a foreign country, the vessel would forestall such a schism with the future. It would "form an indisputable connecting link," he explained to the board of supervisors a year earlier, "between the people of San Francisco of the day 1878 and 1978."[21] His decision to deposit that time-binding leaden box in a granite monument out on the sidewalk rather than within a building, as Mosher and Deihm had done, may have been motivated by experience. No fewer than six great fires and three major earthquakes struck the city between 1849 and 1868, causing substantial loss of buildings, including his own office (with all its instruments and valuables) in 1851. Deeding that monument to the city and its contents to the Historical Society removed another threat to any collection that remained in private hands: that of being sold and dispersed in the event of bankruptcy.

Cogswell's preservation of the present (and recent past) for future historians, specifically, represented a new departure. Mosher and Deihm had addressed their vessels not to scholars but to officials—the future president of the United States or the mayor of Chicago—and to the signatories' descendants. Their purpose was clearly memorial rather than archival. Except for the addition of a few pamphlets and other ephemera, they made little attempt to convey the broader social, cultural, and political conditions of the period. Cogswell, by contrast, was apparently motivated by a vision of 1870s San Francisco as the subject of future historiography. Literary theorist Fredric Jameson describes this projection of historicity onto the present as a "process of reification whereby we draw back from our immersion in the here and now (not yet identified as a 'present') and grasp it as a kind of thing—not merely a 'present' but a present that can be dated and called the eighties or the fifties."[22] The historian Peter Fritzsche has traced this tendency to view one's own present as a distinct historical epoch—and even to give it a unique name—to the rupture of the French Revolution and the Napoleonic wars, which appeared to leave witnesses "stranded" in time.[23] Yet the subsequent economic and social upheavals of urban modernity deepened that impulse, prompting Baudelaire in 1863 to stress what he called "the memory of the present" and Mark Twain and Charles Dudley Warner in 1873 to christen their own moment the Gilded Age (although historians did not embrace the term until the 1920s).[24] Such a reification of or detachment from the present opened up the further possibility of encapsulating it through some kind of representative sampling of objects.

The term *antiquarian box* and the targeting of "the Historical Society" indicate that Cogswell was specifically addressing the amateur, local historians who dominated the practice before the professionalization of the 1880s. Beginning with Massachusetts in 1791, state historical societies (initially privately funded) had sprung up across the

United States attracting a membership of male patricians.[25] Cogswell's vessel, however, was as much an encouragement as an endorsement of those local amateurs. A California Historical Society was indeed founded in 1871, but it had attracted only twenty-five members, who were mostly too "busy with the daily affairs of life" to commit themselves to the "undertaking" or even attend the meetings. By 1874 it was defunct, one of no less than four abortive attempts between 1852 and 1886 to found such a society in California. (Its ghost entry in the city directory may have led Cogswell to assume it still operated.)[26] Moreover, that society had sought antiquarian rather than post–gold rush or contemporary sources—a priority it shared with other libraries such as Hubert Bancroft's.[27] As late as 1887, the professor and future president of the University of California Martin Kellogg complained of the lack of any systematic effort to accumulate historical sources of the present and recent past—a failing that would invite the "reproaches of future generations of Californians." He criticized a bias not only toward the state's early years but also toward great events—"battles and conquests" and "a few notable crises, like that of the Vigilance Committee of 1855"—rather than "the daily life of men." He therefore called for the documentation of "every decade since the first settlement. . . . We want the particulars of the changes that have taken place; of the transition from surface to quartz and hydraulic mining . . . ; of the chronic 'hard times' from which Californians suppose themselves to have suffered; of the incoming of the Chinese, and all the relations of labor to capital."[28]

As if anticipating Kellogg's call, Cogswell solicited messages and artifacts from ordinary rather than leading San Franciscans and accumulated materials that reflected the prosaic rather than the epochal, the deep currents of everyday life rather than what French historian Fernand Braudel later called the momentary, "surface disturbance" of "event history."[29] Books such as *The Bonanza Mines of Nevada* and *Ship Building on the Pacific Coast* document the region's industries; railroad timetables display the speed of mechanized transportation but also the complexity of local times in the decade leading up to the introduction of standard time; while a price list for a local laundry reveals the costs of cleaning clothes as well as local fashions. There are publications on the city's water supply, quarantine laws, and fire department ordinances, as well as a description of the cable car, recently invented by Andrew Smith Hallidie. Elite culinary tastes are documented in a menu from the Palace Hotel; recreational and touristic trends in an advertisement for the North Beach Bathing House, a prospectus for the

Santa Barbara Hot Sulpher Spring Company, and a guidebook to California's geysers; religious customs in an array of hymns, sermons, and tracts; and medical practices in a homeopathy manual and an advertisement for a patent medicine, Dr. A. Zabaldano's Syrup of Eucalyptus. Meanwhile, cultural preferences range from an anthology of Romantic poetry and a copy of the libretto from Verdi's newest opera *Aida* to a program for the popular Hutchinson Family Singers.[30]

Among all this printed matter, periodicals predominate. Where cornerstone deposits typically included one or two newspapers, Cogswell assembled a more comprehensive archive of the city's news media. There are complete copies of as many as twenty-one newspapers that were printed in or around San Francisco along with four weekly and seventeen monthly or quarterly magazines and journals from across the nation. Rather than limiting itself to a single day, as cornerstones did, the vessel included newspapers dating back to 1877 and magazines from the 1850s onward as well as clippings of articles on specific issues. Cogswell's privileging of the newspaper reflected its centrality to San Francisco's daily life. Although the nation's ninth largest city, it ranked third in the number of dailies (twenty-one) and the per capita circulation rate (0.61 copies for each inhabitant), based on 1880 census figures.[31] Circulation rates do not do justice to San Francisco's appetite for newspapers, as each copy was read by multiple people, often out loud in public spaces. To keep track of the ongoing feuds between editors, San Franciscans typically read more than one newspaper each day. The newspapers' economic and cultural dominance of the city was already visible in the office buildings they erected during the 1870s.[32]

In accumulating his archive of newspapers, Cogswell went beyond historical practices of the period. With the notable exception of Josiah Royce, who wrote a history of California that relied heavily on old newspapers, American historians were slow to recognize their value as primary sources until well into the twentieth century.[33] One reason for this was newspapers' partisanship, which was especially pronounced in San Francisco. A second, cited by another early pioneer of history from newspapers, was their "ready accessibility." Historians fetishize "manuscript material because it reposed in dusty archives and could be utilized only by severe labor and long patience."[34] There was also the problem of their sensationalism. Even Kellogg, who commended Royce's efforts and recognized that newspapers are "invaluable in their way, and give a life-like impression of many past events in our State," regretted how they could confuse and distort the historical record.

> The political strifes of the day were magnified out of all proportion. The passing sensation took the lion's share of space; . . . a prima donna . . . a loud-mouthed demagogue or a prize fighter . . . a murder or an embezzlement . . . a fire or a flood. Meantime the persistent forces, the really important events, were slighted or overlooked. Such forces are quiet in their working: such events come and go without bluster or blare of trumpets. . . . The genuine historian would starve, if the newspapers were his sole resource.[35]

This suspicion of newspapers persisted even among innovative, twentieth-century historians such as Marc Bloch and Fernand Braudel of the Annales School, who identified them with the surface froth of event history.[36] It took marginal figures such as Lucy Maynard Salmon, author of *The Newspaper and the Historian* (1923), to realize that newspapers, even (or especially) their most "unauthoritative parts" such as their advertising sections, could reveal the deeper "spirit of a time or locality."[37]

Cogswell supplemented his newspaper holdings with numerous donated artifacts. These included banal, and in some cases mass-manufactured, objects such as a box of breath sweeteners, a mechanical pencil, a souvenir pen and bud vase from the Centennial Exposition, a silk bookmark, a wooden puzzle, a paper weight, fashion accessories (buckle, lapel stud, lace collar, flag pin, and buttons), coins of various denomination (including gold and silver dollars), and, presumably Cogswell's own contribution, some false teeth. This wide-ranging accumulation of contemporary artifacts broke new ground not just for the time-vessel tradition but also for the collecting and preservation of material culture.[38] Eighteenth-century museums and collections such as Charles Willson Peale's, following the Renaissance tradition of the "cabinet of curiosities" or *Wunderkammer*, had prioritized ethnographic exotica, natural history specimens (shells, fossils, etc.), or rarities (unusual animals, ancient coins and medals, etc.) and represented national heroes through painting and statuary.[39] During the nineteenth century, the historical relic—believed to foster an affective, sensorial relationship to the past—became the dominant object in the private collections of figures such as John Fanning Watson and in the public collections and displays of state and local historical societies and museums. These could include ordinary objects (even mass-manufactured ones), but only if they had a connection to—or had been touched by—some historic individual, such as a glass from which George Washington supposedly once sipped. Later in the century, collections came to include relics connected not to historical figures but to epic historical events, as in the bloodstained bible or bullet-torn uniform of an ordinary soldier killed in the Civil War.[40] Yet few, if any, institutions collected the material relics of *present*-day, *un*celebrated individuals, as Cogswell did.

Similarly, while numismatic collections remained limited to ancient coins, those in Cogswell's vessel were contemporary. These coins might be viewed anthropologically, as votive offerings embodying a wish, like those that Romans and Greeks deposited in boxes, foundations, or rivers, or that we might throw into a fountain; as gifts to the future, cementing a relationship across time or perhaps competitively flaunting their own magnanimity to posterity; or as a ritual destruction of wealth, echoing the American Indian potlatch ceremony. Indeed, together with other precious objects, such as a silver spoon and plate and a chunk of silver ore from a local mine, they amounted to an estimated fifty dollars (or twelve hundred dollars today).[41] But they may also have been intended as archaeological evidence of the present in accordance with the new science of numismatics, institutionalized only fourteen years earlier with the incorporation of an American Numismatic and Archaeological Society. The plethora of coins and absence of paper money in the vessel, for instance, confirms Californians' preference for specie, in defiance of the Legal Tender Act of 1862 and the Legal Tender Cases of 1871, which rendered the new "greenbacks" constitutional.[42]

Cogswell's vessel did, to be sure, hark back to the cabinets of curiosities in its eclectic embrace of natural specimens and exotic objects. A small silk drawstring pouch contained forty-seven water-polished pebbles and a shell and was accompanied by pressed flowers, dried leaves, a berry, and a pinecone. Further objects such as seeds, acorns, and legumes, sealed in a glass vial, serve not just as specimens of natural history but as embryonic forms of life possibly intended to blossom a century hence; indeed, a two-thousand-year-old date seed recovered from an excavation at Masada, Israel, successfully germinated in 2005 and will generate its own seeds by 2022.[43] They thus stand in for the time vessel's larger quest to bring Gilded Age San Francisco back to life, or perhaps for the notion that actions in (or historical knowledge of) the present may bear fruit for posterity. Moreover, just as the *wunderkammern* manifested the global reach of the merchants and rulers who owned them, so did Cogswell's vessel contain souvenirs of his (and others') travels overseas, such as coins from Chile, Turkey, and Japan, a four-thousand-year-old piece of mummy cloth from Thebes, and a molar tooth supposedly "extracted . . . from the skull of Robespierre, the Great Republican of France."[44]

Yet whereas the *wunderkammern* claimed to microcosmically represent the entire world and all of time, Cogswell's aim was primarily to encapsulate a single city at a single moment. Exceeding Mosher's and Deihm's more modest goals of memorializing the social or political elite, Cogswell sought to encompass San Francisco as a whole. He included not just the newspaper to which he himself subscribed but all the newspapers of the city. The Republican perspectives of the *Chronicle* and the *Bulletin* are balanced by the Democratic viewpoints of the *Examiner* and the independent but conservative outlook of the *Alta*. The city's diverse religious communities are similarly represented by the Catholic, Jewish, and even atheist newspapers, as its ethnic communities are by the French, German, Italian, Scandinavian, and Chinese newspapers.[45] The inclusion of the *San Francisco China News*—"the first newspaper in the Chinese language ever published outside of the Chinese Empire," according to its founding editor's accompanying letter—constitutes an unusual acceptance of an immigrant group that encountered widespread hostility and violence ever since arriving in the 1850s and 1860s to work in the gold fields and on the construction of the transcontinental railroad.[46] The recognition of all immigrants also diverges from the post–Civil War trend (emphasized by historians such as Mary Ryan) toward the ethnic fragmentation of the civic public sphere in the form of separate parades, festivals, and societies for each group.[47]

Cogswell also exceeded Mosher and Deihm in his inclusion of women. Whereas the earlier vessels admitted the wives of businessmen and politicians and the occasional advocate of women's rights, Cogswell's recognized a much broader range of roles. The "domestic arts" are endorsed in the form of a prospectus for *The Housekeeper's Encyclopedia* and a copy of the "home culture" magazine *Cottage Hearth* as well as in needlework artifacts such as a crochet sample, a pin case, swatches of fabric, and spools of thread; in items of clothing such as a lace collar; and in craftwork such as flower pressing.[48] But the vessel also sanctioned more public roles, such as moral reform. Along with a copy of *The Gem: Ladies Monthly Reform Journal* were two letters and a poem submitted by leading kindergarten advocate Sarah B. Cooper.[49] More controversially, the vessel celebrated women's imminent admission into the male preserves of commerce and the professions. A pamphlet from the California Institute of Bookkeeping and Penmanship advertises "Business Education for Ladies" who are "rapidly asserting" their "business qualities." The guidebook on California's geysers cited above was by Laura de Force Gordon, the legendary trance speaker and outspoken suffragist, who—together with Clara Foltz—had just successfully lobbied for a state constitutional provision guaranteeing women the right to pursue any "lawful business, vocation or profession" and was preparing to argue before the California Supreme Court for women's right to be admitted to a college of law.[50] In an inscription on the frontispiece of *Great Geysers of California*, she testified to her activist struggles:

> If this little book should see the light after its hundred years of entombment, I would like its readers to know that the author was a lover of her own sex and devoted the best years of her life in striving for the political, equal, and social and moral education of woman.[51]

Although it is unclear whether "lover of her own sex" signifies her sexual orientation or her feminism, the recovery of this inscription led to Gordon's posthumous incorporation into the history of lesbianism in San Francisco, thus demonstrating how time vessels can secure affective and not just genealogical kinships across time.

The inclusiveness of Cogswell's antiquarian box with respect to religion, ethnicity, and gender was consistent with his own apparent cultural pluralism. As he declared in a note deposited in the box, he had founded Cogswell Dental College as an institution upholding his beliefs in equal opportunity for men and women.[52] Eight years later, he would found Cogswell Polytechnic College, a nonsectarian institution open to students regardless of religion or gender.[53] His sympathy for woman's rights may also be deduced from his hiring of De Gordon's suffragist partner, Clara Foltz, as his attorney in 1882. Indeed, the latter lived in his house for two years, until their friendship broke down over unpaid legal bills.[54] It is thus telling that Cogswell inscribed his time vessel to boys *and* girls.

In its apparent embrace of all groups within the city, Cogswell's vessel was the archival equivalent of photographic panoramas. The latter were increasingly popular and sophisticated by the 1870s, especially in San Francisco; indeed, the vessel included a book of panoramic photographs of that city.[55] Efforts to encompass whole cities, to enclose them within a single frame by ascending to an elevated vantage point, were a middle-class response to their increasing illegibility and unintelligibility. Vastly overgrown, architecturally cluttered, spatially fragmented, socially complex, and sensorially overwhelming, cities could nonetheless be transformed photographically into consumable, digestible panoramas. Most famous were Eadweard Muybridge's 360-degree views of San Francisco, photographed from the tower of an unfinished mansion on Nob Hill in 1877–1878. Like Cogswell's time vessel, Muybridge's panoramas purported to freeze the city in time, allowing his peers to subject it to rational analysis.[56] Both are comparable to Gilded Age newspapers, which, by organizing news, editorials, and advertisements into separate sections, imposed a verbal order on urban life.[57]

Yet despite scholars' assumptions, Muybridge's panoramas were not entirely for contemporaneous consumption.[58] Echoing time-vessel architects, he envisioned his photographs as a precious resource for future generations, especially the monumental 1878 panorama, which he reserved for a limited edition of nine. Mounted on canvas and bound in leather, the latter were to be locked away in a millionaire's private gallery or library as "sealed vessels within which the heady liquor of the view aged and mellowed."[59] He subsequently wrote that "I hope in the far distant future it will be valued more possibly than it is at present, as a memento of what San Francisco looked like at the end of this present century," especially if "the names of the principal buildings and places" were inserted, as he did on another version. Like a time vessel, he urged it to be "kept from the light as much as possible," so that it remained in "pretty well as good a condition *a hundred years hence* as it is to day."[60]

Cogswell was not alone, then, in seeking to preserve a totalizing "view" of the city for posterity. There may even be a direct connection between time vessel and panorama. As an earlier commentator wrote, "a bird's-eye view of New York" positively invites "contemplative and philosophical" speculations about its destiny: whether it will "become the most populous city [in] the world" or fall victim to "wars . . . , pestilence or famine."[61] This affinity between perspective and prospective visions, between looking down and looking into the future, can be traced back further, to Herodotus's famous account of Xerxes's tears. The Persian king's panoramic view of his army from a hilltop prompted his bitter realization that "of these multitudes not one will be alive when a hundred years have gone by."[62]

At first glance, the initial response of Cogswell's fellow San Franciscans suggests they endorsed his all-inclusive, panoramic portrait of their city. The crowd at the dedication reportedly included men, women, and children; the middle and working classes; and native-born and immigrant residents, with a German and a Mexican American invited to add their own blessings. Viewing it as an occasion to assert ethnic as well as civic pride, San Franciscans had decorated "all the buildings in the vicinity . . . with the flags of different nationalities," reported the *Alta*.[63] Afterward, Cogswell reflected in his diary on the "grand success" of the ceremony. According to "all the daily eve[ning] papers," he proudly wrote, "this was the largest [peaceful?] street gathering . . . seen in San Francisco for nearly 30 years."[64]

Upholding Franklin

Despite this impression of magnanimity, ulterior motives underlay Cogswell's monument. The fountain was not just to quench people's thirst, cool them down in the summer (with the aid of Cogswell's self-invented, "patent refrigerator" system that allowed ice to be added to the water), or cure their stomach complaints (through the supposed healing properties of Vichy water). More subtly, it was to inculcate the doctrine of temperance by offering an alternative to the temptation of alcohol. In his letter to 1979, he alluded to the difficulty of obtaining a free glass of water in San Francisco, which still lacked a municipal supply and thus remained at the mercy of Spring Valley Water Company's rates. "If a person goes into a saloon and asks for a drink of water and did not spend some money for [alcoholic] drink or cigars," he complained, "this person would be looked upon as a mean or parsimonious person." Moreover, to visit those saloons and taverns was to expose oneself to moral corruption.[65] Although Cogswell officially disavowed any direct affiliation to temperance organizations, he divulged his aims at the unveiling by selecting as his master of ceremonies Francis Murphy, a famous revivalist-style temperance orator from New England.[66]

The Franklin monument was not Cogswell's only temperance fountain. He donated five to San Francisco and fifty to towns and cities across the country—including Washington, Boston, and New York—in the last three decades of the century, most of them with cornerstone boxes containing documents and objects.[67] A disciple of Rev. Lyman Beecher's creed of "total abstinence" and a trustee of the temperance organization Boys and Girls Aid Society, whose journal *Our Head Light* he included in the vessel along with temperance songs, Cogswell allocated much of his fortune to these ornate monuments.[68] His ultimate goal was apparently to erect one for every hundred saloons in the nation—which would have amounted to a thousand.[69] Initiated and underwritten by a single benefactor, they differed from conventional monuments, which were overseen by monument associations and paid for by public subscription.

Although Cogswell dictated the monuments' design, the recipient cities could decline planning permission or refuse to pay for the pedestal, shipping, installation, water supply, or upkeep. The San Francisco Board of Supervisors delayed approving the Franklin fountain for several months and then rejected the roof and awning he had included in the design.[70] Even after the unveiling, the supervisors balked at having to pay for its maintenance and did not formally accept the gift until 1881.[71] Cogswell attributed this uncooperativeness to their corruption by the liquor interests. His letter to the historical society of 1979 leveled a charge of "moral" "lax[ity]" against the Populist mayor Andrew Bryant and his largely Democratic supervisors. "Formally . . . engaged in the sale of intoxicating liquors" and now "largely elected and held in power by this class of their constituents," they discourage "by all the means in their power . . . anything that will not" enrich those interests.[72] The brewery-owned "saloon system" did indeed dominate the economic, cultural, and political life of Gilded Age San Francisco, offering the laboring poor free meals, shelter, entertainment, camaraderie, and even cashing their checks in exchange for political loyalty at elections.[73] Its power—together with the preponderance of foreign-born and Catholic residents, and the cultural acceptance of inebriation as an inevitable by-product of the city's dynamism—conspired to thwart the local temperance movement, which failed even to secure state regulation of the sale of alcohol in the early 1870s.[74] Realizing the power of demon rum in San Francisco, Cogswell decided publically to downplay the temperance rationale for his fountain and time vessel. "If I were at present to disclose <u>my true object</u>," he confided to the historical society of 1979, "I am doubtful if it would succeed."[75] Yet through the medium of a time vessel, he could expose those politicians to a *future* public. In contrast to his peers, that future public would be grateful for the gift, especially if it had succeeded in eliminating alcohol—and thus crime, vice, vagrancy, indigence, and every other social evil it allegedly caused.

Cogswell's temperance scheme in turn concealed a deeper class agenda. Not simply a moral and religious crusade, the antisaloon movement was mobilized by urban capitalists to instill bourgeois values of self-mastery, discipline, independence, and frugality among the working class and thus ensure productivity on the factory floor and lawfulness on the streets.[76] Cogswell presented himself as an exemplar of middle-class individualism, a role model, in his own words, "worthy of imitation" by the "young."[77] He emphasized his Horatio Alger–like rise from poverty and orphanhood to riches and respectability not through reckless speculation in mining stocks or other risky enterprises but through sheer exertion, moral fortitude, and self-education in addition to the "kind[ness]" of the "the great and good god who protects the fatherless" and the "benefit of Sabbath School instruction."[78] He posed as the embodiment of Protestant frugality, living not in an expensive mansion or fashionable hotel but in the more modest Commercial Hotel, a "second class" establishment offering "good accommodations at very moderate prices."[79] It is not surprising, then, that he chose Franklin, high priest of self-reliance, thrift, and other virtues associated with the bourgeois-capitalist ethic, to stand guard over his vessel.

Franklin's eighteenth-century virtues were decidedly out of step, however, with 1870s San Francisco. With the advent of industrialization—delayed in California by labor shortages, correspondingly high wages, and the expense of importing raw materials, but eventually triggered by the Civil War, which broke the city's dependence on eastern products—San Francisco had evolved from a mercantile port into a major manufacturing metropolis.[80] While in 1860 industry lagged behind other sectors with only 1,546 employees (2.6 percent of the population), by 1880 it was the largest, with 28,442 (12 percent), elevating the city from twenty-second to ninth in the nation in manufacturing output.[81] Although Cogswell's vessel included statistics celebrating "American Industrial and Commercial Triumphs," the implications for San Francisco's social structure were disastrous.[82] As the city divided into a wage-earning working class and a propertied elite, it presented fewer opportunities for social and occupational mobility. Whereas 16 percent of the city's elite in the 1850s had begun as blue-collar workers, only 5 percent had by 1880—a statistic resembling that of older, eastern cities.[83] There was a similar decline in rates of downward mobility, as the elite shored up their wealth in real estate and other assets. These solidifying class boundaries were increasingly visible at street level.[84] "Behind the palaces" of Nob Hill, wrote the Mexican poet Guillermo Prieto in 1877, "run filthy alleys, or rather nasty dungheaps without sidewalks or illumination, whose loiterers smell of the gallows."[85] Cogswell's vessel thus upheld (literally) the self-made man just as industrial capitalism was rendering it an illusory ideal. In choosing Franklin as its messenger to the future, Cogswell reaffirmed the continued relevance of the Protestant work ethic.

The effect of industrialization on social mobility was compounded by the depression of the 1870s. That depression was particularly devastating and long lasting in California, where local speculators had overinvested in real estate in the expectation that the transcontinental railroad would boost land values; its completion in 1869 brought only a glut of eastern products and laborers.[86] An even greater speculative mania in mining stocks, the Comstock Lode silver boom, collapsed in 1875–1877, bringing down the Bank of California and exposing the incompetence and (according to one contributor to Cogswell's vessel) "gross frauds" of mining executives.[87] These financial setbacks coincided with a severe drought in 1876–1877, which crippled agricultural production and hydraulic mining. Numerous businesses were bankrupted, including the German immigrant Charles Otto's hardware company, a failure solemnly recorded by his son for Cogswell's vessel.[88] Even worse affected was the city's working class. By 1877, wages had plummeted and more than a fifth of white males were unemployed, while thousands of Chinese Americans, discharged by the railroad and mining companies, migrated to the city in search of work. Many of the unemployed ended up as vagrants, dependent on benevolent associations and workhouses for survival.[89] Younger men and even boys who resorted to mugging and theft acquired the epithet *hoodlum*, a neologism coined by the local press in the 1870s. Loafing around street corners and store entrances in ragged clothes, they represented the negation of Cogswell's cherished ethic of work, self-reliance, and self-improvement.[90] In his letter to the future, he complained of this "dangerous class of young men called hoodlums" who "have become a terror to good citizens." Those "caught out at late hours in the night . . . are frequently garroted[,] rob[b]ed[,] knocked down[,] shot[,] slingshotted[,] or otherwise maimed."[91]

Capitalist individualism came under more direct attack during the flashpoints of class and labor conflict ignited by these economic conditions. San Francisco, with its relatively high wages, weak unions, and glutted labor market, escaped the kinds of strikes seen elsewhere.[92] Yet there were signs of emerging militancy. In 1868, an impoverished San Francisco typesetter and union member began to articulate a radical critique of capitalism that eventually attracted a national following. Henry George accurately predicted that year that the railroad would not enrich all, as promised, but widen the gulf between rich and poor by making real estate prohibitively expensive.[93] He elaborated this critique into his magnum opus *Progress and Poverty*, completed three months before Cogswell's time-vessel sealing. Although encompassing the political economy of all civilizations since antiquity, he cited his own city's growing class divisions and unemployment as evidence that it was not exempt from the problems associated with older and larger cities. George's dismay was all the deeper for having once envisioned San Francisco as a city of equal opportunity for eastern migrants like himself.[94] "For most Californians," writes historian Alexander Saxton, "there was no place further to go. Literally and symbolically, San Francisco was the end of the line."[95]

Further signs of incipient dissent arose during the troubled year of 1877. Although California's railroad workers did not join the Great Railroad Strike, many San Francisco citizens—and all but one of its major newspapers—were sympathetic to the cause or at least critical of the railroad corporations.[96] The socialist Workingmen's Party of the United States held nightly demonstrations in the "sandlots" opposite the unfinished city hall. At one such mass meeting, on July 23, violence finally erupted in San Francisco. A local gang dissatisfied with the orators' neglect of the so-called Chinese problem interrupted them with cries of "To Chinatown!" Two days of mob violence ensued against the Chinese and their property and against companies that transported or hired them. Only by resorting again to vigilantism did the bourgeois elite restore order.[97]

In the aftermath of the "July Days," a new threat surfaced. An Irish American "drayman" or beer deliverer, Denis Kearney, emerged as a leading soapbox speaker in the sandlots and formed his own Workingmen's Party of California (WPC). Despite its name, the WPC was neither socialist nor allied with organized labor; its platform largely consisted of denunciations of corrupt politicians and complaints about the Chinese "problem."[98] Yet Kearney's occupation of downtown streets and even Nob Hill, with his band of workingmen and his calls to "hang, shoot, or cut the capitalists to pieces," provoked fear in the elite.[99] By the time Cogswell unveiled his Franklin fountain, the WPC had parlayed its support into electoral votes, winning all thirty of San Francisco's delegates (and a third of all delegates) to the convention for rewriting California's constitution.[100] With another round of elections just weeks away, Cogswell's vessel was sealed against the backdrop of ongoing street skirmishes and rancorous political campaigning. Endorsements of Kearney's famous slogan, "The Chinese Must Go!," even found their way into the vessel.[101]

Although Kearney proved a paper tiger and the WPC dissolved the following year, documents in Cogswell's vessel reveal the extent to which the elite feared a repeat of the Paris Commune and New York draft riots. Kearney's rise appeared to confirm English historian Lord Macaulay's earlier prophecy—reprinted in the latest issue of the San Francisco literary journal *The Argonaut* and deposited in the vessel—that "[economic] distress everywhere makes the laborer mutinous and discontented, and inclines him to listen with eagerness to agitators who tell him that it is a monstrous iniquity that one man should have a million while another cannot get a full meal." Americans would be faced with a dilemma: allow "some Caesar or Napoleon" to restore order at the expense of liberty, or let civilization be "plundered and laid waste by barbarians"—barbarians "engendered within your own country by your own institutions." The editor who reprinted Macaulay's words viewed the recent strikes, riots, and rise of Kearney as evidence that the "pernicious doctrines of Socialism" had already taken root in the United States.[102]

The more conservative newspapers deposited in the vessel also testify to the apocalyptic fears provoked by the WPC's dominance of the constitutional convention. Adoption of the new constitution, the *Alta* warned, "would be little short of a calamity," as its taxation and anticorporate provisions would strike a fatal "blow" to the "prosperity of the state" and usher in a "reign of the brutal and ignorant." San Franciscans should "prepare for the worst," for a fate "infinitely worse than any [natural] disaster she has ever experienced in the past."[103] Such apocalyptic rhetoric, in George's view, rendered "calm discussion . . . impossible."[104] Although Cogswell abstained from addressing these controversies in his own messages to the future, he vented his political views in his diary. On May 7, 1879, as Californians headed to the polls, he declared that if they voted to ratify the constitution, this would be regarded "by all the better class . . . as an evil day . . . & so says HDC." Although he might have approved of the clauses guaranteeing women the right to enter the professions and universities—the result of successful lobbying by his lawyer Foltz and time-vessel contributor Gordon—he appears to have shared the *Alta*'s view of the yes vote as a victory for the "revolutionary schemes" of a "miserable pack of ignorant and degraded scoundrels."[105]

While some advocated repressive responses to the specter of class insurrection—strengthening the police, deploying troops, constructing armories, curbing immigration, and hanging anarchists—Cogswell aligned himself with those favoring a gentler, ameliorative approach.[106] The only "radical cure for revolutions," declared a *Catholic Monitor* issue he deposited in the vessel, was to inculcate industriousness and faith among the masses.[107] This idea of uplift motivated him to found a "home" for working women in the Barbary Coast and a dental college (with relief fund for "indigent" students) at the University of California—both documented in the vessel. The need to counter the temptations of "gin shops" and "revolutionary brigades" was a rationale his contributor Cooper also employed to advocate kindergartens.[108] Like Mosher's and Deihm's safes, Cogswell's vessel thus embodied reformist hopes.

The refashioning of hoodlums into respectable citizens was a task Cogswell expected his Franklin fountain itself to perform. He dedicated it to "our boy and girls" so that "they have some stock or ownership in it which will create in them a kind of pride which will cause them to patronize and protect the fountain from mutilation or distruction [*sic*], and eventually cease visiting these liquor saloons and gradually make better associates and I hope become good citizens, as the power of example even on some of the worst subjects is usually for good."[109] To further this goal, he planned to appoint as the fountain's official custodian "some unfortunate who might be on the street begging or stealing." In return for polishing its brass rails and faucets, cleaning the surrounding sidewalk, and perhaps also protecting the vessel from thieves, this boy would be permitted to sell his "wares or [polish] boots etc and thereby get an honest living without begging and would make one of his class self reliant and an example to other boys."[110]

Cogswell's time-vessel fountain, like his women's shelter and Cooper's kindergarten, was located in the city's most notorious slum and vice district precisely to redeem that underclass.[111] The intersection of Kearny Street and Montgomery Avenue was the very epicenter of the Barbary Coast, described by a contemporaneous city guide as a "stagnant pool of human immorality and crime" contaminating the "surrounding blocks" like a malarial swamp.[112] Cogswell knew the neighborhood well; his "second class" hotel was less than a block away. To help decontaminate that slum, he designed lamps to encircle Franklin. If, as the city guide claimed, hoodlums were known to hang out on "dimly-lighted street corners" and in "dark alley-ways and nooks," those lamps would lure them out into the light, rendering them visible and thus governable.[113] The fountain, then, was part of a bourgeois-capitalist struggle to reclaim the city's public spaces from the "dangerous classes," which it would literally enlighten.

Cogswell's time-vessel messages did not refer explicitly to Kearney or the WPC, but their frequent references to "hoodlums" can be read as implicit allusions. Both mainstream parties used that term as a pejorative for the insurgent workingmen.[114] By characterizing the 1877 riots as the "malicious mischief" of "a few hundred vicious boys," Democrats absolved their affiliates, the Anti-Coolie Clubs, and insinuated that the Republicans had exaggerated the crisis to increase their powers.[115] For Republicans, the hoodlum theory deflected attention away from their own share of responsibility for the unrest; they had spurred on the WPC in the hope of dividing the Democratic vote.[116] As articles in the vessel pointed out, both parties were complicit insofar as their corruption and chicanery drove San Franciscans into Kearney's arms.[117] Above all, blaming a few indolent hoodlums enabled the bourgeois elite as a whole to disavow the deeper social and economic forces that had generated the unrest. The discourse of hoodlumism transmuted a class problem into a moral and generational one.

Demonizing the WPC as hoodlums served, furthermore, to conceal the elite's implication in the racial politics of Sino-phobia. Although historians have emphasized working-class animosity toward the Chinese, a pamphlet included in the vessel reveals the involvement of the city's leading citizens. Published in 1869 by the California Immigrant Union (CIU), it responded to the state's economic "stagnation" by condemning the Chinese on both economic grounds ("their earnings all leave the country[,] [t]hey consume very few articles of American manufacture," and then they "return to China"), moral-cultural grounds (their supposed resistance to Christianization, their failure to learn English, and the prostitution of their women), and political grounds (their ignorance of the American tradition of liberty).[118] The CIU sought instead to promote respectable, skilled immigrants from Northern Europe, "those liberty-loving races *whose descendants we are ourselves,*" by dispatching agents and pamphlets to those countries and facilitating land purchases in rural California.[119] The pamphlet's author was not a Kearneyite workingman but a leading businessman and promoter of regional railroads. His CIU was "composed of responsible and prominent business men" and even boasted four governors, including Leland Stanford.[120] Their ulterior aim may have been to undermine organized labor by bringing in new workers to dilute its bargaining power and by co-opting its Sinophobic rhetoric. Cogswell's vessel thus followed Deihm's and Mosher's in endorsing white middle-class hegemony and the capitalist status quo even while conveying reformist sympathies for the laboring masses. Just as the centennial safes concretized a faith in the durability of the nation and its institutions by addressing its officials a century hence, so did the Antiquarian Box imply a faith that San Francisco (and its "historical society") would survive the current troubles.

The Fate of the Franklin Fountain

At nine a.m. on July 3, 1879, Cogswell visited his local police station. Concerned, perhaps, that celebrations of Independence Day or disagreements over the new constitution, which came into effect that day, would spill over into street violence, he sought to "engage [a] policeman" to protect his new monument.[121] Only three weeks after its dedication, the Franklin fountain and time vessel had already attracted verbal and physical abuse, which Jerome McCarthy, the street boy he had appointed as its custodian, was powerless to prevent. In an undated letter to the board of supervisors, Cogswell complained of "cups and chairs" being "stolen by vandals that have no fear of detection in the night," the supervisors having denied his application to install lamps.[122] Far from remedying the situation, the supervisors allowed the fountain to suffer further deterioration and neglect. Finally, in June 1904, four years after Cogswell's death, it was dismantled and banished to a secluded spot in Washington Square Park—an act that, according to one theory, precipitated the great fire and earthquake two years later, as Cogswell had allegedly placed a kind of pharaoh's curse on his vessel, wreaking a natural disaster on the city if anyone disturbed it before its target date.[123] In its new location, it continued to be troubled by vandals, and its pipes were soon capped.

The Franklin monument was not alone in suffering such mistreatment. A host of indignities were heaped on another statuary fountain Cogswell installed four years later on Market Street, two blocks from the Ferry Building (**fig. 2.3**). There were reports of boys inserting dead rats into the pipes to startle unsuspecting drinkers, wrenching off the faucets, and assaulting the statue with cobblestones or peashooters.[124] Vehicle collisions inflicted further damage, causing its lamps to lean so much that they reminded one observer, ironically, of lurching inebriates.[125] In the early hours of New Year's Day, 1894, a crowd of bohemian writers and artists put it out of its misery, pulling it down with a rope, decapitating it, and amputating its limbs.[126] Cogswell fountains in other cities were also objects of iconoclasm during this period. One in Rochester, New York, was attacked with crowbars, and another in Rockville, Connecticut, was thrown into the local lake.[127] Elsewhere, city governments preempted such attacks by removing the offending fountains. Of the more than fifty Cogswell donated, only a handful still stand today, including one opposite the National Archives in Washington, thanks ironically to an exclusive drinking club, the Cogswell Society, whose members meet to toast the good doctor and fabricate stories about his great accomplishments.[128]

While all monuments, as Georges Bataille and other theorists have argued, are liable to incite popular violence—physical and symbolic—Cogswell's clearly provoked an unusual degree of animosity.[129] To some

FIG. 2.3 Anonymous, *Monument, statue, and drinking fountain, to [sic] Henry D. Cogswell, ca. 1888. Corner of California, Market, and Drumm Sts*, photograph, ca. 1888. Roy D. Graves Pictorial Collection, BANC PIC 1905.17500 v.10:9—ALB, Bancroft Library, University of California, Berkeley.

extent, this can be attributed to public antipathy for the cause they promoted. The temperance movement remained unpopular, especially in cities dominated by the economic and political power of the breweries and their saloons, and by the antiascetic values of Catholic immigrants from Ireland and Germany. Concerns about the cleanliness of the water supply and the hygiene of a shared spout together with belief in beer's health benefits further contributed to the rejection of temperance fountains and a preference for public amenities such as toilets.

Cogswell's decisions regarding the fountains' design did not help their cause. Many were topped with an anonymous temperance orator holding out a glass, presumably of water. Duplicated across the country, the "silent orator" was widely criticized on aesthetic grounds. The Brooklyn *Eagle* branded it a "monstrosity," a "hideous creation," and a "monumental eyesore"—"absolutely destitute of a single element of [artistic] merit . . . a standing contradiction of every principle of art and a libel on American workmanship."[130] Mark Twain issued a similar judgment when he saw the version in Rochester, New York, in 1884: "The man looks as if he'd been nine days drowned. It has a putrid, decomposed sort of a look that is offensive to a delicate organism." He advised locals to form a mob and throw at least the top half of the statue into the Erie Canal.[131] A further problem was the "silent orator's" unfortunate resemblance to Cogswell. Newspapers across the country denounced him for using his fountains to aggrandize himself.[132]

But the hostile reception of Cogswell's monuments, including the Franklin fountain and time vessel, also betrayed deeper currents of class resentment. Local communities may have objected to the notion of bowing to take a drink, as if in humble gratitude to the munificent millionaire. San Franciscans certainly objected to the expectation of gratitude toward the Spring Valley Water Company, which had offered to provide the free water for the Franklin fountain. At the dedication, the mention of that infamous, monopolistic company—then under attack from the WPC for exploiting a common, natural resource—aroused loud boos from the crowd, and a "decided Sand-lot yell: No! no!"[133] By paying its high rates, ordinary San Franciscans were effectively underwriting Spring Valley's supposed benevolence. Henry George aimed a similar argument against Cogswell's philanthropy. As the latter financed his putative gifts with endowments of land, it was the city's poorer residents who were funding them through rent payments. He was thus merely a "philanthropist by proxy."[134] George also exposed Cogswell's guiding doctrine, temperance, as a capitalist folly. "You cannot . . . root out [in]temperance, or prevent destitution," he declared in an 1878 speech at the workingmen's stronghold, the San Francisco Metropolitan Temple, "until you . . . stop the tendency of wealth to concentrate in a few hands." The appearance and disappearance of those "admirable virtues" of "prudence, economy, and temperance," he continued, are directly proportionate to the rise and fall of wages.[135] The only real and lasting solution to social problems was to redistribute the land.

George would likely have objected, too, to the conceit that the Franklin fountain would preserve the memory of contemporary civilization. Cogswell's initial letter to the board of supervisors had argued for the vessel on the grounds that its contents would demonstrate to future San Franciscans that their predecessors were "tolerably well civilized," no doubt to refute the stereotypes of a frontier boomtown but also, perhaps, to counter the newspaper reports of a city overrun with marauding workingmen. Like Muybridge's 1878 panorama, whose fifteen-minute exposures depopulated the city's streets and emphasized the grandeur of its architecture, the vessel ultimately played down the disorder.[136] In an essay on Kearney, George repudiated such assertions of civilization, claiming that they merely papered over the imminent threat of class apocalypse.

> It is true that we have the public school and the daily paper. . . . But, "scratch a Russian and you have a Tartar." Look at your civilized man when fired by that strange magnetic impulse which passion arouses in crowds, and you read in his eyes the blind fury of the Malay running amuck. You will understand how handkerchiefs hemmed with the sewing machine might be dipped in blood, and hearts carried on pikes through streets lit with gas![137]

In the event of such a catastrophe, George argued in *Progress and Poverty*, efforts to preserve civilization—such as the practice of burying some "mementos" in a "corner stone" of public buildings in preparation for "the time when our works shall be ruins and ourselves forgot"—would prove futile. The coming class war, waged on the vast and treacherous battlefield of industrial cities and with modern weapons such as "petroleum, nitro-glycerine, and dynamite," would exceed all prior wars in destructiveness.

> It is startling to think how slight the traces which would be left of our civilization [were] it [to] pass through the throes which have accompanied the decline of every previous civilization. Paper will not last like parchment, nor are our most massive

buildings and monuments to be compared in solidity with the rock-hewn temples and titanic edifices of the old civilizations.

The aftermath of American civilization would thus resemble earlier dark ages, when bands of "squalid barbarians" lacked "even the memory of what their ancestors had done."[138]

Such objections indicated that for all Cogswell's efforts to invigorate and expand practices of civic memorialization, his approach had significant limitations. In employing a traditional statuary monument as a vehicle for his time vessel, he rendered the latter vulnerable to acts of egalitarian iconoclasm. In inscribing his own name on the pedestal, establishing himself as the sole overseer of its "antiquarian" contents and eschewing any collaboration with established institutions or municipal officials, he undermined its claim to be a collective effort of commemoration by the community. And in harnessing the monument and vessel to his cherished political cause of temperance, he provoked indignation from working-class immigrants and condemnation from radical critics like George. The zinc figure of Franklin thus became a lightning rod for various competing futures: a capitalist utopia of individualistic teetotalers, a Sinophobic utopia of white workers, and a Single-Tax utopia of classless abundance.

THREE

Annals of the Present, the Local, and the Everyday: The Centurial Time Vessels as Heterodox History, 1900–1901

If Charles Mosher, Anna Deihm, Henry Cogswell, and others jointly but independently conceived of the time vessel, the negative responses—ranging from indifference to denunciation—indicated a difficult birth. The only vessels launched during the remainder of the century were modest and underpublicized affairs: a copper box deposited in Boston's Faneuil Hall by the Ancient and Honorable Artillery Company of Massachusetts in 1881; a "memorial chest" dedicated by freemasons in New York City's Masonic Temple in 1894; and a series of "century boxes" for the centennial of three Ohio cities: Marietta and Cincinnati in 1888 and Cleveland in 1896.[1] Depositing materials for a specified interval of time had yet to ignite the public imagination.

Yet during those years, the idea was fleshed out in utopian fiction. Transported through hypnosis, dreams, or some mystical or technological device to a utopia now located in another time (often fifty or a hundred years in the future) rather than another place (the original meaning of *utopia*), protagonists invariably took material belongings with them. The rooms in which they time traveled thus served as de facto time vessels. In Edward Bellamy's *Looking Backward* (1888), the novel that spawned so many of these utopian fictions, Julian West's subterranean sleeping chamber preserves certain objects: a gold pendant locket containing a photograph of his fiancée, Edith Bartlett, which moves her great-granddaughter Edith Leete to kiss it, and a Boston newspaper reporting labor unrest, which amuses her father, Dr. Leete. It also preserves a secondary vessel within it: a safe containing "several thousand of dollars in gold, and any amount of securities" along with a packet of private letters from his fiancée, as his hosts discover in the sequel *Equality* (1897). These ancient relics function not simply as romantic relief from the lengthy expository passages about utopian institutions but also as devices to emotionally transport the reader into that utopia.[2]

Whereas West's transmission of objects to the future was inadvertent, the result of his mesmerist accidentally inducing a deep trance, other utopian novels featured protagonists who *chose* to travel in time and were therefore more deliberate in their depositing of objects. In Alvarado Fuller's *A.D. 2000* (1890), Junius Cobb, a disillusioned soldier, fills an iron box with "papers and valuables which I hope to again see and use" and a valise with "a few articles" before locking himself in his time machine located inside the Statue of Liberty in San Francisco's Ashbury Heights. The inventor whose crystallized ozone powered Cobb's time machine placed another vessel within it: a "hollow rod of copper" containing a letter.[3] Indeed, time vessels proliferate in Fuller's novel, cropping up at critical moments in Washington, DC, on remote islands, and even at the North Pole.[4]

Time vessels also materialized in dystopias. In Mark Twain's short story "The Secret History of Eddypus, the World-Empire," written in 1901–1902 but not published until 1972, a thirtieth-century historian narrates the rise to power of Mary Baker Eddy's Christian Science movement. After the collapse of the American republic, Eddy imposed a theocratic state, which eventually colonized all other countries except China. Now, under the Eddymanian Popes, religious freedom is suppressed, books are destroyed, and the historical record is altered. One of the only links to the past is a manuscript that a certain "Mark Twain" had written early on in the struggle and deposited for five hundred years in a vault in Washington, DC, before he was hanged. A thousand years lapsed, however, before "shepherds digging for water came upon the vault after piercing through a depth of thirty feet of ancient rubbish, and they broke into it and brought up the relic, all the volumes complete, and all sound."[5] Usually dismissed as an anti-

Christian Science screed, "Secret History" was acutely concerned with the possibilities and limits of communicating across time.[6]

At this point, interest in depositing actual time vessels reemerged, stimulated by the century's turn but perhaps also by the outpouring of utopian fiction. In spring 1900, vessels were sealed at two colleges: a box of diaries, essays, and photographs by professors and students at Harvard, conceived by geographer William Morris Davis and organized by head librarian William Coolidge Lane, in consultation with President Charles W. Eliot; and a box filled with "college things of interest" by Mount Holyoke's graduating class and sealed at its commencement ceremony. Several months later, three further vessels were launched to mark the new century. In Kansas City, Judge Charles E. Moss invited revelers to contribute materials to a "century box"; in a private ceremony in Detroit's city hall, Mayor William Maybury consecrated another "century box"; and at their annual dinner, the *Chicago Tribune*'s staff sealed messages of greeting and hope in a similar box.[7] On August 4, 1901, citizens of Colorado Springs contracted the time-vessel bug, gathering at Colorado College under the aegis of businessman Louis Ehrich to deposit their "century chest." In the subsequent months, the practice spread to Massachusetts, where boxes were dispatched at the First Congregational Church of Rockland and the Peabody Institute Library.[8] Finally, Twain himself developed a time vessel of sorts during these years: a long, autobiographical manuscript whose publication he planned to defer, initially until 2000, as the president of Harper and Brothers suggested in 1900, and ultimately for a hundred years after his death.[9]

Although the architects of these centurial vessels (as I call them) followed in Deihm's, Mosher's, and Cogswell's footsteps, there was no acknowledgment of precursors. "The conception of the 'Century Chest' originated with Mr. Louis R. Ehrich of Colorado Springs," claimed one of its contributors, "and this community was the first to carry it into effect. All others are copies."[10] Even Twain's "pet scheme" of postponing publication was heralded by the London *Times* as "unparalleled in the history of literature."[11] (In fact, it had been in use since at least 1792, when the steamboat's inventor John Fitch recounted indignities and denounced adversaries in six volumes of manuscript, depositing them in the Philadelphia Library and stipulating that they be opened and published thirty years after his death "so as not to embarrass" his children—instructions that were duly carried out).[12]

Despite their denial or unawareness of precursors, these figures reinforced an ongoing expansion, already initiated by Cogswell, of the time vessel's functions. They sought not merely to memorialize politicians, the business elite, or themselves but also to convey information documenting a larger social whole—the city or the college campus—in part for the future historian's benefit. If Deihm's and Mosher's vessels had evoked the funerary practices of embalming and death masks, these new ones aspired to the epistemological status of an archive. They thus included materials such as typed or mimeographed documents that did not bear the material trace of any human body. Embracing photography and even phonography for its documentary rather than auratic value, they sought to capture the typical and to record urban and domestic environments. They used essays and journals to communicate diverse aspects of the urban or collegiate community, from its governing institutions to its quotidian activities. And to impart the feel of everyday life, they welcomed material artifacts too. They imagined recipients using these sources to reconstruct their community's political, social, and cultural life at the dawn of the twentieth century.

In so doing, the centurial vessels implicitly challenged contemporaneous historical practices. Few historians went beyond political, military, and "great man" history, and those who did tended to be amateurs, whose influence was by then waning.[13] The historical profession was coalescing around the model of a "scientific" history of the nation that drew on written records, particularly government documents, to trace the evolution of laws and institutions. By contrast, local history, cultural beliefs, and material artifacts were largely spurned as the province of "unscientific" antiquarians, relic hunters, dilettantes, and, worse, female historians.[14] Few contributors to the centurial vessels were historians, but through their writings and offerings they implicitly expressed beliefs about what would and should count as history. They thus contributed to the expansion and democratization of historical practice.

While constructing local archives of the present, however, these vessels did not renounce the earlier function of memorialization altogether. They echoed their predecessors' faith in extending the temporal reach of the human body, in particular by articulating a fantasy of physically touching or embracing their future recipients. With the elaboration of features such as the sealing of secret messages, the centurial vessels stimulated notions of an affective relationship across time. They thus articulated an "embodied future" that countered the "abstract future" so confidently projected by scientists and social scientists of the period. This yearning for intimacy across time generated, in turn, a sense of ethical duty to posterity. The time vessel itself could serve as a kind of pledge and reminder to work actively toward a better future.

Reaching Out to the Millennium

If the deferred sealing of Deihm's and Mosher's vessels in 1879 and 1889 proved anticlimactic after their initial success at the Centennial Exposition, those of 1900–1901 were discharged amid much greater fanfare because of their incorporation into turn-of-the-century celebrations. Detroit's, for instance, was part of a larger series of spectacles Mayor Maybury organized. The ceremonial sealing in his office was scheduled to be completed on the stroke of midnight, at which point city hall's exterior would be illuminated with electric bulbs, the national flag raised from its cupola, its four-ton bell chimed twenty times, and a "flight of rockets" fired from the top of a nearby skyscraper. Simultaneously, the city would burst out in light and sound, as the Mayor ordered "every . . . home, . . . every church, factory and business building" to be lit up and "all church bells . . . rung." As it turned out, the mayor mistimed his dedicatory speech, and so all these celebrations—combined with the unofficial pyrotechnics and pandemonium of carousing Detroiters—drowned out the vessel ceremony. Nevertheless, before dispersing, all who witnessed the sealing posed on city hall's snowy steps for a flash-lit photograph.[15]

Even more spectacular, and certainly more successfully orchestrated, was the Kansas City vessel's backdrop. The Century Ball featured historical costumes (for the dancing guests) and took place in a vast convention hall ornamented by ten thousand colored light bulbs (**fig. 3.1**). "The most important . . . feature" of the ball, the vessel was placed at the southern end under a sign. "All evening the box stood receptive," reported one newspaper, and "there poured into it a steady stream of matter." At one point, "a grand march was formed by the dancers and as they filed past . . . they dropped into it any token they chose."[16] On the last stroke of midnight, "dead silence" descended on the fifteen thousand guests, followed by "the bursting out of a cheer, . . . [and] the flashing of a myriad of brilliants across the polished arena."[17]

The timing of these ceremonies accorded with the prevailing belief that the century began on January 1, 1901, rather than a year earlier, on the logic that the first millennium began on AD 1. Although this issue was hotly debated across the United States and Europe through the 1890s, as at every century's end since the 1690s, a consensus eventually settled on 1901, the most notable holdouts being Kaiser Wilhelm's Germany and his colonial possessions, a few loyal German Americans in regions such as the Upper Midwest, and Mary Baker Eddy's Chris-

FIG. 3.1 *The Minuet in Convention Hall Just before the Stroke of 12,* illustration for "The End of the Century in Kansas City," *Kansas City Star,* January 1, 1901. Newspaper clippings related to the Century Box, Century Box Collection, Kansas City Museum, Kansas City, MO.

tian Scientists.[18] In fact, one contributor to the Detroit vessel made the claim (echoed by modern scholars) that neither date was correct; because of an error in the calculation of anno Domini in the sixth century, the celebrations were four years late.[19] Aware that this controversy was lingering from "corner drug store" to "street car platforms," the committee organizing Kansas City's celebrations "appealed to a board of experts" before opting for 1901.[20] Regardless, the postponement only heightened the anticipation.[21] Detroit's and Colorado Springs' vessels were further augmented by the coincidence of local anniversaries: the twenty-fifth anniversary of Colorado becoming a state, the thirtieth anniversary of Colorado Springs' founding, and the two hundredth anniversary of the city of Detroit. The vessels were thus part of a larger cluster of celebrations.

For all its arbitrariness as a temporal marker, the passing of the nineteenth century stimulated, in the United States as elsewhere, not just celebration but also introspection. This often took the form of a collective look backward as critics weighed the social, political, cultural, and technological developments of the previous hundred years. Typically, these journalistic reviews—such as "A Century's Progress" and "The Story of the Nineteenth Century," deposited in the Colorado and Kansas City vessels, respectively—affirmed the idea of progress even while acknowledging some unfavorable or unfulfilled developments.[22] Yet the retrospection inspired by the century's turn could also take a more despondent, nostalgic form: "This dreadful gruesome New Year, so monstrously numbered," Henry James wrote to a friend on January 1, 1900, "makes me turn back to the warm and coloured past and away from the big black avenue that gapes in front of us."[23]

The new century engendered anticipatory visions too. That Americans felt they were on the cusp of a new era is evident in the range of phenomena to which they applied the adjective *new*. The coming epoch was described as that of the New Woman, New Negro, New Negro Woman, New South, New Ethics, or New Democracy.[24] When not busy writing obituaries for the past century, newspapers and magazines were christening the next one, prophesying the tasks and struggles to come in articles such as "The Burden of the Twentieth Century" or "The Time-Spirit of the Twentieth Century."[25] Some articles looked simultaneously backward and forward, as in "The Nineteenth Century: A Review, an Interpretation, *and* a Forecast."[26] Cultural historian Hillel Schwartz terms this bidirectional orientation *janiformity*—after Janus, the two-faced Roman god of beginnings and endings. Although evident in earlier ends of centuries, this threshold experience of being suspended between past and future was particularly intense in the 1890s. It was "an epoch-in-waiting . . . at once an endtime and a betweentime, years of nervousness, decadence, boredom and thrill-seeking, suicide and Ferris wheels, faithlessness and occult philosophies, anarchy and artificiality."[27] It also differed from earlier centuries in its heightened self-reflexivity regarding such cultural phenomena; the term *fin de siècle* was a neologism of the period.[28]

Insofar as all the vessels (except Harvard's) programmed 2000 or 2001 as their target date, they followed another end-of-century tendency: a fixation on the millennium. More than just another centurial event, the year 2000 (like 1000) was overdetermined by Christian eschatological hopes. In the wake of Bellamy's novel, at least five other utopian authors (including Alvarado Fuller and John Jacob Astor IV) published prophecies of that year.[29] In the dying days of 1900, newspapers across the country ran articles—or whole sections—offering similar descriptions and visions of their cities a hundred years hence.[30] One such issue, deposited in Kansas City's vessel, took the form of a newspaper from the future. Dated the 43rd of Saturn, 2000, it was otherwise identical to the "yellow newspapers" of 1900 in its liberal use of banner headlines, exclamation marks, and cheap rag paper and in its reporting of local gossip and scandal alongside "serious" news.[31]

The collecting of predictions was a new feature. Whereas Deihm's and Mosher's vessels contained none and Cogswell's just one, the centurial ones were brimming with them.[32] Mayor Maybury instructed his contributors to summarize the previous hundred years and predict the next hundred, futurology thus proceeding organically from historiography. His choice of contributors covered a wide range of topics, from Detroit's finances, labor relations, and police department to its parks, streets, vehicles, schools, churches, and newspapers.[33] Predictions even extended to trivial and personal topics. In his contribution to the Harvard chest, Professor Adams Sherman Hill speculated whether Sunday newspapers would exist when it was opened (he hoped not), whether meteorological forecasts would be more accurate, and whether his grandson would be chief justice of Massachusetts or "a common laborer or a farmer."[34]

These vessels were in fact part of a larger boom in prophecies. Although divination had enjoyed a privileged, religious status in ancient Egypt, Greece, and Rome—with quasi-divine powers conferred on oracles, sibyls, and seers—it fell into disrepute thereafter. Associated with insanity, folly, witchcraft, and paganism, divination was prohibited in Judaism and Catholicism and survived only on the social and geographical periphery in folk cultures, fairgrounds, and among itinerant fortune-tellers.[35] The emergence of modern science fiction and utopian fiction in the late nineteenth century finally legitimated, rationalized, and secularized the practice.[36] The prediction of future worlds even claimed the status of a science—subsequently termed *futurology*. In *Anticipations of the Reaction of Mechanical and Scientific Progress upon Human Life and Thought* (1901), H. G. Wells disavowed the loose, extrapolative prophecy of his 1890s "scientific romances" for a more rigorous, inductive method modeled on geology and archaeology and involving the extraction of "operating causes" from recent history and the calculation of a kind of "average outcome."[37]

Utopian and science fiction authors were not alone in advancing "scientific" prophecy. The rise of Darwinian evolutionary theory and modern statistics inspired scientific "experts"—from sociobiologists and demographers to biometricians and eugenicists—to project various social and biological futures. William Morris Davis, the conceiver of the Harvard vessel, was the founder of geomorphology (the study of how landforms have evolved over thousands of years through the "cycle of erosion") but also a key proponent of environmental determinism, the notion that civilization's future can be predicted from topography, climate, and other physical facts.[38] Positivist methods similarly emboldened economists, sociologists, and criminologists to forecast financial fluctuations, social pathologies, and crime rates.[39] These scientists and social scientists arguably sought to contain the threatening indeterminacy of the future and the radical singularity of any future event. They represent what theorist Elizabeth Grosz calls "the attempts of the old to contain the new"—to nullify its threat by incorporating it into its "already existing frameworks."[40]

The forecasts in the centurial vessels, however, often diverged from those of "experts." Not only were they authored by people of diverse social backgrounds and political viewpoints (as we shall see) but they were also sealed away rather than publicized, thus disclaiming any intent to become self-fulfilling prophecies. Many also acknowledged the doubts and contradictions inherent in any prediction. Whereas Wells claimed in his 1902 lecture, "The Discovery of the Future," that the future (of societies, not individuals) is just as "fixed," "determinate," and thus knowable as the past, time-vessel contributors were typically more skeptical.[41] The president of Detroit's chamber of commerce began with the disclaimer that "one can but speculate" about the future and that such speculations "may fall far from that of the reality."[42] In beginning his aforementioned predictions with the words "I wonder whether," Professor Hill similarly hedged them as vague conjectures or daydreams. The centurial prophecies were thus more akin to utopian wishes. Detroiters wrote them, according to one contributor, "with hesitation and doubt, but yet in hopefulness," and the Kansas City box contained numerous calling cards inscribed with wishes such as "May the dreams and ideals of the closing century be realized in the new."[43] These time vessels were, in effect, modern wishing wells or secularized prayer boxes promising to grant any wishes dropped into them.

According to the Marxist "philosopher of hope" Ernst Bloch, such wishes need to be taken seriously as repositories of utopian longings that project a better future.[44] Unlike the closed, deterministic future of the scientific experts, the future imagined in time vessels was an open-ended realm of potentiality—although not a nebulous one, as the time span oriented the wish to a definite due date. "My mind turns," wrote Ehrich in Colorado, "on all the *possibilities* which these hundred years may bring, on the various stations in life in which these lines may find you."[45] The French phenomenologist Eugène Minkowski sheds further light on this radical openness of hope. Whereas everyday activity and expectation keep us fixated on an immediate future, he wrote in the 1930s, hope (and desire)—mediated by other persons, by things or institutions, or, in our case, by a metal box—can project us into a future beyond our lifetime, giving us "an intimation of all that could be in the world." Rather than fearing the future (as Henry James confessed to), messages such as Ehrich's exemplify Minkowski's notion of affirming it.[46] They thus contradict historians' assumption that Americans responded to the social upheavals of modernity merely by seeking refuge in nostalgic pasts.[47]

These vessels sought, moreover, to forge an intimate and personal relationship with those distant, mediated strangers. Rather than reducing the future to abstract forecasts and statistical extrapolations, contributors apprehended it in more direct and embodied ways. "A number of persons," wrote the *Kansas City Star*, "have availed themselves of the opportunity, through the medium of the century box, to communicate with the shadowy beings of the Twenty-first century."[48] As with their precursors, this intimacy was instilled by the sense of remote presence that time vessels, like other one-to-one communication devices, afforded. Yet while addressed on their exterior to some general collectivity, such as the town's future citizens or the college's class of 2000, the centurial boxes also contained many documents addressed to specific individuals. Those addressees were sometimes blood descendants, reflecting the democratization of genealogy, especially among Anglo-Americans and those who had migrated westward.[49] But more often they were professional or official successors, such as the president of Colorado College or of the local chapter of the Young Women's Christian Association (Colorado Springs); the chief of police, the pastor of the Westminster Congregational Church, and the president of the New Boys' Union (Kansas City); and the city controller and the mayor (Detroit).[50] Others addressed themselves to subgroups: Colorado College's male students, professors' wives, and any "native Americans . . . or the descendants of such who may be living in the year 2000"; Kansas City's Christian Scientists; and Detroit's Scots, Jews, working classes, and surgeons.[51] (Only one contributor was so bold as to address, as Deihm and Mosher did, the future president of the United States.[52]) Although etymologically linked to sexual reproduction, posterity could be given bodily form without presuming a biological or "blood" relation.

Each of these inscriptions was animated by a fantasy of miraculous communication akin to that of "spiritual telegraphy" (conversing with the dead through a spirit medium) and "thought transference" (telepathy). This kinship would explain why these vessels attracted so many Christian Scientists and spiritualists and why the idea interested spiritualist-inspired authors such as Bellamy.[53] Twain, too, had a long-standing fascination with paranormal phenomena. While consistently exposing fraudulent spiritualists, he championed telepathy in an 1891 article and engaged a spirit medium to contact his daughter after her death in 1896.[54] His antipathy toward Christian Science arguably stemmed not from its mystical and spiritualist notions but from how Mary Baker Eddy interposed herself. Time-vessel recipients would not require a mediating individual to hear the words of dead souls. Communication would be direct. Some even imagined it as two-way: "If your science," beseeched a Mount Holyoke student, "shall have taught you . . . the power of communication with the unseen world from which we may possibly be overlooking your destiny—we beg you to reply to this message of ours."[55]

So strong was the yearning for communion with the millennial generation that several contributors expressed a desire to attend the unsealing—whether through reincarnation, resurrection, ghostly visitation, or some kind of time travel. "I wish that I might be present," wrote a Colorado Springs doctor to his "dear friends" of 2001, "when you open the iron box and read this letter. . . . Be careful how you criticize it for my spirit may be hovering around, or I may be present in my 'astral body' listening to what you are saying."[56] A Detroit newspaperman merely hoped "to be resurrected long enough to glance over the head lines" of 2001—"if," that is, "the twenty-first century has newspapers and those newspapers are reliable"—while a Kansas City couple submitted a calling card inscribed with a "Happy New Year" to "their reincarnated selves." [57] Even those resigned to remaining buried relished the thought of their voice issuing forth. Twain began his autobiography with the words, "I am literally speaking from the grave."[58] The trope recurs in his alter ego's preface in "Secret History": "I am come from the grave, where I have mouldered five hundred years—look me in the eye!"[59] Some imagined their recipients' reactions. A "curious gaping crowd" will stand "awe struck" over the century box, speculated a Kansas City journalist, "feeling the presence of those who for many years have been peacefully slumbering." The crowd will "hear a voice from out of the indefinable realms of space shouting that good old Missouri rallying cry"—already the unofficial slogan of that state—"Show Me!"[60] Such fantasies testify to time vessels' power to grant not just a sense of personal immortality but also a vicarious experience of time travel. To understand the attraction of this practice, we need to attend not just to its archival function but also to its experiential seductions, above all, that of manipulating time.

For all this talk of mysterious voices and ghostly presences, contributors expressed their affection for their millennial recipients most often through tropes of physical touch, such as the handshake across time, as evoked in Mosher's earlier engraving (**fig. 1.2**). "I stretch my hand across the coming Century," intoned Mayor Reed of Kansas City, Missouri, "to clasp the shadowy fingers of the unborn mayor of 2001"—a sentiment also uttered by his counterpart across the state line.[61] Christian Scientists were especially fond of the image. "Knowing as we do that there is no space or time in 'divine Mind,'" wrote one, "I reach across these hundred years and clasp your hands."[62] Sometimes the paper itself served as a connective tissue between present and future bodies. The sealed manuscript in Twain's "Secret History" begins by imagining that the reader's hand, resting now where "mine rested last," will "feel some faint remnant of [its] warmth."[63] Such tropes of tactility exemplify the embodied future that time vessels could engender.

Expressions of physical contact sometimes evoked deeper emotional or even romantic intimacy. Several contributors addressed their recipients as "dear friends," and one offered "a loving kiss."[64] In her epistle to Mount Holyoke's future students, college magazine editor Margaret Ball expressed "a bond of sympathy between us" stemming from their shared "natures."[65] This sense of intimacy may have been fostered by the knowledge that they were writing in secret, for future eyes only. Some of the centurial vessels allowed contributors to seal their messages in envelopes, thus extending the earlier analogy with mailboxes.[66] Privately addressing the Colorado College girl who "stood highest" in the class of 2000, one man fantasized about being "preserve[d] for a century" by "a fountain of youth" so that I could "stroll with you to some quiet place out of doors and talk of Real Things."[67]

This intimacy and privacy prompted confessions. Some confided love, such as the Kansas Citian who wrote about "Clara Rieke a beautiful girl whom I hope to some day make my wife," or the student at Harvard's sister college who declared her "love" for her English professor, who "really knows about everything" but "doesn't take a bit good care of himself."[68] The Harvard vessel was well stocked with confessions. Most were minor, such as revelations of their frustration at the slow progress of their books or the shortcomings of their lectures. Archibald Coolidge—scion of a prominent Boston family, assistant professor of history, and later director of the Harvard Library, cofounder of the Council on Foreign Relations, and founding editor of *Foreign Affairs*—acknowledged that he gave one lecture that was "rather muddled" because he had forgotten "part of my notes," delivered another on a subject "I did not know much about," and on yet another occasion switched the subject from Germany to Turkey because of "nervous[ness]" induced by a colleague's presence. He even admitted to cutting a lecture because he was "not properly prepared." Aspersions were also cast on other lecturers. "Not wildly exciting" was his verdict on a visiting French poet's lecture. Finally, Coolidge revealed his efforts to exploit family connections to secure an ambassadorship.[69]

While Coolidge's confessions were somewhat trivial, others bared their souls to the future. Deputy Keeper of the University Records and temporary lecturer in history, William Garrott Brown revealed his unhappiness at Harvard. He was not "best fitted" for the archival position, his salary was "too small," and his senior colleagues were condescending.[70] An Alabaman, he aspired to become a writer of southern history and fiction and was working on a biography of Andrew Jackson. Yet he wondered whether "anybody [would still] know of the book" when the chest is opened and confessed grave doubts about his writing abilities. Even more candid are his disclosures regarding his private life.

> [I] have never had a serious love affair, + it does not seem likely that I shall ever marry. My deafness + natural shyness with women keep me from any sweet relation of friendliness with girls. This is perhaps the chief source of my frequent spells of intense melancholy. . . . I should like to have children about me, but it is not customary for the nice sort of people in Cambridge to take unmarried men into their homes.[71]

Nor did Brown have the consolation of a supportive family. His parents had died by the time he was twenty and his brother and sister a year earlier, leaving his "family sadly broken." Brown concluded with a poignant wish: "May love, and faith, and a good work to do, prove easier achievements [for my future recipients] . . . than I have found them."[72] He did not know that he was in fact on the verge of great success. His Jackson biography was well received, and he published six more books—including the influential *The Lower South in American History*—over the next five years. In 1906, however, he contracted tuberculosis and died seven years later, aged only forty-five.[73]

Few contributors to the Harvard vessel were as candid as Brown. More typically, their confessions took the form of complaints, particularly about the students. "The boys . . . received my announcement that there would be no lecture on Thursday," grumbled Professor Hill, "with more enthusiasm than could have been desired."[74] His English colleague Barrett Wendell also noted that his "class did not seem much interested" in his lecture on early American literature. They were also ignorant of British history, not knowing who was on the throne when Cotton Mather published his *Magnalia Christi Americana*.[75] Wendell's and Hill's courses are generally considered to have pioneered the teaching of composition in American universities (hitherto under the grip of classics and rhetoric), paving the way for creative writing programs.[76] Yet after twenty years teaching "Stylish Composition," Wendell admitted to his future reader that he found it "tedious work," and that it had "utterly failed" to "breed enthusiasm for literature." Perhaps the "old-fashioned classics" provided better "training" after all. Moreover, the heavy instructional workload left his "head in a whirl" and was "utterly destructive to continuous thought."[77]

Professors complained about colleagues, too. Coolidge did not relish meetings and reported evading one committee assignment.[78] Some singled out certain professors for special criticism. The famous psychologist Hugo Münsterberg, according to his colleague Paul Hanus, has an "intense eagerness to be an authority on everything" and "to create a sensation by posing as the opponent of all 'shams.'" Hanus was looking forward to the next faculty meeting for an "opportunity" to challenge him.[79] Even the renowned president Charles W. Eliot came in for criticism. "Everybody admires him," wrote Wendell, "but he lacks the faculty of inciting 'affection.'"[80] Conversely, some professors were quite complimentary about themselves. Of all the speeches made the previous night at a social function, Münsterberg boasted that his elicited "the greatest applause."[81]

While the centurial vessels prompted contributors to imagine their recipients as admirers, confidantes, friends, or even lovers, they also provoked anxieties about ever reaching them. The temporal gulf seemed at times too vast. "It is a far cry to you of the year 2000," intoned a Colorado Springs chorus director; "May my ink be black enough to bridge the space!"[82] Although addressing specific individuals may have helped to negate the unknowable otherness of the future, the experience of writing to someone "as yet unborn" stirred up "mingled feelings" in Ehrich's daughter, Leah.[83] It "seems a rather curious thing," she wrote, for those of her generation to address their descendants when they have "no idea of having any" and have not even "found our mates."[84] As the mother of only one child, Dora Noxon wondered whether "you [her descendants] are there"; to "dare write a letter to my posterity" seemed an act of "extreme optimism."[85] When our name is called out at the opening ceremony, worried Clarence and Louise Arnold, "will there be someone to claim these pictures and letters or only a great silence?"[86] This anxiety may have been a further impetus behind others' decision to address future members of their profession or social group.

Despair also stemmed from the apparatus' inherent unidirectionality. Leah Ehrich found it "inexpressibly sad . . . that we can never reach you, never know how you receive our words sent in this manner, never know how much or how little you sympathize with us and feel gratitude for our small achievements." She even wondered whether the chest's contents "will be as interesting and valuable to you . . . as we hope."[87] Such uncertainty left the local newspaper editor at a loss as to what to write and led another Coloradan to begin, "If some of you only could appear to me now and tell me what you would like to know of this day and age."[88] After all, added yet another, "it is not always easy to say anything worthy of even *contemporaneous* consideration."[89]

The presumed intellectual superiority of their future readers further intimidated writers. "We are haunted by premonitions," Margaret Ball of Holyoke College divulged, "that you will have learned in the kindergarten as much as we achieve in college. We greatly respect our posterity, as is suitable for those to whom Darwin has taught evolution."[90] This mental burden may have discouraged some from writing altogether. One Coloradan offered a litany of excuses—including his "tendancy [*sic*] to procrastinate"—for failing to write a "carefully preparred [*sic*]" article in time for the sealing.[91] Several contributors seemed to feel undeserving of the future's attention, implying that the time vessel conferred an unnatural, artificial fame unlike that acquired through great accomplishments. Even Colorado Springs' mayor J. R. Robinson felt he had "done so little that is worthy"—and had written this letter so "crude[ly]" and "hastily"—that when the chest is opened, "my grandchildren or great grandchildren . . . may prefer to conceal their identity, or the fact that they are my descendants."[92] Concerns such as these perhaps prompted one Kansas Citian to inscribe just three words on his calling card to the future: "What's the use."[93]

Yet the difficulties and limits of communicating with posterity only intensified their yearning for intimacy with it. "As I write," Leah Ehrich pined, "I find myself wondering . . . if your heart tingles with the thought of we who are dust as mine does with the thought of you who are as yet nonexistent."[94] Indeed, she and others without children, such as William Garrott Brown, were among the most fervent contributors to time vessels, which served perhaps as surrogate connections to posterity. Moreover, there are perhaps no more powerful declarations of affection than those that cannot be reciprocated. They are comparable to those love letters in Julian West's safe in Bellamy's *Equality*. Written by his fiancée before his 113-year trance left her effectively bereaved, they sparked a "quite indescribable emotion" in him and a "silent sympath[y]" in his hosts.[95]

"A Stimulus to Better Things": The Time Vessel as Ethical Spur

Despite all their uncertainty about communicating with the future, contributors were convinced their enterprise exerted a palpable influence on their present. Far from a passive receptacle, the time vessel was an active agent provoking a host of affective and ethical responses in its participants. In particular, it induced meditations on mortality. To communicate across a century was to confront one's death, even if those fantasies of reincarnation testify to the impossibility (as Freud believed) of fully imagining one's own nonexistence.[96] Two Detroiters signed off their messages with the tribute Roman gladiators supposedly made to Emperor Claudius: *nos morituri te salutamus* (we who are about to die salute you).[97] A number of Kansas Citians were moved to copy onto their calling cards various poems about human transience, such as the *Rubaiyat of Omar Khayyam*, or to pen their own verses:

Remember, thou are Mortals
As we who pass, tonight, these Portals
And a hundred years from now
Gently pause and silent bow
For us—who have gone Before.[98]

Even in Colorado Springs, a spa town or "hospital city" (as one contributor called it), where disease and death were particularly common, the vessel heightened the awareness of mortality. "Writing these letters to you," Leah Ehrich explained,

> makes us realize as we have never realized before, how fleeting is life and how short and precious are the few years allotted to us. In this busy hurried life of ours we have but little time to think of Death—inevitable as we know it to be—we never think of it, never grasp that it can come and take us into the unknown, but lately [during the preparation of this time vessel] we have felt it as never before.[99]

Time vessels did not merely evoke one's own death. One Detroiter ruminated, like a latter-day Xerxes, about how "an entire generation will have come and gone."[100] Thoughts also turned to the death of their children and grandchildren, including those not yet born. An elderly woman in Colorado Springs realized even her "dear grandchildren," Alice and Edith, will be dead by the time her letter is opened.[101] As a memento mori, the centurial vessel countered the cultural "denial" of death that was emerging in the late nineteenth century.[102]

Such thoughts of mortality kindled varying emotions. Some merely expressed a sense of uncanniness. For one Colorado Springs lawyer, writing "a letter which you know will be read, if read at all, by people to whom you have ceased to be even a memory and to whom your whole life and surroundings are only the subject of antiquarian research," was "a trifle eyrie [*sic*]."[103] Others, such as the aforementioned grandmother, found the task "ghastly," while Mary Noyes Shaw, organizer of the Rockland vessel, "shudder[ed] to think of writing this for a hundred years away. There is a grievesome feeling that makes one sad."[104] Yet there was also a "melancholy pleasure" to the task, as one Harvard professor put it.[105] By promising to preserve one's written (or photographic) testimony, the time vessel may have alleviated the death anxiety that it incited.

While suggesting that time vessels rouse contributors from their immersion in the quotidian, Leah Ehrich also believed they precipitate a deeper engagement with their present. "We find ourselves unconsciously . . . wondering if this or that will interest you as it does us."[106] In phenomenological terms, the time vessel lifted the veil of habitual perception enveloping their everyday experiences, allowing them to view their lifeworld afresh, through future eyes. Objects normally overlooked—or unlikely to endure—suddenly gained visibility and significance. Harvard librarian William Lane noted "an old and rather tumble-down building" on Boston's busy Washington Street that "will surely have disappeared" by the time his message is opened.[107] Contributors thus imagined their material environment as an archaeological site. They perceived their present as the future's past.

Contemporaries embraced the time vessel also to inspire action in the present. Describing one's activities for some future reader, wrote one Harvard student, "causes me to ask myself the question, 'Am I accomplishing the purpose for which I am living?'"[108] More often, the vessels were described as spurring action for the sake of future generations. Ehrich's century chest arose out of his efforts to inculcate a sense of duty to posterity, or what he called *posteritism*. He first explored this "quality of looking, with beneficent regard, far into the distant future," in an 1889 article in the *Magazine of Western History* calling for the preservation of a portion of northwestern Colorado as a national park, "before it is defaced by the saw-mill and the mining shaft." The decimation of the buffalo and forests is due to our "reckless, improvident disregard of the future."[109] But posteritism was in the ascendant, he believed, as religious superstition declined. Mankind is turning from "dreamy, nebulous" notions of heaven, he declared in his time vessel dedication, to a "practical duty" of improving the here and now and from ancestor worship to "worship of the unborn."[110] Anticipating Carl Becker, he suggested that posteritism might even serve as a substitute or secular religion, which would cultivate that "highest human virtue," namely, a "sacred regard" for future generations.[111] Indeed, Ehrich may have drawn on the Judaism he had earlier renounced. His conception of the vessel as a reminder of generational obligations evokes the legendary Ark of the Covenant, the gold-plated chest in which the Israelites allegedly preserved the stone tablets of the Ten Commandments and other sacred relics for hundreds of years.

Although Ehrich's neologism failed (so far) to enter the lexicon, some contemporaries picked it up. *Out West* magazine declared posteritism a "laudable idea" and applauded its "formal christening" at the time-vessel ceremony. The world is "full of people" too "nearsighted" to realize that regard for posterity is the "magnet of civilization" and the "incentive" of humanity's continuance. "If every city in America would bequeath its Century-chest to the future, what a civilizing influence it would have! Not on the future, maybe—but on us."[112] Time-vessel contributors also invoked the idea, if not the term. "We live not alone for ourselves," wrote one Coloradan, "but for you."[113] Some of these invocations were merely rhetorical and self-serving, motivated more by narcissistic desire for future recognition and gratitude, a "dying hope" that posterity "will generously appreciate what [we have] . . . done for [it]."[114] But others deployed the idea as a genuine spur to altruism, an ethics of posterity. For the editor of the *Colorado Springs Gazette*, the prophecies in the chest were not merely utopian wishes but concrete pledges to be fulfilled over the coming years.

> In committing to the chest the picture . . . of what the city is today, and of what we hope it will be in the future, we owe it to ourselves and to those who will disclose our record, to do our full part towards the fulfillment of these lofty anticipations.

He hoped, moreover, that the chest would continue to serve as an ongoing "promise and a pledge."

If in the hundred years to come dissensions threaten the safety of our political fabric, if immorality indicates decay, or if selfishness and avarice attack the foundations of our social and civic establish[ment], let the presence of the chest act as a stimulus to better things.[115]

The editor thus imagined society as a continuing process to which each generation pledges itself. In fostering an affective bond with posterity, the time vessel promoted a sense of duty to it.

Records of the Contemporaneous Everyday

Among the unborn addressed by centurial messages was the future historian. The year 1900 marks a watershed, when the time vessel's funerary function of preserving individuals' material traces was subordinated to its archival one of recording and transmitting information about the present. While Mosher and Deihm did little to convey sociocultural conditions and Cogswell's efforts to do so were inhibited by his own urge to self-commemorate, the centurial vessels actively espoused that task. Contemporaries imagined the ecstasy such troves of untouched sources would induce in historians. "A treat is in store for the antiquary and students of history who will pursue their investigation around Kansas City one hundred years hence," reported the *Star*, imagining "the expressions of wonder on [their] faces when they rummage through the mass of papers that box will contain" (**fig. 3.2**).[116] Believing that historical value is directly proportional to the lapse of time and hence extending his target date to five hundred years, Twain's alter ego in "Secret History" pictured his manuscript rising

like a lost Atlantis out of the sea; and where for ages had been a waste of water smothered in fog, the gilded domes will flash in the sun, the rush and stir of a tumultuous life will burst up on the vision, the pomps and glories of a forgotten civilization will move like the enchantments of an Arabian tale before the grateful eyes of an astonished world.

Such sources would be particularly valuable as their encapsulation safeguarded them from what Twain called "meddling scholars": those archaeologists, philologists, and editors who amend, annotate, abridge,

Scene at the Opening of the Century Box, Rooms of the Kansas City Commercial Club, January 1, A. D. 2001.

FIG. 3.2 Anonymous, *Scene at the Opening of the Century Box, Rooms of the Kansas City Commercial Club, January 1, A.D. 2001*, cartoon in *Kansas City Times*, January 6, 1901. Future officials pore over the time vessel documents. Newspaper clippings related to the Century Box, Kansas City Century Box Collection, Kansas City Museum, Kansas City, MO.

or interpret texts. These documents would reach the future historian "complete, undoctored . . . no word missing, no word added."[117] Two of the centurial vessels even contained an inventory, a kind of finding aid for future researchers.[118]

Those researchers were assumed by then to have taken up local history. Whereas Deihm's and Mosher's projects were national in scope, the centurial ones focused on a single college, city, or community. In Ehrich's words, "stress was laid almost entirely upon the local surroundings."[119] This placed them at odds with contemporary historians. Both professional, "scientific" historians and leading amateurs dealt with the nation and its gradual triumph over particularist attachments to region or section.[120] Although a generation earlier leading professionalizers, such as Herbert Baxter Adams, had encouraged their students to research the local, by 1900 they had effectively abandoned the subject to the civic boosters, gentleman dilettantes, and women who populated the historical, antiquarian, and preservation societies. This decline in the status of local history was not reversed until the 1930s.[121]

Another way in which centurial vessels departed from their precursors was in restricting themselves to a single moment in time. Whereas Deihm, Mosher, and Cogswell included materials from the recent past and added further material all the way up to their sealing, the centurial vessels defined the present more narrowly as a single year or, in Harvard's case, a single month. Their attempt to historicize and archive the here and now also put them at odds with "scientific" historians, who believed in the necessity of temporal distance to determine the significant facts and sources. The latter asserted their objectivity by repudiating the very notion of contemporary history.[122]

While contracting their temporal and geographic scope, centurial vessels expanded thematically to include diverse aspects of the community. Rather than just memorializing the accomplishments of leading individuals, they sought to capture a larger social whole even if those efforts were constrained (as we shall see) by class, racial, and ethnic biases. In particular, emphasis was on everyday activities. The Harvard circular signaled this microcosmic intention to "bring together for the benefit of our successors . . . as complete a record as possible of the present daily life of the University." In addition to teaching, research, and administrative work, it suggested covering one's "domestic occupations, . . . social relations, entertainments, etc."[123] Contributors responded with detailed descriptions of, for instance, the dress of a typical Harvard student ("the sack-coat . . . cut rather short; the waistcoat double-breasted. The trousers . . . turned up . . . to display the fancy stockings, often crimson in color.")[124] In one of the prefaces he drafted for his autobiographical manuscript, Twain emphasized how it "will furnish an intimate inside view of our domestic life of to-day not to be found in naked & comprehensive detail outside of its pages."[125] Similar motives led Margaret Ball to deposit her account book, not (as Cogswell had done) to illustrate her financial acumen but to document the "expenses of an *average* student at Mount Holyoke at this time."[126]

Bound in pigskin, Ball's account book doubles as a material artifact, one of many deposited in 1900–1901 to further convey the texture of everyday life. The Colorado Springs chest included such items as metal hairpins, badges, a silk fan, and a local department store's fabric samples, while Kansas City's contained a clock, a coin, and a two-cent stamp.[127] Some of these may have been offered (like those in Cogswell's vessel) as relics so that future recipients could experience an affective, sensorial relationship to their original owners. But most seem intended to be representative items documenting a broader social experience. As such, they anticipate the decline of the historical relic in the early twentieth century as museum curators came to view artifacts as general specimens of social-historical knowledge.[128]

In their archival orientation toward the material, the typical, the quotidian, and the contemporary, the centurial vessels thus projected a new kind of historiography. Academic historians, including seven of the eight history professors who contributed to the Harvard vessel, established their professional authority by focusing on the "higher" realm of the state and its diplomatic, military, political, and constitutional history. They thus elevated themselves above the everyday lives of the masses, which they coded as vulgar, corporeal, feminine, and thus ahistorical.[129] They also distinguished themselves by denouncing amateurs' and antiquarians' fetishistic attachment to relics and artifacts and instead elevated the written document. Thus, unlike ethnology or biology, history emerged as a text-centered rather than an object-centered discipline. Although by the late 1890s the pioneering historian Lucy Maynard Salmon was writing on such quotidian subjects as domestic servants and incorporating the vernacular, local evidence of train schedules, college catalogs, or laundry lists into her essays and classes along with mundane objects such as games, tools, kitchen appliances, and women's fans, she came up against her college president and academic publishers.[130] Professors elsewhere did not incorporate material culture and the decorative arts into US history courses until the 1930s.[131] The vessels thus gambled on a historiographical revolution by 2000.

That historiographical revolution was anticipated by the work of the one cultural historian who contributed to the Harvard chest: Charles Homer Haskins, the founder of medieval history as a professional field of study in the United States.[132] Just as Lane's chest would shed future light on the modern university, so did Haskins (in an 1898 article in the *American Historical Review*) seek to elucidate Europe's medieval universities. Historians, he wrote, have addressed "what may be called the anatomy of the medieval university—its privileges and organization, its relations to king and pope, [etc.]," but they have overlooked "its inner life . . . the daily life and occupations of its students." The latter necessitated neglected sources such as rhetoric manuals, which included samples of letters students wrote (or might write) to their families. But he warned against viewing any source as a transparent window onto the everyday past. For one thing, students' missives were motivated by ulterior motives—persuading parents to send additional funds or permit them to extend their studies—and were thus silent about the "wilder side of university life." Second, the manuals included only letters that were typical and applicable to other students; idiosyncratic ones were not copied and were therefore lost. Finally, low literacy rates and insufficient archives meant that "very few such letters could have been written, and, if written, . . . preserved." "What a pity," Haskins lamented, "that out of such a mass of letters there are none that tell us in simple and unaffected detail how a young man studied and how he spent his day!"[133]

These archival limitations were what centurial vessels sought to overcome. To convey the substance of the everyday, they strove to create documents undistorted by ulterior motives. They relied less on existing printed sources such as newspapers, pamphlets, and books and more on "intentional sources." Individuals in Colorado Springs and Detroit were commissioned to write essays on preassigned topics. These essays—numbering about sixty for each vessel, with some as long as forty pages—covered a broad range of economic activities, institutions, technological networks, spatial environments, and cultural phenomena (including contemporaneous slang). Maybury described them as "testimony from living witnesses" designed "to give you as clear an insight as is possible into the social, religious, moral, commercial and political affairs of Detroit and of the times in which we live."[134] Unfortunately, he gave his writers just two days, a deadline several considered inadequate.[135]

To register the local and the everyday, Harvard librarian William Lane opted for a different device. Throughout March, each contributor was to keep a journal, "recording faithfully, and in as much detail as he can, all that goes on . . . including his college work, his professional interests, his family relations, his amusements." Lane further suggested carrying a notebook so as to "watch [one's] days and . . . occupations, and to note down [matters] as they occur," which later can be written up in the journal.[136] The plan yielded seventy-two journals that range from events at work and in the newspaper to monetary concerns, errands to town, illnesses, and recreational hobbies. Professor Hanus, adopting the term Haskins used in his *American Historical Review* essay, called it a documentation of the university's "inner life."[137]

Hearing of the project, the *Washington Times* agreed it would be "valuable" to future historians. Earlier generations wrote and preserved journals as well as long letters, which now allow us to make "conjectures" about the past. But today "few" keep diaries, and "the art of letter-writing is falling into disuse." The letters that are written are unlikely to be preserved—not enough "storage room"—and "are apt to be filled with matters which will throw no light on our civilization." The Harvard scheme thus promised to revive a lost art, reminding all families of their duty to future generations to keep a journal and to make it a "lively, well-written, and thoughtful record" rather than a "mere escape-valv[e] of sentimentality." It may seem "uninteresting" now but will fascinate "readers of history" a century hence. "For time is a sort of alchemist and changes the common iron and pebbles of everyday life to rare treasures."[138]

Accurately observing and evocatively recording everyday life was a skill included in the Harvard composition curriculum. Hill and Wendell believed the modern university should prepare students for careers in science and business by focusing less on the rhetorical arts and more on expository writing. For them, good writing demanded "constant practice" (and red-ink correction) rather than learning abstract rules from textbooks. They thus required students to submit short, daily (or weekly) essays called "themes."[139] In a few pages—to be deposited "in a box at the professor's door not later than ten-five in the morning"—students had to narrate some incident or perception they had experienced that day (or week). Collected over the year, they constituted a kind of journal of "fleeting impressions."[140] The instructions to draw directly on the senses (rather than internal thoughts) accorded not only with President Eliot's belief in the "powers of observation" and "inductive" investigation and with William James's and Hugo Münsterberg's studies of perception at Harvard but also with contemporaneous artists' and writers' explorations of everyday experience.[141] According to an alumnus, the task was designed to cultivate something the instructors termed "the daily theme eye"—that "optic" sensitivity to experience possessed by "Henry James and the police reporter of the New York *Sun*" alike.

It became needful, then, to watch for and treasure incidents that were sharply dramatic or poignant, moods that were clear and definite, pictures that created a single clean impression. The tower of Memorial [Hall] seen across the quiet marshes against the cool, pink sky of evening; the sweep of a shell [rowing boat] under the bridge and the rush of the spectators to the other rail to watch the needle-like bow emerge. . . . By training the daily theme eye, we watched for and found in the surroundings of our life, as it passed, a heightened picturesqueness, a constant wonder, an added significance.[142]

A belief that such writing would allow future readers to penetrate to the marrow of contemporary experience appears to have prompted Wendell, Hill, and their colleague John Hays Gardiner to submit 382 unread student themes for March 1900 to Lane's chest.[143] Given Wendell's complaints about workload, it may also have been to avoid grading them.

The vessels sought to circumvent another tendency of historical sources noted by Haskins—that of harboring ulterior motives and consequently skirting or distorting the truth—by providing confidentiality. The Harvard journals and themes were to be submitted unread, the former "placed flat in stout envelopes . . . and sealed by the writers."[144] Contributors in Colorado and Detroit similarly submitted their messages in sealed (some with wax) envelopes.[145] Harvard went one further by establishing complex rules of access. The chest could be opened in 1925, but none of its records could be viewed or cited until 1960 unless the writer had died, his family or literary executors requested it, and the library council voted to allow it. Lane allowed contributors to specify further restrictions on the future use of their journals, such as that they be read first by certain individuals. Thirteen years later, he permitted one contributor to withdraw his journal altogether.[146] The quarantining of individual messages was not without precedent. By the late eighteenth century, private individuals (such as Fitch) who donated their papers to manuscript collections—whether of historical societies, university libraries, or the Library of Congress—were occasionally permitted to place a temporal restriction on access. But those policies remained loose and uncodified well into the twentieth century, and those institutions remained interested more in distant (colonial and revolutionary) figures than in recent individuals. Moreover, by 1899 there had emerged a public archives movement, led by the American Historical Association, that denounced such absolute, timed restrictions on access as undemocratic. Time vessels' insistence on sealed messages thus marked yet another divergence from official archives.[147]

Not all the centurial messages, to be sure, were kept completely confidential. Mayors and other dignitaries read theirs out loud at the ceremony.[148] Some contributors showed theirs to relatives or dictated them to their secretary, wife, or other scribe.[149] Lane himself wrote his journal as a series of dispatches to a distant friend who collated and returned them at the end of the month—an idea he recommended to others as a means to write "without reserve" and as engagingly and "vividly as possible," without any "posing."[150] In Kansas City, there was no guarantee of confidentiality at all; guests at the ball freely removed others' calling cards and read their messages.[151] Even at Harvard, the lingering fear of one's message falling into the wrong hands led some to moderate their views.[152] Nevertheless, many contributors indicated that confidentiality liberated them to speak their minds. The centurial vessels thus yielded—as we shall see—more candid assessments of their nation, cities, and institutions than those found in earlier vessels and certainly more so than other commemorative endeavors of the period, such as local histories, museums, or historical pageants.[153]

The most vehement advocate of such an approach was Twain, who was looking for a way to write an autobiography that told "the truth, the whole truth and nothing but the truth, without malice, and to serve no grudge, but, at the same time, without respect of persons or social conventions, institutions, or pruderies of any kind."[154] He who writes his memoir to be read during his lifetime, Twain wrote in his preface, "shrinks from speaking his whole frank mind." Only "if I knew that what I was writing would be exposed to no eye until I was dead" could "[I] be as frank and free and unembarrassed as [in] a love letter." His models were the unreserved, confessional writings of Giacomo Casanova, Jean-Jacques Rousseau, and Samuel Pepys.[155] In response to his friend William Dean Howells's skepticism, however, Twain abandoned this belief in the possibility of writing the full truth about oneself and devoted the autobiography to *others* he had encountered in his life.[156] By deferring publication for a hundred years, he could describe them without fear of offending them "or their sons or grandsons."[157] In the essay "Privilege of the Grave," written in 1905 but not published until 2008, Twain recommended this model for others. We all suppress certain "unpopular opinions," so why not "reveal these secrets of ours . . . from the grave . . . ? Why not put these things into our diaries . . . and leave the diaries behind, for our friends to read?"[158] He reaffirmed this idea in "The Secret History." The manuscript rediscovered in Washington was valuable because its author had written it "for a distant posterity who could not be hurt in their feelings by it." His pen was thus "the freest that ever wrote."[159]

The Work of History in the Age of Mechanical Reproduction

The centurial vessels' emphasis on their future archival function was also evident in their appropriation of modern media of inscription. Each of them did, to be sure, contain traditional handwritten documents. Harvard requested contributors to write at least "a portion" of each journal by hand, presumably for evidence of penmanship rather than to transmit their characters or souls.[160] But they also embraced mechanical writing. Nineteen of the Colorado Springs messages and twenty-two of Detroit's were typed—and five of the latter were mimeographed, a process only a decade old. Invented to transmit individuals' unique aura through precious photographs, autographs, or relics, the time vessel could now also convey generic and reproducible documents.

A prominent champion of such machines was Twain. Often cited as the first to submit a book in typewritten manuscript (*Life on the Mississippi*, 1883), he now developed a new method to crank out his autobiographical time capsule. Instead of writing it out by hand first (as he usually did), he dictated it directly to his secretary and had his daughter, a trained typist, type it out.[161] Even the fictional manuscript in "The Secret History," originally hand drafted to ensure physical contact between author and first recipient, was "secretly printed by a member of [Twain's] family on a machine called a type-writer," an indication that Twain continued to celebrate modern writing machines even after his disastrous investment in the Paige typesetter.[162] While stimulating Twain's dreams of mass dissemination, such technologies also evoked a nightmarish loss of authorial control and rights. He therefore drew up precise instructions regarding how and in what form his autobiography was to be published: expurgated at first, and in progressively less expurgated editions every quarter century thereafter. Twain's publisher suggested adding a clause allowing it to be disseminated via the "phonographic" and "electrical" media that will have replaced the traditional book by then.[163] In "Secret History," Twain's alter ego issued similar instructions for his manuscript, granting the government (rather than commercial publishers) the right to "publish it at the rates current at said remote time," with proceeds to go to the drafting of a "sane" copyright law.[164] Twain's interest in time vessels was thus related to his campaign for an act to preserve author's rights in perpetuity.

This turn to the typewriter may have been motivated not just by its newness—although introduced at the Centennial Exposition, it became standard in offices only during the 1890s—but also by concerns about the future legibility of handwritten documents.[165] While inscribing her epistle, Leah Ehrich found herself "wondering if you also use a pen."[166] In Bellamy's *Equality*, Edith struggles to read the love letters in West's safe, as "the progress of invention" has made "handwriting, and the reading of it" virtually "a lost art" by AD 2000; people send messages only by telephone or phonograph. "For our important records," Edith adds, "we still largely use types, of course, but the printed matter is transcribed from phonographic copy, so that really, except in emergencies, there is little use for handwriting."[167] The vessels' conscription of the typewriter and mimeograph thus signaled their emphasis on a uniform, businesslike transmission of information, or (in Martin Heidegger's words) a debased view of the word as merely a "means of communication."[168] At this time, typewriters were largely for business correspondence. To type a letter to a friend was "almost a personal insult."[169] The new archival paradigm thus appeared to conflict with the older ideal of physical intimacy with the future.

Mechanical media, to be sure, did not necessarily diminish the affective bond with future (or past) generations. The first commercial typewriter was the invention of an ardent spiritualist who was "interested . . . in communicating with the other world." Subsequent models, wired invisibly to their operators, were used in séances by the 1890s to contact the dead.[170] The phonograph generated fantasies of communicating from the grave too. Although initially used for dictating ephemeral business letters and, by the 1890s, for listening to commercially recorded music, it was believed (as it was for the camera a generation earlier) to extend one's physical or even spiritual presence across time. Several messages were cut for future generations, including one by British politician Joseph Chamberlain, recorded at Alexander Graham Bell's laboratory in 1888 and deposited in the Smithsonian for reproduction in "perhaps 100 years."[171] Through wax-cylinder recordings of fight songs deposited in their vessel, Colorado College students similarly sought to rouse future fans of their football team.[172]

Yet by 1900, when the first phonographic archives were founded in Europe and the United States, the medium was beginning to be touted for its documentary possibilities. Recordings of speeches would be more reliable historical sources than written transcripts, the journal *Phonoscope* insisted in 1896, because they preserve "every modulation and inflection of voice" and thus prevent "disputed readings."[173] Also preserved were cultural forms, such as Native American chants, regional dialects, and folk tales or songs. In 1900, *Phonoscope* proposed creating a complete "graphophonic record of life at the end of the nineteenth century for the benefit of posterity."[174] Few such proposals, however, were actually enacted in these early years. By including phonographic recordings, the Colorado Springs chest thus stood in the vanguard of this archival movement.

The time vessel's archival turn was most visible in its reenvisaging of photography as a documentary medium. Each centurial vessel contained numerous photographs, with Colorado's boasting as many as 415. Some were conventional portraits that contributors attached to their messages, thus evoking the memorial function of the earlier vessels. However, those portraits were unaccompanied by any claim about conveying individuals' souls, and they typically depicted ordinary people: students, churchwardens, a fireman, a five-year-old girl.[175] Several lacked captions identifying the sitter, while some indicated representative types, such as "College Girl, Senior Class" (**fig. 3.3**).[176] There were no portraits of President Eliot in the Harvard chest, but there were photographs of a Cambridge postman doing his rounds (**fig. 3.4**) and an Irish-born fruit vendor whose donkey and cart had graced the neighborhood for forty years (**fig. 3.5**).[177]

Most photographs, however, were of buildings, streets, and interiors assembled into collections to encapsulate the social and built environment of their locality. One Mount Holyoke student used an examination "blue book" as an album for snapshots of activities around campus, from climbing down dormitory fire-escape ropes (**fig. 3.6**) to performing in the college theater. The Harvard chest contained a specially commissioned collection of sixty-two photographs by eight photographers depicting Cambridge streets, university buildings, students' rooms, and professors' houses—places "which though not interesting in themselves," wrote Lane, "are sure to change or pass away in the course of a few years, + of which no photographic record at present exists" (**figs. 3.7, 3.8**).[178] For the Colorado chest, a professional photographer took ninety-three photographs of Colorado Springs' streets and parks; its civic, educational, and religious buildings; and its commercial and domestic interiors—all eerily depopulated, like some movie set or a mysteriously evacuated city (**figs. 3.9, 3.10**). Carefully staged photographs of living spaces—especially college dorms, with their pennants and sporting paraphernalia—were by 1900 a popular genre employed by the middle class to transform familiar but ephemeral settings into stable documents of personal memory and family history. One photographic contributor to the Harvard vessel apparently made a lucrative business out of capturing his fellow students' rooms (**fig. 3.8**).[179]

FIG. 3.3 Anonymous, *College Girl, Senior Class*, photograph, ca. 1901, from a collection of "Photographs of Citizens of Colorado Springs" submitted to the Century Chest. Folder 150, Colorado Springs Century Chest Collection, 1901, Ms 0349, Colorado College Tutt Library, Special Collections.

In addition to preserving ordinary people and spaces, the photographs were largely the work of undistinguished photographers. Whereas earlier vessels transmitted photographs by prominent professionals (Mosher and Brady), the centurial vessels relied heavily on amateurs, thus attesting to the mass democratization of photography through new technologies (the dry plate, roll film, and the box camera) and new institutions (journals and clubs).[180] Lane asked "everyone" at Harvard "to contribute what photographs he can, particularly pictures of his home, both interior and exterior views." He also enlisted the Harvard Camera Club (then administered by T. S. Eliot's brother, Henry Ware) and organized

FIG. 3.4 Walter Babcock Swift, *Cambridge Postman*, photograph, 1900. Harvard University "Chest of 1900," HUA 900.13, envelope 7, Harvard University Archives.

a photographic competition for the "best collection of pictures of buildings and college scenes" with himself as one of the judges.[181]

In their documentation of built environments and reliance on amateur photographers, these vessels evoke—and were perhaps influenced by—the British survey movement. Formalized under the National Photographic Record Association (1897), this movement felt it was their duty to posterity to preserve vernacular landscapes and customs (predominantly rural, but also urban) threatened by modernity.[182] Rather than leave memory to accident or fate, the association self-consciously memorialized the present. The founder called on amateurs across the nation "to fix the fleeting features of the epoch in which we live and hand down to posterity the 'outward and visible' state of things as

FIG. 3.5 Chauncey Blair Borland, *John the Orangeman and Donkey*, photograph, 1900. Harvard University "Chest of 1900," HUA 900.13, envelope 6, Harvard University Archives.

FIG. 3.6 [Louise W. Dodge?], *Favorite Sport, Going down the Fire Escape Ropes*, cyanotype photograph, 8.8 cm × 8.65 cm, ca. 1900. Dodge inserted this and other photographs of college life into a blue book she submitted to the Mount Holyoke Class of 1900 Box. Mount Holyoke College, Archives and Special Collections.

they existed." Photography was considered ideal for this. As the London *Times* observed in 1900 (anticipating Walter Benjamin's theory of photography's optical unconscious), an "apparently irrelevant and unnoticed detail" accidentally captured in a photograph might be what is most interesting to a future generation. At the same time, amateurs could legitimize their hobby by giving it a larger purpose. Yet photography's radical contingency and stylistic variability was also a threat to the archive. Amateurs were thus instructed to adopt a straightforward, factual approach free from pictorialist aestheticism or darkroom manipulations—a documentary or "archive style" that also characterized the Harvard and Colorado photographs.[183] They were to further contain the photographs' mutability by labeling and captioning them and filing them in boxes, cabinets, or envelopes.

Through the transatlantic circuit of lectures and exhibitions, such ideas were taking root in the United States too. The Boston Camera Club carried out a survey of Boston's historic sites around 1888.[184] That same year, the photographer George Francis urged American libraries and historical societies to systematically accumulate photographic archives of the present to enable future scholars to "trace the process of evolution" of rural and urban environments. "To do our full duty to posterity," he pleaded, "we must not fail to hand down to them the best possible picturing of our lands, our buildings, and our ways of living," including tenement houses and industrial mills that are "fast passing away." Yet not until the 1930s did institutions implement such surveys or did professional historians recognize photographs as sources.[185] The Harvard and Colorado vessels thus represent rare early attempts to systematically document the contemporary landscape for posterity through carefully staged and labeled photographs.

FIG. 3.7 Walter Babcock Swift, *College House from Corner of Dunster Street*, photograph, 1900. Harvard University "Chest of 1900," HUA 900.13, envelope 7, Harvard University Archives.

FIG. 3.8 Julian Burroughs, *untitled* (interior view of a student room), photograph, ca. 1900. Harvard University "Chest of 1900," HUA 900.13, envelope 8A. Harvard University Archives. In one theme essay, a student describes how "Room-mate and I took photographs of our room. Went through the usual process of piling everything up on one wall to impress folks at home" (W. H. Mearns, weekly theme, Friday March 2, 1900, box 55, HUC).

Prophylactics against Decay

The time vessels' increasing reliance on the modern media of photography, phonography, and typography raised problems of durability. Early, tinfoil phonographs only preserved sound for a short time or a few playbacks, and removing the foil from the cylinder effectively erased them. Second-generation wax cylinders, such as those in the Colorado vessel, were removable but not much more durable; they "would not stand the ravages of time," predicted the Kansas City *Star* in 1902, "being necessarily of soft, impressionable material." The idea of phonographically preserving voices for posterity was thus more fantasy than immediate possibility.[186] Similarly, the low-quality, acidic paper used on mimeograph machines was much less durable than older stocks, causing many of the mimeographed messages in the Detroit vessel to suffer yellowing and deterioration.[187] Concerns about photographs' longevity

FIG. 3.9 F. P. Stevens, *Interior: F. L. Gutmann's Drug Store, Southwest Corner of Bijou & Tejon Streets, 1901*, photograph, 1901. Photograph 61, folder 160, Colorado Springs Century Chest Collection, 1901, Ms 0349, Colorado College Tutt Library, Special Collections.

FIG. 3.10 F. P. Stevens, *Modern Plumbing*, photograph, 1901. Photograph 84, folder 160, Colorado Springs Century Chest Collection, 1901, Ms 0349, Colorado College Tutt Library, Special Collections.

also persisted through the turn of the century despite improvements in techniques and materials.[188] Futuristic fiction reflected on this perishability of modern materials. In *A.D. 2000*, the objects found in a sealed chamber are "ready to fall in pieces at the least shock."[189] Future societies also continue to employ ephemeral media. In a rare concession of imperfection in Bellamy's utopia, West's host acknowledges the fragility of their own phonographic records relative to the written records of the nineteenth century.

> The riper civilization has grown, the more perishable its records have become[.] The Chaldeans and Egyptians used bricks, and the Greeks and Romans made more or less use of stone and bronze, for writing. If the race were destroyed to-day and the earth should be visited, say, from Mars, five hundred years later or even less, our books would have perished, and the Roman Empire be accounted the latest and highest stage of human civilization.[190]

Time-vessel contributors nevertheless made efforts to mitigate this loss of durability by drawing (like Deihm) on older as well as newer materials. The handwritten messages in Colorado were written in India ink on bond paper and inserted, unfolded, in large cloth-lined envelopes, while pure linen stock was used in Detroit. For photographs, platinum (or bromide) prints, far more stable than the albumen prints used by Mosher and Deihm, were preferred—or required in the case of the Harvard chest. And for good measure, Colorado's official photographer wrapped his in tinfoil.[191] Moreover, to deposit a phonograph or photograph in a vessel was to shield it from repeated use and exposure to light and air.

A further response to the perishability of modern records—whether recorded, photographed, typed, or handwritten—was to ensure the integrity of the container. Breaking with Mosher and Deihm, none of the centurial vessels used a bank safe. Not only did bank safes have weak points such as ornaments, hinges, and locks, they also potentially attracted future thieves by appearing to contain valuables. In *Equality*, Bellamy indirectly suggested a further drawback to safes: their possible obsolescence by 2000. Although West remembered the combination after 113 years and the chamber's low humidity kept the locks rust free, the mechanism was broken—and there is not "a man in the world who could pick [it]." In a utopia of equally distributed wealth and guaranteed allocations of the nation's capital, there is no impulse to steal. Nor is there demand for safes' fireproofing capacities, since the abolition of insurance has led to the eradication of fires. Consequently, making or fixing safes—like handwriting—has become a lost art. "What a ridiculous little box it turns out to be for such a pretentious outside!," exclaimed Edith's mother when they finally opened it.[192]

Instead of safes, the centurial vessels consisted of uncomplicated boxes ranging from the size of a shoebox (Detroit) to that of a large trunk (Colorado Springs; **fig. 3.11**). The latter was made of plate steel

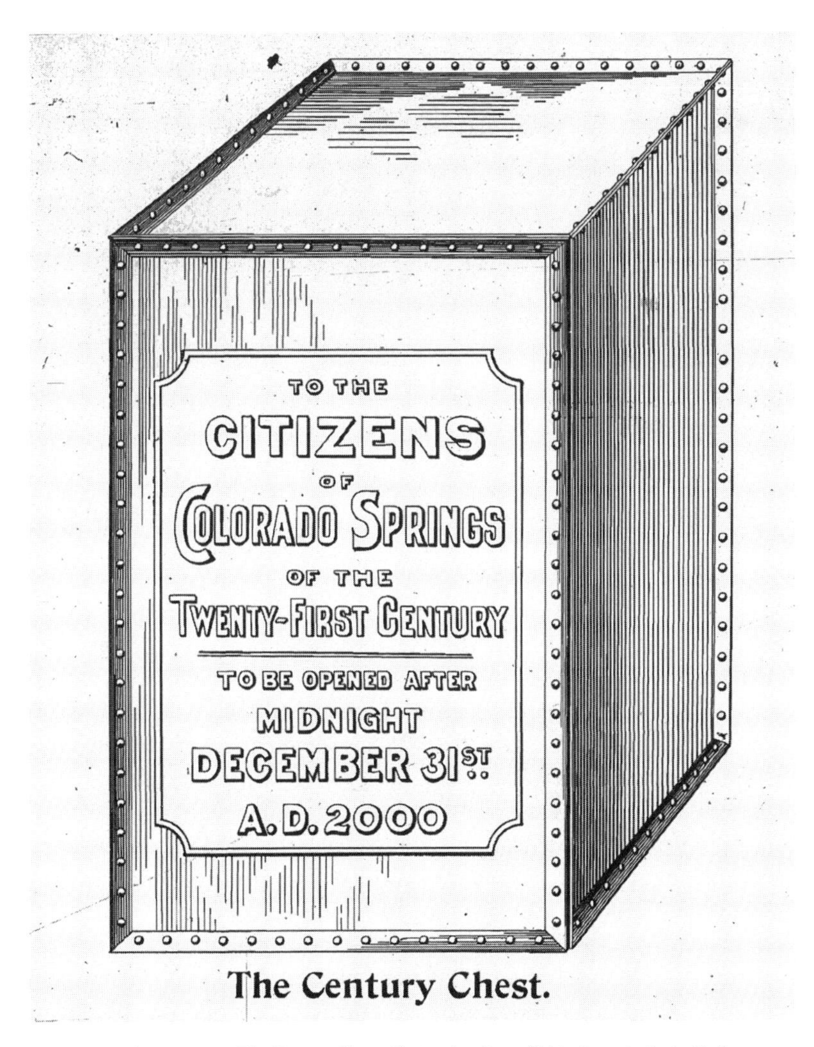

FIG. 3.11 Anonymous, *The Century Chest*, illustration from "A Heritage to Posterity," *Colorado Springs Gazette*, August 5, 1901, 7. Folder 3, Colorado Springs Century Chest Collection, 1901, Ms 0349, Colorado College Tutt Library, Special Collections.

FIG. 3.12 The Kansas City Century Box with some of its contents: a silver horseshoe, a commemorative clock, silk ribbons, and a city directory. Kansas City Museum, Kansas City, MO.

with lead lining and joints, and with its quarter-inch-thick walls, it weighed two hundred pounds. Kansas City's (**fig. 3.12**), Detroit's, and Rockland's were made of copper (the former using heavy copper plates). Harvard and Mount Holyoke used wooden boxes, the former lined with zinc, the latter nested within a larger metal one. All of them were given a secure, airtight sealing either with steel rivets (Colorado), solder (Mount Holyoke), or a combination of solder and "heavy sealing wax" (Detroit).[193] Like Deihm's safe, these boxes served as advertisements for their manufacturers as well as evidence of their donors' largesse.[194]

A final precaution was to deposit the boxes out of sight. In a further departure from their predecessors, which were installed in conspicuous locations, the centurial vessels were hidden away: in bank vaults in Detroit and Rockland, in the entrance to Kansas City's Convention Hall, or in college libraries. Such measures were motivated not just by fear of looting or tampering but arguably also by concerns about some larger cataclysm. The "center of this great American continental republic" was the safest place for such a deposit, wrote one Coloradan; no "foreign invasion or . . . domestic turbulence" could endanger it there, even for five centuries.[195] Removal from public view, however, rendered the vessels more likely to be forgotten. Thus, as a fail-safe some were entrusted to office holders—Detroit's city controller or Harvard's president and librarian—to be passed down to their successors. Rockland's clerk was to verify the chest's presence in the vault annually, while Mount Holyoke's class of 1950 was to appoint three "trustees" to perform a midcentury checkup.[196] The Ancient and Honorable Artillery Company had devised a similar tactic, enclosing their century box within a larger one to be opened after just fifty years and attaching a request that the inner vessel be "forwarded."[197]

Participatory Ventures

The centurial vessels sought to document not only ordinary citizens' living environments but also their testimony. Women were better represented than in earlier vessels, contributing half of Colorado Springs' articles and a third of Detroit's, in their capacities as librarian, housekeeper, poet, needleworker, gardener, teacher, or student (Colorado), or as city official, YWCA manager, charity worker, suffrage advocate, or mother (Detroit).[198] These vessels also solicited contributions from a reasonable (although, as we will see, not all-inclusive) range of ethnic groups.[199] Moreover, although the bourgeois elite were still overrepresented—the heads of businesses, clubs, or city departments wrote many of the essays—working people were included. A streetcar conductor, printer, and two members of the Council of Trades and Labor Union contributed to the Detroit vessel, and a builder and union leader contributed to Colorado's.[200] Similarly, Twain refused to devote his manuscript to all the famous people he had encountered. Deferred publication obviated the need to appeal to contemporaries' fascination with celebrities, allowing him to include "bootblacks and shoemakers" or anyone who has "keenly excited my interest"—a goal shared by his alter ego in "Secret History," who sought to represent people of all "ranks," "grades and callings."[201] For ordinary, nonpublishing contributors such as the Coloradan Arthur Taft, a time vessel offered the "only chance to be read in the future."[202]

This greater inclusiveness applied even to the elite institution of Harvard. Although the original proposal was limited to faculty, Lane decided to open it up to staff and students with a view to representing all ages and "all grades of the life of the university."[203] To encompass "different interests and classes" and "make the picture more complete," he carefully selected specific students—graduate and undergraduate, male and female—some of whom went on to have distinguished careers, such as Lucy Sprague (Mitchell), the future pioneer of progressive children's education and founder of Bank Street College.[204] He also included the perspective of dining hall stewards and janitorial staff and entertained the idea of "Mr. Jones the bell-ringer" contributing "some account of his life + his many years of service."[205] After one professor complained that the plan still omitted one's family members ("even a prose instructor or assistant often has a family and their position, views, and duties are a large influence on the life of the university, the force behind the throne in many cases") and that they must have "their say . . . independently of any supervision from husbands or parents or brothers," Lane extended the invitation to them too.[206]

For all these concessions, each vessel's architect was, to be sure, from society's upper echelons. Moss had made his fortune as an iron manufacturer before becoming a prominent judge and alderman in Kansas City.[207] Maybury rose to the mayoralty of Detroit as a lawyer (usually representing capitalists and corporations), a member of Congress, and a leading Mason.[208] Lane was descended from a leading New England family and belonged to numerous historical, library, and literary societies.[209] While from more modest Jewish immigrant beginnings, Ehrich attended Yale thanks to his father's successful clothing company, which he inherited and expanded, eventually employing some seven hundred workers and generating two million dollars in annual sales by his own estimates. He sold his stake, invested in real estate and bank stocks, converted to Christianity, and in 1885 moved to Colorado Springs, building himself an opulent mansion to house his extensive library and impressive collection of Dutch and Flemish art works, which he claimed was the largest in private hands in the country.[210]

Unlike Mosher, Deihm, and Cogswell, however, none of these architects emblazoned their names on the plaques affixed to their vessels. While retaining some control over their projects—Moss apparently "decide[d] what shall be put in and what left out," and Ehrich's family submitted the most letters, photographs, and objects—they generally devolved their authority to the community.[211] Ehrich formed a committee to assign topics to the appropriate citizens.[212] He also, crucially, allowed unsolicited submissions. Almost two hundred people—far outnumbering the sixty commissioned writers—contributed letters, photographs, and other materials to him in the days leading up to the sealing. "I have taken too much liberty to volunteer my letter," wrote one, "but I feel it is the duty of every intelligent man."[213] Unlike the preceding vessels and those that followed in the 1930s, which required the contributions in advance, the sealing ceremony provided a further opportunity for unauthorized items. Approximately eighty individuals or families, unlisted in the official inventory, brought messages and materials, while as many as two hundred and seventy slipped in their calling cards, many of them inscribed.[214] At that point, as there was "still considerable room in it," the chest was only "temporarily sealed," giving citizens one more day—announced the *Morning Gazette*—to take advantage of free postage to posterity.[215] An even longer extension of the official deadline was allowed in Kansas City. For sixteen days after the ball, as Moss continued to collect material, the box remained open and on display in the window of Jaccard's jewelry store, where "all day a curious throng . . . gathered about [it]."[216] Meanwhile, the newspapers informed "anyone who has not left his name therein" that they "may yet do so, if on the list of those in attendance at the ball."[217] Perhaps the most lax of all was Lane, who permitted one journal that was eighteen days late.[218] The vessels thus evolved into collaborative projects involving numerous individuals—some officially appointed, others self-appointed.

Opening the vessels to diverse contributors also exposed them to a flood of unsolicited feedback. While many viewed the schemes favorably and were flattered to be asked to contribute, others, especially in Cambridge and Colorado Springs, were more critical.[219] Numerous Harvard professors agreed to participate only grudgingly, complaining to Lane about this burden on their schedule at a busy time of year.[220] Insufficient time was also a common excuse for declining the invitation altogether. The famous historians Edward Channing and Albert Bushnell Hart abstained, the former stating he did not have "thirty-one hours" in the day, the latter accusing Lane of "adding a new penalty to human life."[221] Busyness (or illness) may also account for the large number of professors who accepted the invitation but failed to follow through or who submitted journals that had numerous missing days or otherwise disregarded the instructions.[222] "The cares of the world" prevented even Lane from completing his diary. Falling "ingloriously . . . by the wayside" midway through the month, he resolved to complete the missing days over summer recess, drawing on his rough notes. Those notes accompanied him on a recuperative trip up the Nile that winter and on another vacation the following summer. Finally, after seventeen months, he conceded defeat, confessing sympathy for others who failed the task.[223]

These petty complaints about the strain of keeping a journal were accompanied by more substantive critiques. One was that the time vessel was unnecessary because future historians wishing to reconstruct the world of 1900 would have plenty of other sources at their disposal. "When this scheme was proposed," wrote an assistant professor of history, "hardly any one . . . [thought] it would be seriously undertaken. . . . I regard it as a waste of time. Whatever difficulties students of our college history may encounter two or three generations hence, . . . lack of documentary materials will not be one of them."[224] The distinguished English professor and dean of the college LeBaron Russell Briggs concurred. Rejecting Lane's invitation, he declared that "the twentieth century will not be the loser."[225]

A more serious critique was that time vessels failed, for various reasons, to attain their goal of documentary objectivity. First, unless one already kept a journal, the task of spending "a couple of hours" writing one each day would paradoxically "prevent the day from being a normal one."[226] Second, contributors questioned whether their daily life was representative of others. "It really isn't right for me to presume to give a diary, expository of the average life here," wrote one student. "With most fellows I have absolutely no tastes in common. . . . I live in a little coterie of eccentrics."[227] Nor was the promise of confidentiality deemed sufficient to ensure objectivity. Sealing one's letter allowed one to write without regard for the opinion of one's contemporaries, but it could not—as Twain eventually realized—suspend the awareness that one's words would, or could, eventually be read by someone. (Only perhaps by retaining one's anonymity or destroying one's document could one truly avoid self-censoring; indeed, Twain turned to the "scheme" of writing letters he never intended to send, which permitted greater "frankness & freedom").[228] One female graduate student thus dismissed the Harvard chest as "futile" because the core "elements of our lives" would be removed (as "we do not trust ourselves to record [them] where they shall *ever* be seen"), leaving only the external record or "shell." The future reader "will, as it were, sit among husks."[229] Furthermore, the fact of being invited to submit a journal fostered a certain "self-consciousness" that might lead one to distort the truth.[230] Munsterberg did "not believe in the scheme" because it "puts no premium on posing while it decidedly invites to pose"—and he clearly needed no invitation.[231] Finally, even if time vessels' custom-made intended sources obviated the bias of ulterior motivation that hampered traditional sources such as medieval students' letters, they nevertheless harbored a longer-term motive, that of influencing posterity's view of the present. A Colorado Springs judge considered it an attempt to forestall the opinion of the next century in regard to those of today, and . . . to juggle the evidence. I fear that the best face upon all matters will be incorporated amongst the treasures of this chest and I am inclined to think that there may be, unconsciously, an attempt to conceal some matters which it would be necessary for the citizen of the next century to have presented to him in order to give an unbiased opinion of our importance.[232]

An even more devastating charge was that the vessels would simply bore their recipients. According to the judge, it was pointless to describe the conflicts and issues of the day, as even those that "caused a slight blaze at the time" will "not be of enough interest to the next century to note." Nor would a prophecy, noted a fellow attorney, much "interest . . . those who live when . . . [it] shall be fulfilled or disproven."[233] Recording everyday life also seemed senseless to those with unvarying routines, such as the astronomy professor whose work life "would consist exclusively of the continuation of a long job of computation" and his domestic life of "reading the newspaper" and feeding the furnace.[234] Professor Hill seemed bored by his own journal: "Little to record. I have had the usual amount of work and rest . . . the usual downs in health. . . . I have been to my clubs at the usual intervals, and have enjoyed the usual amount of conversation." Rereading it, he found it "even duller," and briefly considered throwing the whole "wasted" effort into the fire.[235] His colleague Wendell was also struck at the end of the month by his journal's apparent "insignificance" and hoped instead to be remembered by his books.[236] And by the eighth page of his journal, Lane himself began to wonder whether the reader would even have the "persevere[nce]" to reach that page and "whether the simple details of private life will really have any interest after all sixty or a hundred years hence."[237]

Above all, these criticisms and refusals to participate—especially from several Harvard historians—indicate that the time vessels may have posed a threat to the authority of professional historians. Since the introduction in America of the research seminar in the 1880s, the recovery of sources had become a central aspect of the profession. A romantic and gendered mystique had grown up around the (implicitly male) historian's heroic struggles to reach out-of-the-way archives and to penetrate the dust and disorder to unearth the decisive, virgin source. In purporting to furnish the future historian with all the sources he would need, the vessels thus denied him the opportunity to showcase his research prowess. Scientific history further asserted that documents become "sources" only after the fact, through the historian's expertise in extracting from an archive what is significant, which in turn depended on his trained ability to perceive hidden causes and truths.[238] From this perspective, the centurial vessels made the cardinal error of presuming that sources could be identified a priori, that they could preexist the historian. As unofficial, local depositories, they also challenged the American Historical Association's post-1895 initiative to systematize the collecting and cataloging of manuscripts in federal and state archives.[239]

These very concerns appear in a letter William James wrote to Lane explaining that he was on sick leave in France and would be unlikely to contribute. "[It used to be] enough for a generation to make raw history. Now we must half cook it for the future historian, and soon we shall have to compose it for him in advance. What an easy time he'll have in the year 2000 simply unsealing & transcribing a finished work!"[240] From the standpoint of James's radical empiricism, the time vessel must have appeared a futile attempt to freeze the flux of life, to grasp and preserve a specious present—while the idea of viewing that present from the mediated perspective of a future date implied an inauthentic mode of existence, a detachment from the immediacy of experience.

As a result of all these problems, some concluded that the only thing time vessels would reveal to the future was "how much energy and time we wasted"—or, in the judge's view, how preoccupied some of us were with the "importance of the present." Instead of sealing the "doings of the present time" for posterity, he added, we should entomb them "*forever*"—what we might call an anti–time capsule.[241] Yet they also revealed how diverse contemporaries, not just historians, debated the potential value and biases of collecting intended sources for the future—the question of whether one could create a "Pepys . . . by order."[242] Noting that it was "interesting" to see how "different men" viewed the scheme, Lane decided to deposit in the chest all letters responding to his request, whether they approved or disapproved, accepted or declined.[243] The Harvard chest is a rare instance of an archive that calls attention to its own absences and limitations.

Toward a Secret History

Exemplifying this autocritical tendency, Twain offered the most stringent appraisal of the time vessel. Even while writing his deferred autobiography, he exposed the inherent contradictions of such undertakings in his "Secret History of Eddypus." The specifications of his fictional time vessel can be read as correctives to other vessels of the period, including his own. Instead of adopting the customary hundred-year interval, his alter ego "Twain" had chosen the deeper future of five hundred—a time remote enough for the waning of historical knowledge to have rendered his manuscript invaluable, like a "lost Atlantis" rising out of the sea, yet not so remote that its language would have become unreadable by the "common people." And instead of describing the surface phenomena and acknowledged facts of society, he had sought to reveal the underlying factors of *"circumstance and environment."* The greater precautions "Twain" took to ensure his vessel's survival and recovery also implied the inadequacy of other vessels. He wrote not on paper but on vellum and deposited it not in a visible location but underground, "in a vault constructed especially for it below the foundations of the new Presidential palace" in Washington, DC. And whereas others relied on local office holders to ensure their vessels were not forgotten, "Twain" deposited a record of its existence and location in the public archives, presumably a less fallible custodian.[244]

Even these modifications, however, could not prevent the vessel from overshooting its target. Covered in "thirty feet of ancient rubbish" in a city long forgotten, the manuscript was discovered accidentally by shepherds ten rather than five centuries after its depositing.[245] (A similar fate befell the century safe that the hero in *A.D. 2000* had deposited in the US treasury's archives in 1888, with written instructions for its opening to be passed on to each successive treasury secretary. The instructions were soon "mislaid," the safe buried "behind huge piles of papers and boxes of documents," and the key and combination code forgotten. It was twelve years past its target date before a file clerk came across it.)[246] By the time "Twain's" manuscript is recovered, the language in which it was written is obsolete. Its value as a pristine document is thus diminished by the fact that only a few philologists could "spell out its meanings." Those meanings are further obscured by the lack of other texts from the period (which have been almost entirely destroyed by the Eddymanian Popes) and by ignorance of the author (who is believed to have been a bishop rather than a novelist). Oblivious to Twain's comic reputation, the future historian mistakes his absurd burlesques for literal insights, resulting in a comic muddling of names: Columbus apparently arrived on the Mayflower, and yellow journalism was "invented by Ralph Waldo Edison" during the glorious reign of the Black Prince, son of George III. Time vessels are not only subject to the vagaries of forgotten locations, linguistic extinction, and cultural misinterpretation; they are also, Twain suggests, susceptible to deliberate suppression. To avoid the censorship of the Eddymanian church, which prohibited the excavation of archaeological sites and the translation of ancient texts, the manuscript cannot be published and disseminated but has to remain concealed even after its rediscovery. Twain's story thus implicitly impugned time-vessel advocates who, preoccupied with the vulnerability of documents to material decay, overlooked the possibility of their intentional destruction and suppression.[247]

For all the problems of suppression and misinterpretation, the forbidden manuscript does at least enable the narrator—a dissident, underground historian—to write a "secret history" in invisible ink countering the official master narratives of the Christian Science popes. Twain may even be arguing that a time vessel opened on its target date is ultimately of little value, as it suggests there has been social and cultural continuity. Paradoxically, only by being forgotten, as a result of some historical rupture, can a time vessel enable the future to remember. Furthermore, its success will remain dependent on the hermeneutic skills, and indeed bravery, of its future readers. Rather than passive recipients—as William James presumed—they would need to be active interpreters of the recovered materials so as to construct a counter history out of the gaps and silences of the archive. "One of the most admirable things about history," explains Twain's narrator, is "that almost as a rule we get as much information out of what it does not say as we get out of what it does say."[248] In the next chapter I will thus sketch such a counterhistory drawn from the materials in all these centurial vessels. As Professor Haskins recommended in his essay on medieval sources, it has been "necess[ary]" to examine "each collection as a whole before utilizing any of the documents it contains."[249]

FOUR

Seeds of Hope: "Posteritism" and the Political Uses of the Future, 1900–1901

The centurial vessels departed from traditional memory sites and institutions and from the growing orthodoxies of professionalized history in their embrace of the history of the present, the local, and the everyday; their inclusion of new kinds of sources; and their diversity of voices. Did they also represent a departure in how they responded to social and political issues? What did they have to say about the industrialization and class stratification of their cities, their nation's emergence as an imperialist power, the encroachment of technology on everyday life, or the possibilities of challenging the solidifying hierarchies of gender, race, and class? How did the vessels' conceit of speaking directly to posterity color their contributors' views of the present and future?

The centurial vessels certainly offered an occasion, not unlike a ceremonial parade or exposition, for Americans to affirm faith and pride in their city or nation. Some contributors did transmit reports of civic accomplishment and unprecedented commercial growth that betray the influence of dominant discourses of boosterism and nationalism. Yet the option of sealing messages unread evidently emboldened others to critique their city's or nation's cultural, moral, political, and socioeconomic flaws. Belying the public celebrations of Dewey's victory in the Philippines, their private messages expressed concern about the drift toward imperialism. They also divulged genuine fears of impending class warfare, thus calling into question our retrospective assumptions about prevailing confidence in the capitalist order.

In the process, the vessels became the focal point of a larger contest over the memory of the present. Would the working class be remembered as "happier, more prosperous and contented" than ever, as one contributor asserted, or as "virtual slave[s]," as another claimed? Would labor unions be recalled as fomenters of "disloyalty and treason" or as champions of the people's "rights and liberties"? Would future historians eulogize William McKinley as the president who triumphed over Spanish tyranny or criticize him as the "greatest criminal" of his time? And should Harvard be remembered as an ivory tower of distinguished, male professors or as a broader community in which women and black people struggled for recognition, and janitors and servants labored for low wages?[1]

The centurial vessels were also arenas for competing predictions about 2000/1, from visions of unrivalled power and prosperity to premonitions of anarchy, upheaval, or despotism. More fundamentally, they differed over the question of the future's legibility. Some contributors followed the ascendant paradigm of (social) scientific prognostication in viewing the future as closed or predetermined, the logical outcome of assumed laws of economic and technological progress. In extrapolating current trends across the coming century, they effectively colonized the future, that is, reduced it to the terms and assumptions of the present. Even the new technologies they envisioned were merely extensions of current technologies catering to current needs and organized along current lines. Such closed futures typically mark an investment in perpetuating existing hierarchies of race, gender, and class.

Other contributors, however, retained a sense of the openness or otherness of the future—its contingency on multiple, unknowable factors. By 2000/1, their college, city, or nation might have diverged from its current course. A sense of living through an era of profound transformations—in which cities were expanding at unprecedented rates, technologies were extending the human body, industrial workers were becoming a political force, the nation was acquiring its first colonies, and women and blacks were claiming new rights and liberties—led many to refute complacent assumptions of continuity. Several spoke of being "at the turning point" or "threshold" of a new age, of living on the cusp between a century of "transition" and a century of "fulfillment," or of their period being remembered as a dark age preceding an enlightenment. The open future was a nebulous, uncharted realm of possibility in which certain institutions, racial categories, and ethnic groups—or indeed one's city—might no longer exist. Time-vessel messages even envisioned—enthusiastically or fearfully—the displacement of capitalism or the republic by some other system, socialist, anarchist, or theocratic. Indeed, to deposit artifacts and photographs of their middle-class milieu was to anticipate a time when they will be relics of a defunct era.

In tracing these conceptions of the future, we may recover the politics of hope. The vessels stimulated contributors to imagine the ultimate, posthumous fulfillment of their hopes—not just for their own city, community, or progeny but also for larger movements. Eugenicists dreamed of the biological purification realized by their proposed restrictions on breeding. Christian Scientists envisioned becoming the "universal religion." And pacifists longed for a world without war. Similar fantasies of vindication were deposited by homeopathic doctors, progressive reformers, promoters of parks and wilderness, temperance advocates, suffragettes, civil rights activists, and above all labor leaders and socialists. For each group, the invoked future was a powerful motivating force. The Progressive Era's social and cultural conflicts were thus between competing futures.

Desires for a better future, Ernst Bloch reminds us, can veer off into "abstract" utopias, which are merely idealist and compensatory. The vessels, however, allowed contributors to fashion "concrete" utopias by designating a target date by which they would materialize, thereby endowing the utopian wish with an anticipatory vector. Contributors also included a history of their group or movement, thus rooting their hopes in the preexisting, or what Bloch called the "real-possible."[2] In Elizabeth Grosz's terms, those groups mobilized the "unactualized potential" of the past to produce the "resources for multiple futures."[3] They thus embraced historiography without necessarily subscribing to that historical determinism that projects dominant trends into the future. They sought instead to activate those suppressed aspects of the past that harbored *alternative* possibilities.

Contributors could also root their hopes in an embodied relationship with posterity. Indeed, Bloch believed that for utopianism to become concrete, it must involve a "venturing beyond" oneself and thus an affective dimension.[4] By addressing their sealed messages to specific individuals and invoking some kind of physical contact or bond of solidarity across time, they linked their hopes to actual workers, women, or black people of the future. They also concretized utopia through tropes of bodily absence, as when one clubwoman suggested that gender equality will have rendered clubs obsolete, leaving no one to receive her message. These contributors, in other words, wrote for a future that was something other than the mirror image of the present.

Finally, we will see how these vessels' capacity to project open, contingent, concrete, and embodied futures led to their deliberate deployment as devices for instilling a sense of responsibility to posterity. Although sometimes inciting self-congratulatory celebrations of the present, they could also convey the possibility, indeed necessity, of acting for the future. Louis Ehrich originally conceived his century chest precisely to popularize this ethic, which he called *posteritism*. The scheme had some success, as the idea was taken up elsewhere in the country and applied to unintended agendas, including the protection of labor.

Metropolitan Dreams and Doubts

At first sight, the centurial vessels appear to celebrate urban progress. The invitation to write to the future certainly prompted some to trumpet their city's miraculous commercial and demographic growth. In just a hundred years, proclaimed one Detroiter, a cluster of "some three hundred small houses . . . [and] a fort" has mushroomed into "a city of 300,000 souls"—a growth attributed by the president of its board of trade to the "unparalleled" expansion of transcontinental commerce (**fig. 4.1**).[5] There was similar pride in Kansas City's rise within "three decades" from a "struggling hamlet of cabins" to a twin city of "nearly a quarter of a million" with a livestock market and railroad hub to rival Chicago.[6] Locals testified to these economic forces by depositing a directory of the city's livestock exchange, an account of its largest grain trading company, and commercial and railroad ephemera.[7] Its selection to host the Democratic National Convention of 1900 was welcomed as recognition of its urban ascendancy, as was the rapid rebuilding of Convention Hall after it burned down three months before the delegates were to arrive. To consecrate that instant monument to the "Kansas City spirit," the century box was deposited a year later in its new entrance.[8]

The Colorado Springs century chest also appears to express civic pride. Contributors spoke of how over the previous decade the arrival of new railroad lines and the establishment of the nearby springs as a popular destination for tourists and convalescents had doubled the

FIG. 4.1 Anonymous, untitled photograph of downtown Detroit, ca. 1900, deposited in the Detroit Century Box. City hall stands on the left, and the Majestic Building on the right; Fort Street and Woodward Avenue intersect in the foreground. A description was typed on the verso: "This picture was taken in the central business portion of the city of Detroit and from the window of a [building?]." Detroit Century Box Collection, Detroit Historical Society.

city's population to more than twenty thousand, qualifying it as a "city of the first class."[9] As in Detroit and Kansas City, such evidence was touted as proof that it had weathered the depression of the mid-1890s and that its wealth rested on sounder foundations than those of western boomtowns.[10] They credited these blessings to its visionary founder William Jackson Palmer, who was still around to contribute to the time vessel—an indication of how western urbanism compressed historical time.[11]

These centurial vessels also extolled the industrialization of their cities. Almost overnight, Detroit had morphed from a mercantile to "essentially a manufacturing and industrial city," remarked the Michigan Stove Company's president.[12] Although beyond the manufacturing belt, Kansas City had emerged as a center not just for processing and packaging meat but also for producing goods for farmers in the hinterland as acknowledged by a catalog of harnesses and a yearbook of the local Manufacturers Association.[13] Even the spa town of Colorado

Springs was in the throes of industrialization. Initially, the process had been retarded by the "high price of labor" and elite distaste for factories' "disagreeable odor," wrote the iron manufacturer who made the century chest. "All this is changing just now" as a result of the discovery of gold twenty miles away at Cripple Creek on October 20, 1890, "& we are at the turning point."[14] By 1900, Colorado Springs boasted several mining companies (**fig. 4.2**) and the nation's "leading" exchange for mining stock along with ore-reduction plants, smelters, railroad yards, and ancillary factories—mostly shunted to the poorer, adjoining town of Colorado City.[15]

To negate the stereotype of burgeoning, industrializing cities as crude and unsophisticated, the vessels drew attention to their cultural flowering. The director of Detroit's art museum demonstrated "What Detroit Has Accomplished in Art"; its leading organist reported the "wonderful transformation and growth in all departments of music"; and the manager of its opera company described the "exquisite decorative effect[s]" of its newly completed opera house, with similar claims registered for Colorado Springs and Kansas City.[16] The latter's vessel attested to its cultural sophistication by memorializing the Century Ball, reportedly the nation's grandest New Year's celebration.[17] Editors cited (and deposited) their newspapers as further evidence: the Colorado Springs *Gazette* was a purveyor of "clean, decent, conservative, progressive and able journalism" and a buttress against the spread of

FIG. 4.2 Anonymous, *Curtis Coal Mines—Immediately North of City*, photograph, ca. 1900, submitted to the Colorado Springs Century Chest. The town's founder, Colonel William Jackson Palmer, complained in his time-vessel message about the pollution from nearby factories and mines, including this one, which burned coal to provide electricity for Colorado Springs. Folder 152, Colorado Springs Century Chest Collection, 1901, Ms 0349, Colorado College Tutt Library, Special Collections.

"Yellow Journalism" to this "home of culture," while Detroit's *To-Day* stood for "permanency and prestige" rather than "partisan newspaper-making."[18] Lastly, cultural growth was manifest in these cities' built environment. The president of Detroit's Common Council spoke of "a beautiful city, well paved and sewered, with broad avenues and handsome homes, pretty parks and drives," while a Colorado Springs realtor waxed lyrical about "cottages and mansions of the highest types of architecture . . . [m]assive business blocks, most modern in their construction and convenience, beautiful churches, and school houses."[19]

Several contributors boldly predicted even greater glories by 2000/1. Drawing on the rhetorical tradition of urban boosterism, they adduced their city's climate, resources, and rail communications to prove that the coming century would be one of unabated and unprecedented commercial growth.[20] Detroit, wrote the stove manufacturer, "will surely be of wonderful prosperity and industry" and "in the front ranks of the commercial and manufacturing world."[21] Assuming a steady annual increase of one percent, the president of the Merchants' and Manufacturers' Exchange calculated there would be five thousand factories employing half a million workers by the time his letter was opened.[22] Charles Freer, the railroad-car manufacturer and subsequent founder of the Smithsonian's Freer Gallery of Art, added that Detroit's products would be renowned for their "superiority of workmanship and quality of materials."[23] (None of these prophets, however, anticipated the economic supremacy of the automobile; although Henry Ford had recently founded the Detroit Automobile Company, with legal and financial support from Mayor Maybury himself, it was dissolved the same month the vessel was sealed.)[24] Visions of abundance were especially vivid in western towns like Kansas City, which had a reputation for boosterish histories predicting urban greatness.[25] Following those boosters, a treasury clerk imagined "a railroad center that will be the pride and admiration of the World with a Union Depot covering 2 or 4 blocks . . . monster freight sheds . . . miles of horse and mule market stables—prodigious packing houses—and manufactories galore."[26] Such visions rested on a conception of the future as the logical and predictable outcome of current trends and as governed by some variant of determinism (environmental, economic, technological, etc.). According to the temporal logic of linear progress, those cities would ascend to ever greater wealth.

Contributors offered concrete predictions of demographic and spatial growth too. The Kansas City clerk estimated it would be home to "upwards of two millions" and extend "8 or 10 miles Eastward &

Southward," while another Kansan wagered that "palatial homes" would crowd the road all the way to Excelsior Springs, twenty-five miles away.[27] Population estimates for Detroit ranged from one to four million, presuming "no war, pestilence, or other destructive influence."[28] Sprawling "for long distances in every direction," it will have absorbed suburbs like Grosse Pointe and even the outlying towns of Pontiac and Ann Arbor. In fact, with business districts stretched out along local and suburban railway lines, "the marked distinction" between cities and countryside will have "broken down."[29]

These contributors' faith in sublime, ceaseless growth induced imperial dreams of their city's dominance of its hinterland. Detroit's dominion, according to some, would extend across the border into Ontario.[30] The boldest prophet of its future metropolitan status, newspaper publisher James Scripps, pointed to its superior climate (relative to other cities on the Great Lakes), perfect harbor, and unrivalled access to vast raw materials—all signs that its continental sway would exceed that of Constantinople over the Byzantine Empire.[31] Mayor James Reed of Kansas City, Missouri, invoked even grander historic precedents—the Holy Roman Empire and the Kingdom of the Pharaohs—as proof that "when the sun of the twenty-first century shall rise it will fall upon the spires and minarets of a mighty American city, and gild them till they glow with glory." Whereas earlier boosters had promoted Kansas City as a regional metropolis—the "Athens" or "Queen" of the West—Reed and others now articulated its destiny as the national metropolis.[32] The time vessel's architect Charles Moss presumed his letter would find the US president residing in Kansas City, the new capital.[33] Even more boldly, the mock newspaper from the year 2000 that was deposited in the vessel reported Kansas City's selection as headquarters of all the amalgamated nations, thanks to the efforts of McKinley's great-nephew George. A precondition for that glorious ascent appears to have been the merger of the twin cities.[34] Between 1855 and 1879, there had been four attempts to incorporate the entire city into Kansas, which would have prevented the inefficiencies and petty rivalries that hindered later development.[35] Reviving those schemes, Mayor Marshman of Kansas City, Kansas, addressed his message to the future mayor of a unified metropolis. "Municipal government" will have been "simplified" and tax dollars saved by having "one mayor, one council, one police department, one fire department."[36] A rare instance of such collaboration, the vessel itself and the unified telephone directory deposited in it concretized that hope for future consolidation.[37]

Fantasies of millennial glory, however, mingled in the vessels with doubts or even anxieties. One rector suggested Colorado Springs "may be a great and flourishing city" by 2001 yet acknowledged that it may just as easily have "passed utterly out of existence"; after all, "a hundred years is a long time."[38] Contributors cited various threats: a great fire (one had almost "swept [the city] out of existence" three years earlier); the depletion of the local mines (causing it to "lose its commercial importance, just as the Massachusetts seaboard towns did . . . a half century ago"); or the moral erosion of social and political institutions by "selfishness and avarice."[39] Kansas Citians similarly acknowledged the possibility that "some great convulsion of nature [a tornado perhaps?] may wipe our beloved city with all its fond associations and revered memories from the map." Some were old enough to remember the uncertainty of the 1850s and 1860s, when political antagonisms, sectional skirmishes, and railroad construction delays nearly dashed their urban ambitions.[40] For these contributors, the future remained a realm of uncertainty. Contingent on unforeseeable factors and events, it could hardly be predicted with any confidence.

Perhaps emboldened by the centurial vessels' assurance of confidentiality, contributors also expressed criticisms of their city. Alongside claims of cultural sophistication were complaints of cultural backwardness. The opera house manager admitted that Detroit still lacked "a large music hall, a concert organ, a permanent endowed orchestra and a more efficient chorus."[41] Its newspapers also came under question. "The *News* is unreliable," declared an editorial writer at the Detroit Free Press, the *Tribune* "colorless," the *Journal* a "bigoted partisan newspaper," the *Today* "a puny monstrosity . . . [with] no merit except cheapness," and his own newspaper "poorly edited."[42] Such candid assessments were especially common in Colorado Springs. "In the matter of public art our town is decidedly remiss: there is not a statue within its borders, not even a single fountain of artistic design," admitted Samuel Caldwell, a local doctor, while "another of our crying wants is a Public Library building."[43] Its architecture was also found wanting. The town's leading architect declared that the vast majority of houses are stylistically deficient and "turned out by the dozen" by uneducated builders—some so eclectic that "the architects of the Parthenon . . . [would] turn in their graves!" A visitor's first impression of the town is "very disappointing," agreed another resident. "I thought it such the ugliest place I had ever seen."[44]

Even more alarming were the moral threats to their cities. Alcohol was a particular concern in Kansas, the cradle of the temperance movement and the first state (in 1881) to enact constitutional prohibition.

Residents of Kansas City, Kansas, however, could merely patronize their sister city's thriving taverns and saloons. If one response was to raid them and smash bottles—as the Kansan temperance radical Carrie Nation did in 1901—another was to entrust to the century box a wish that "Kansas City have . . . no saloons" by 2000, thus reinforcing time vessels' long association with teetotal utopias.[45] Intemperance was also the scourge of the middle classes in Colorado Springs. Palmer had founded it as an antisaloon town to avoid the vices associated with western boomtowns, and by 1900 it was the largest "no license" community in the state, a fact crucial to its reputation as a health resort.[46] But alcohol was still readily obtainable from its less respectable twin Colorado City and, as Dr. Caldwell complained, from social clubs. According to another resident, "there is more intoxicating liquor sold and drank in this so called 'temperance town' than in any other city of its size." These and other contributors feared alcohol was leading to a general breakdown of morals and the spread of urban vices.[47] Liquor consumption was thus a grave problem—"the cause of more harm and sin than all other evils combined," lamented a local minister.[48]

This moral corruption had apparently spread to city government. Colorado Springs "has had a very bad set of men to manage its affairs," one contributor alleged. "Nearly all the Mayors have been more or less mixed up in some irregularity," and three of the last four treasurers "have been short in their accounts"—most recently, to the tune of $30,000.[49] Charges of mismanagement and graft (or "boodling") were also widespread in Kansas City, Kansas, while across the state line Tom Pendergast had already established a powerful political machine—raising doubts in both cities about the mayor-council system.[50] As the president of Detroit's chamber of commerce confirmed, municipal corruption was a major issue of the day.[51]

More than just passively recording the present, time vessels represented an active pledge to solve these and other municipal problems for posterity. Rather than imagining utopias in the abstract, they sought to bring them about through concrete action in the present. At the Colorado Springs ceremony, Ehrich argued that "posteritism" should be "applied" at the local level to nurture a "passion in each community to make itself more desirable, more beautiful, and more uplifting as a home for the generations that are to follow." This ethic of care for a city's future inhabitants needed to be "re-kindled," because the "intense love for the city which characterized the old Greek and Roman world" has gradually "disappeared."[52] The time vessel thus appears to have been a tool for municipal reform, a movement launched in the 1890s by progressives who were alarmed by the spread of ethnic political machines across urban America. Along with purging corruption and inefficiency from city government, municipal reformers advocated cautious budgetary management. Detroit's city controller declared in his time-vessel message that he avoided fiscal imprudence out of duty not only to "present" but also "future generations." With his thirty-year bonds, "each generation will . . . pay its proportion for the use of . . . permanent improvements" rather than offload the costs onto its successors.[53] Public ownership or regulation of utilities—another goal of municipal reform—was similarly advocated in terms of its benefits to posterity. Detroit's Mayor Maybury, although a pro-business Democrat, acknowledged and continued the progressive legacy of his predecessor Hazen Pingree, soliciting testimonial accounts of two utilities (water and public lighting) that had been successfully municipalized.[54]

Contributors also spoke of a sense of "guardianship" or "trustee-ship," on behalf of posterity, over the local landscape.[55] Municipal reform was closely allied to the City Beautiful movement, a crusade by architects, landscape architects, and urban planners to impose aesthetic—and by extension moral and social—order on American cities. One advocate for a comprehensive civic plan for Colorado Springs, a former forestry commissioner, applauded the creation of parks downtown. The trees were still "small," he wrote, but it "gives promise of future beauty and usefulness." He also reported on the "parkways" or landscaped medians he had introduced to solve the problem of the city's wide streets (in a fit of posteritism, Palmer had endowed the original town plat of 1871 with 140-foot-wide avenues, resulting in complaints by 1900 about the accumulation of weeds and litter and the prohibitive cost of paving). The project remained ongoing—only a few avenues had so far been "parked"—and would "need the fostering care of later generations." So, too, would the ultimate goal of a network of public parks across the city.[56] Detroit's park system, according to a message from its commissioner of parks and boulevards, was "perfectly adapted to extension as the growth of the city may demand."[57] Just as tree planting was a symbolic act of commitment to posterity, so were images of organic growth—plants bearing fruit or seeds sown for future harvests—frequent metaphors for that commitment. Now was a time for "clearing and ploughing," wrote one Detroiter, "to make better the grain of the 20th and 21st Centuries."[58]

Imperialism and the Seeds of National Decline

The centurial vessels' purview did also extend beyond the urban to the national. Again, despite our assumptions about the solidity of late-Victorian faith in progress, the centurial prognosis was colored by considerable misgivings. Certainly they included nationalist fantasies of America's supremacy by the year 2001, such as newspaper magnate James Scripps's prediction that it would be "unquestionably the richest" and "most populous" country "the world has ever known."[59] Others were less certain. They viewed progress not as inexorable but as contingent on certain unknowable variables: the retention of Christian faith, the repudiation of the "pursuit of wealth and pleasure," or adherence to "truth, justice, and righteousness."[60] It was the "earnest wish" rather than conviction of one guest at Kansas City's Century Ball "that a grander, greater and more humane people may celebrate the dawn of the 21st century." Whether national progress in the twentieth century would eclipse that of the nineteenth was, for another guest, a "question" only his future readers could "answer."[61]

Visions of future progress, in any case, could provoke an unsettling sense of the present's comparative backwardness. "In the light of your great achievements," wrote another Kansan, "we wonder . . . what you will think about us."[62] One Coloradan feared his paper would resemble "a relic of a well-nigh primeval age and its crude ideas," while Leah Ehrich despaired how "all that we have done and accomplished must seem very small and meager to you."[63] This sense of the present's inferiority to the future was a feature of what one historian has called "the modern regime of historicity." In earlier periods, most notably the Renaissance and the Enlightenment, the present had been considered inferior to antiquity.[64]

If such statements still affirmed an unbroken line of progress, others questioned even that. One Colorado rector envisaged a future "change" of direction that would lead our descendants to conclude that we had been "on the wrong track." The nature of that historical turn varied. The rector cited the belief "that our Republic . . . [will be replaced by] a form of limited monarchy," while another Coloradan saw "Intemperance, Mormonism, and Anarchy" looming. What was certain to the latter was that the twentieth century would be an ordeal—"a crucial test from which we will emerge tried by fire."[65] Mark Twain presented those coming tribulations with particular acerbity in "The Secret History of Eddypus." He employed the time vessel not simply to impugn Christian Science but to shatter assumptions of progress and ever-expanding liberty and to flesh out his late conception of eternal cyclicality. The reversion to monarchy and papacy exemplified his broader "Law of Periodical Repetition."[66]

For a significant number of time-vessel contributors—above all Twain—the greatest threat to the republic was the recent imperialist activity.[67] After the *USS Maine* sank in Havana harbor in 1898, fifty-one residents of Colorado Springs—thirteen of whom later participated in the time vessel—had signed an open letter to President McKinley. Printed in several newspapers and subsequently deposited in the vessel, it urged nonintervention in Cuba.[68] Ehrich, the instigator of that protest—as far as he knew, the only one at that point—became a prominent member of the American Anti-Imperialist League, chairing its 1900 convention. There, in protest against McKinley and his new vice presidential candidate Theodore Roosevelt, he (unsuccessfully) urged anti-imperialist Republicans to break away and form a third party.[69]

Ehrich conveyed these political convictions in an essay on "national issues" for the chest. Whereas open forms of discourse, such as a letter to a newspaper or dedicatory speech for a time vessel, required a certain deference—especially as Theodore Roosevelt, now vice president, was present at the Colorado ceremony—a sealed message allowed Ehrich to speak more candidly about the Spanish-American War. For evidence of the mentality that precipitated such a "wicked" war, he suggested his future readers glance at the opening pages of Roosevelt's best seller *The Rough Riders* (1899).[70] This suggestion thus subtly undermines Roosevelt's message—which Roosevelt himself read out and deposited in the same box—that celebrated his former regiment, the Rough Riders, as "a splendid set of men, these grim hunters and miners of the mountains, these wild riders of the plains."[71] As for McKinley, to Ehrich he seemed "weak, superficial, pliant, characterless." Ehrich was particularly critical of McKinley's proclamation of Benevolent Assimilation, which had declared America's sovereignty over the newly liberated Philippines not as "invaders or conquerors, but as friends, to protect the natives." That "infamous" proclamation effectively "crush[ed] out the rights and liberties of millions of men," leading to a Philippine insurgency and a deadly, protracted guerrilla war in which thousands of "brave Filipino patriots" and American soldiers were unnecessarily slaughtered. For setting the nation on this course, Ehrich considered McKinley "the greatest criminal living on the earth today" and a grave "dange[r]" to the republic. "Posterity's opinion" of him would be akin to that of King George III.[72] Although the Colorado chest contained some proimperialist messages, others shared Ehrich's sentiments. A local rector lamented that his fellow Christians across the nation had failed to oppose, and had in fact supported, this "intensification of the Martial spirit."[73]

The diaries and essays Harvard professors and students deposited in their vessel the previous summer also evince a surprising degree of opposition to the war, which (as several contributors reported) had cut short some alumni's lives.[74] Paul Revere Frothingham, a preacher at Harvard, blamed the Philippines quagmire on McKinley's "preposterous" proclamation, as did theology professor Joseph Henry Thayer, who was particularly critical of its cynical appeal to the "nobler sentiments—benevolence, religiosity."[75] One student predicted that by the vessel's opening, "the harvest of imperialism" will have been "reap[ed]," but "in this same sowing will be the seeds of ultimate collapse," while another championed the anti-imperialists' "struggle to keep this . . . country . . . free from foreign interests and complications."[76] It was not just America's new colonial ventures that came in for censure but also those of other nations. The latter student, like "most of [his] companions," felt a "deep sympathy" for the Boers despite America's official support of the British. Although the Boers' Siege of Ladysmith had been broken two days earlier, he hoped they "will yet be free from British tyranny." Thayer, too, considered it a "horrible and needless war."[77] Some hoped that not just colonial but *all* wars will be gone by the millennium—that "peace on earth" and "brotherhood of man . . . will have beaten the sword into ploughshares, the spear into pruning hook."[78]

The time vessel thus enabled anti-imperialists to conjure an alternative, open future to their opponents' visions of ever-expanding power, liberty, and abundance. Rather than precisely delineating that future, they fantasized (in the mood of the future anterior) about what would have become of the ideologies and reputations of the present. They could imagine a time when imperialism and war would have become untenable. And they could imagine future generations looking back at Roosevelt's best-selling book as testimony of delusion rather than valor, at McKinley as a King George rather than an Abraham Lincoln, and at themselves as heroic defenders of the republic rather than "unhung traitors," as Roosevelt branded them.[79]

Technological Wonders and Dangers

The centurial vessels also elicited conflicting opinions on the nation's technological advances. Voiced by ordinary Americans, those opinions shed light on popular attitudes toward new inventions. Investigating their effect on society, historians of technology have focused on the inventors themselves, the corporate networks and experts administering them, or the writers and artists who celebrated or critiqued them.[80] Ordinary individuals do figure in social histories of technology, but as users (complicit or resistant) rather than commentators.[81] They are often assumed to have consented to the technological optimism of inventors and utopian novelists or to have embraced technologies for their utilitarian, productive functions.[82] The centurial vessels, however, contain testimony that restores the voices of ordinary people and the complexity of their attitudes toward technology.

Those vessels, to be sure, did contain numerous celebrations of recent innovations and confident predictions of future marvels. Extrapolating from current trends, needs, and technologies, those predictions inhabited a closed future. This was particular evident with regard to transportation. Merchants and manufacturers issued predictable forecasts about the commercial triumph of airships (the first Zeppelin flew in 1900) or interoceanic canals (Roosevelt had called for America to complete the Panama Canal, recently abandoned by the French, in 1898).[83] Assuming further developments in electricity, one woman imagined future residents getting around Colorado Springs by plugging themselves into a transit system. "You will pick up your electric cane, or umbrella from its stand in the hall, then you will step out from your front door, connect with the wire overhead, and go swiftly down town"—a solution perhaps to bourgeois fears of intermixing on crowded streetcars.[84] The fantastical uses of electricity were almost equaled by those of compressed air. Contributors prophesied that pneumatic tubes, already used to deliver mail, would transport humans. No doubt uneasy about being pumped through a pipe himself, Detroit's police commissioner imagined it being used to dispatch criminals to the police station.[85]

Contributors also celebrated how existing technologies had extended, and would continue to extend, the scope of human communication. Telephones were liberating women by enabling them to order goods from home, enthused one Coloradan housewife.[86] Deposited in Colorado Springs and Kansas City, telephone directories served not only as ready-made memorial registers to the middle class (not the elite, who were already opting to be unlisted) but also as evidence of how far the new technology had already spread.[87] Some predicted that by 2000, telephone cables would span the continent and even the globe, while others imagined communication by "sparteograph" with

Martians.[88] One contributor believed that advances in spiritualist communication would allow their future recipients to send a reply.[89] And if the phonograph already allowed the living to speak or sing to the unborn, further enhancements could produce what one Coloradan called a "mental phonograph"—a device that could detect and record people's unspoken thoughts by registering their vibrations. Seemingly oblivious to the dystopian implications, she believed it would reform humanity by purifying people's minds of shameful thoughts.[90]

Besides amplifying the voice and sharpening the ear, technology was also celebrated for extending the eye. X-ray photographs were included in the Colorado chest to show how that latest "instrument of precision," invented five years earlier, could expose "the hitherto invisible structures of the human body."[91] Rather than imagine new uses for such visual technologies, contributors presumed they would serve prevailing functions, such as providing entertainment. "Air yachts," predicted the mock newspaper from 2000, would enhance the spectacle of fireworks by launching them into space from eight thousand feet up, thereby announcing Kansas City's appointment as world capital to everyone living within six hundred miles. Meanwhile, "hectreoscopes," transmitting sound and light waves, would broadcast that display to "all the principal cities of the world."[92]

A common motif in the vessels was the technological conquest of nature. The naked flame, wrote one Coloradan, that "primitive" source of warmth and illumination, was being supplanted by modern systems of steam heating, hot-water radiators, and electric lighting. Millennials will have further eliminated fire hazards by generating steam (as well as electricity) at central stations rather than in kitchens or basement furnaces.[93] Locals also prided themselves on their modern plumbing, a solution to the age-old problem of human waste. The photographer Fred Stevens submitted visual evidence of the fine toilet fixtures (**fig. 3.10**), as well as electric lights and telephones, to be found in middle-class homes.[94] Agriculture, too, will have transcended nature's vicissitudes by 2001. Storms, hurricanes, and cyclones will be defused by chemicals that absorb atmospheric heat, releasing it during cold spells to protect crops, and droughts will be mitigated by vast, government-built irrigation systems or by "scientific appliances" that generate rain. Some predicted liberation from agriculture altogether through laboratories that synthesized fruit and vegetables—but not, alas, steaks, which will "become a banquet rarity."[95] An alternative scenario was the replacement of such produce by "pellets of patent compressed food."[96]

The natural processes most frequently transcended in these technological fantasies were those of human sickness and aging. A flu epidemic gripped the campus during the preparation of the Harvard chest, and a smallpox epidemic had just hit Colorado Springs, where tuberculosis convalescents were a continual reminder of disease and death.[97] But the discovery of the bacillus that caused TB gave Dr. Caldwell hope that "improved methods of sanitation and treatment" would reduce the number and mortality rates of "'lungers,' as we call them, among you."[98] Doctors in Detroit predicted similar advances in surgery, such as more precise instruments and improved anesthetics. Medicine would also become preventative, "eradicat[ing]" germs and unsanitary conditions before they cause illness.[99] Some envisioned the eradication of *all* contagious diseases. Eighty-year-olds would be considered "in the prime of [their] youth," and a few would live beyond the age of 180.[100]

Applying existing technologies to existing needs, these predictions reduced the future to present terms, thus colonizing it. Others, however, were less confident about such extrapolations. The pace of innovation in their lifetime led many to conclude they were on the brink of a new era of unimaginable technological wonders. "No vision of the most imaginative dreamer," wrote a Coloradan, "is impossible of realization to the century, on the threshold of which we now stand." This technological breakthrough (akin to what current futurologists call the *singularity*) would be so profound as to mark a rupture in human history, consigning the America of 1900 to the realm of prehistory or a dark age.[101] Such a great leap forward rendered specific predictions pointless. Given the vast "possibilities" of electricity in so many "branches" of human life, conceded one Detroiter, "all attempts at conjecture [are] worse than futile."[102]

Technological change also prompted ambivalence or outright consternation. Several Coloradans decried the accelerated rhythm of life in their new world of rapid transit and instantaneous communication. "One of the characteristics of our age," remarked Dr. Caldwell, "is the desire to go everywhere in a hurry."[103] For some residents in that health-conscious town, this was a cause for concern.

> Even some of our babies have been obliged to take "a rest cure." I hope your century will not get whirling around so fast that the babies will be thrown off entirely.—I sometimes feel very anxious about you! . . . . Regularity is observed about their meals, but we have not reached the eastern standard in our care of them; baby-carriages in the east are sent out armed with clocks and thermometers.[104]

The application of science and technology to so many spheres of life—including child rearing—raised larger moral and spiritual questions. "Mechanical contrivances will continue to be invented to lessen drudgery," wrote the wife of another Colorado Springs doctor, "but will the moral and spiritual development keep pace with these for the uplifting of mankind?"[105] If the town's future residents wish to recreate the "charity," "geniality," and "good fellowship" that characterized the town's early years, advised one of its longest-standing residents, they must look beyond modern conveniences and "applied science."[106]

There were doubts, too, about technology's reliability. Even as it celebrated the spectacular powers of the hectreoscope, that mock newspaper from AD 2000 contained numerous reports of mechanical breakdown, which exposed an unhealthy dependence on modern conveniences. "A section of the [moving] sidewalk on Central street remained stationary for nearly two hours this morning owing to a slight break in the machinery," the future reporter disclosed, "and people going east on that side had to walk nearly a block." Similar malfunctions caused suspensions of the transatlantic pneumatic mail system and of the solar-powered illumination of a thirty-eight-story office building. Meanwhile, pedestrians continued to be plagued by swarms of aerial craft landing in the streets despite the mayor's "campaign pledge to enforce [an] ordinance prohibiting [it]."[107]

That newspaper drew attention also to the obsolescence and creative destruction that accompany technological progress. It announced the recent archaeological discovery of a "curious machine" called a bicycle and urged the government to prevent the extinction of the horse.[108] There were also prescient fears that future advances in transportation would render entire urban districts obsolete. People will no longer need to live near downtown, predicted a Colorado Springs realtor, and so Cascade Avenue will cease to be "fashionable."[109] Moreover, implicit in the Colorado Springs photographs of telephones, linotype machines, or plumbing and electrical fixtures was a sense of the future obsolescence of technologies themselves. Through a kind of "temporal slippage," the act of photography seems to consign the present object to the past.[110] In anticipating a time in which they will have become historic specimens, these photographs embody what Fredric Jameson called nostalgia for the present—an impulse, however, not specific to the postmodern.[111] Manufacturers, such as D. P. Alderson of Kansas City, were acutely aware of the destiny of our "present styles of implements" to become "relics for your museums."[112] And when our devices and inventions are "supplant[ed]" by new ones, wondered the aforementioned Detroit stove manufacturer, "what will become of all of our great factories and foundries"?[113]

Others queried the assumption of technological progress itself. Believing that automobiles—of which there were only about ten in Colorado Springs—were too expensive, noisy, noxious, and unreliable "to come into general use," Caldwell predicted that the simple bicycle would still prevail in 2001.[114] (Hedging his bets, Stevens photographed both "locomobiles" and bicycles for the vessel; **figs. 4.3, 4.4**). Similarly, the noninclusion of a phonograph to play the recordings in that vessel implies a belief that such a device would remain in use a century hence. Another Coloradan challenged the assumption of unlimited fuel supplies undergirding future visions of technological progress. Well before the year 2001, he warned, "the rapid exhaustion of the coal beds of the world will have become a serious question." Alternative energy sources will be needed, such as "water power," transmitted wirelessly across vast distances.[115] Others also envisioned sustainable sources, from solar and hydroelectric power to municipal recycling plants for converting refuse into "heat, light, and power." The latter

FIG. 4.3 F. P. Stevens, *Locomobile, One of the First in our City[.] The Gentleman is C.E. Palmer, a Mining Engineer*, photograph, 1901. Photograph 87, folder 160, Colorado Springs Century Chest Collection, 1901, Ms 0349, Colorado College Tutt Library, Special Collections.

FIG. 4.4 F. P. Stevens, *Modern Bicycle for Women, 1901*, photograph, 1901. Another contributor to the chest celebrated the bicycle as a convenient and economic tool for "the business woman." Photograph 91, folder 160, Colorado Springs Century Chest Collection, 1901, Ms 0349, Colorado College Tutt Library, Special Collections.

would contain free bathing and laundry facilities for families unable to afford such "modern . . . accommodations in their homes," an acknowledgment that technologies would remain unequally distributed even in 2001.[116]

Finally, predictions of the coming wonders of Western medicine were outnumbered and challenged by even more fervent declarations of faith in spiritual and alternative healing practices. Harvard professor George Herbert Palmer, an expert on natural religion among other things, claimed in his message to have been saved as a child by homeopathy, administered by the wife of Samuel Hahnemann, homeopathy's founder.[117] Despite considerable hostility from mainstream physicians, homeopathy had made substantial gains by 1901. A Detroit practitioner, one of sixty in that city, wrote proudly in his message about a new homeopathic hospital and dispensary that had already treated thirty-two thousand patients, many of them poor. With the establishment of the Detroit Homeopathic College in 1899, he sensed the system's permanent acceptance. Unlike surgical procedures of the period, the mildness of homeopathic remedies ensured that, even if they did not cure patients, at least they would not kill them; he reported a mortality rate of just 6.14 percent.[118] Another alternative therapy touted in the Detroit vessel was biopathy. This now-forgotten practice rejected all medication, viewing the human body as an electric battery (consisting of twelve "elements") that, if kept topped up, could run for 140 years or more. A message from a biopathic healer declared that it would be the "only system of medicine" by 2000, which rendered his explanation of it somewhat redundant.[119]

While biopathy appears to have had only one adherent, many propounded the benefits of Christian Science. More than just a religion, Christian Science offered an alternative healing practice to patients who had lost faith in the medical profession. Mary Baker Eddy claimed to "discover" Christian Science after her miraculous recovery from a serious accident in 1866, a recovery she attributed to her reading about Christ's power to heal. Viewing sickness as derived from mental causes, she "scientifically" elaborated a healing method that rejected all drugs and somatic treatments in favor of prayer and claimed to have healed afflictions from blindness and deafness to tuberculosis and cancer.[120] A similar tale of healing appears in the Colorado Springs vessel, contributed by a woman who publicized Christian Science there, Dora Noxon. After "the best physicians" failed to alleviate her severe peritonitis, Christian Science eradicated it in two weeks—testament to the power of the mind.[121] Contributors expounded Christian Science's active opposition not just to medicine but to scientific materialism in general. Noxon enclosed a longer essay in which she critiqued materialists for reducing man to "a piece of clay," while another follower in the same town declared Christian Science "a savior of men from the miasma of materialism, skepticism and 'Science.'"[122] An essay in the Detroit vessel by Annie Knott, who had embraced Christian Science when it apparently saved her son after he drank a bottle of carbolic acid, described it similarly as part of a larger intellectual "reaction from a period of materialistic speculation in Science and religion."[123] Noting their creed's remarkable growth—arguably the fastest in the nation, increasing 462 percent over the previous decade—they predicted it would be the "universal religion" by 2001, having "swallowed up" everything that impedes "spiritual harmony and divine control."[124] "Though the paper" on which Annie Knott was writing "may be turning to dust[,] the enkindling Truth which dictates these words shall have in a large degree liberated all the nations of the earth from the bondage of superstition."[125] In this light, Twain's Eddymanian writings can no longer be dismissed as the ramblings of a paranoiac; Christian Scientists themselves cherished visions of global domination.

"The Future American": Racial and Ethnic Forebodings

The centurial vessels also elicited conflicting estimations regarding their cities' and the nation's ethnoracial conditions both in 1901 and in 2001. Representatives of older immigrant stock complained of an engulfing tide of aliens from southern and eastern Europe. The secretary of Detroit's Franco-American Club predicted that the city's French population—its first ethnic group—would be negligible by 2001 compared with those from "other nations," while the secretary of the St. Andrew's Society hoped that there would still be "some living Scotchmen" to read his paper.[126] "As the old German population dies off," noted another Detroiter, so do its newspapers.[127] Such groups viewed the time vessel as a last chance to preserve a trace of their culture and customs. With the threads of oral and performative memory wearing thin, they resorted to a disembodied memory device.

Even representatives of growing ethnic groups expressed concerns. While hailing the increasing admittance of Jews into Detroit's commercial, professional, and political life, one Jewish industrialist lamented their continued social separatism—the tendency of the "Jew to confine himself almost strictly to association with his own people." He hoped the younger generation would overcome this, so that by 2001 "almost nothing [will] . . . distinguish the Jew . . . from the non-Jew."[128] Another Jewish industrialist similarly dreamed that his people "will harmoniously blend their lives with those of their fellow citizens of every station and denomination."[129] Voiced by these representatives of Detroit's older, "respectable," German Jewish community, such Americanization rhetoric betrays lingering suspicions about the newer, eastern European Jewish immigrants' capacity to assimilate.[130]

The vessels also tended to exclude certain ethnic groups. By 1900, first- and second-generation Polish immigrants constituted 8.76 percent of Detroit's population, and the Irish made up 5.26 percent, thus outnumbering the French or French Canadian immigrants (2 percent), yet no essay regarding the Polish or the Irish was deposited in the century box.[131] It is possible, too, that they declined to participate in this Anglo-directed performance of pride in their city given their primary identification with their own ethnic communities.[132] The Colorado Springs vessel was also heavily weighted toward Anglo-Americans, as

Ehrich (despite his German Jewish origins) appears to have solicited representatives of Protestant churches, Masonic lodges, business associations, social clubs, and other WASP-dominated institutions, including Colorado College and the newly formed Society of the Children of the American Revolution. Reflecting the late nineteenth-century appropriation of genealogy to assert racial pedigree, several appended family records to their messages, one tracing a lineage back to an English emigrant of 1632, the vessel substituting here for older media for transmitting those records to one's descendants, such as the family bible, needlework sampler, or commonplace book.[133] Another hoped her great-grandchildren will not have forgotten "the truth and steadfastness of purpose of our New England ancestors."[134] For those with a shorter family tree, the vessel offered a compensatory vision of how it might have grown by 2001.

More subtle efforts to safeguard Anglo identities are evident in the Harvard chest. Adams Sherman Hill arguably founded the composition program there out of concerns that the mother tongue was under threat not only from mass-circulation newspapers, with their liberal use of slang, but also from other languages. "Whatever is addressed to English-speaking people," he declared in *Principles of Rhetoric* (1896 ed.), "should be in the English tongue" and "contain none but English words and phrases . . . combine[d] . . . according to the English idiom."[135] Instead of teaching classical languages, he added in *Our English* (1897), colleges should train students first and foremost in English with rudimentary Anglo-Saxon for English majors.[136] As language was viewed as a deep repository—a time capsule, so to speak—of a nation's values, uniting its speakers across time and space, grammatical purity was key to preserving ethnonational purity.[137] The student essays in the vessel were thus more than sources about everyday life. They embodied and affirmed Anglo-American linguistic, and by extension cultural, identity.

A pair of anonymous photographs in the Colorado vessel hints even more surreptitiously at racial agendas **(fig. 4.5)**. Depicting twelve local "gentlemen" and twelve "ladies," they ostensibly demonstrate how a generic face might be produced by superimposing portraits until individual features (or "deviations") dissolve, thus further exemplifying how the centurial vessels deployed the camera for documentary rather than memorial purposes. A hereditarian subtext becomes apparent, however, when we consider how composite photography had been developed in the late 1870s by the British evolutionary anthropologist and founder of eugenics, Francis Galton, to show which groups should

FIG. 4.5 Anonymous, *12 Colorado Springs Gentlemen* and *12 Colorado Springs Ladies*, composite photographs, 1901. Folder 150, Colorado Springs Century Chest Collection, 1901, Ms 0349, Colorado College Tutt Library, Special Collections.

be encouraged to reproduce and which discouraged. He thus contrasted composites of healthy Anglo-Saxon officers, ministers, and scientists with those of criminals, tuberculosis patients, and Jews (**fig. 4.6**).[138] The Colorado photographer may have learned of the technique from Harvard medical professor Henry Bowditch, whose composite photographs purportedly revealed innate class differences between Boston's doctors and its horsecar conductors.[139] He also followed Bowditch in selecting twelve individuals as the ideal number for a composite face. Although the eugenics movement remained marginal in the United States until the 1910s, such photographs—like Mosher's earlier project—lay the groundwork for its popularization.[140]

Beyond distinguishing supposed racial characteristics such as intelligence and beauty, composite photographs had a predictive function. They claimed to show how robust future generations would look if selective breeding were introduced—or how degraded if it were not. The photographic blending of faces directly evoked the notion of "blending" that dominated understandings of inheritance until Gregor Mendel's theory of genetics was posthumously rediscovered in 1900.[141] Galton thus presented his composites of Royal Engineers as "a clue to the direction in which the stock of the English race might most easily

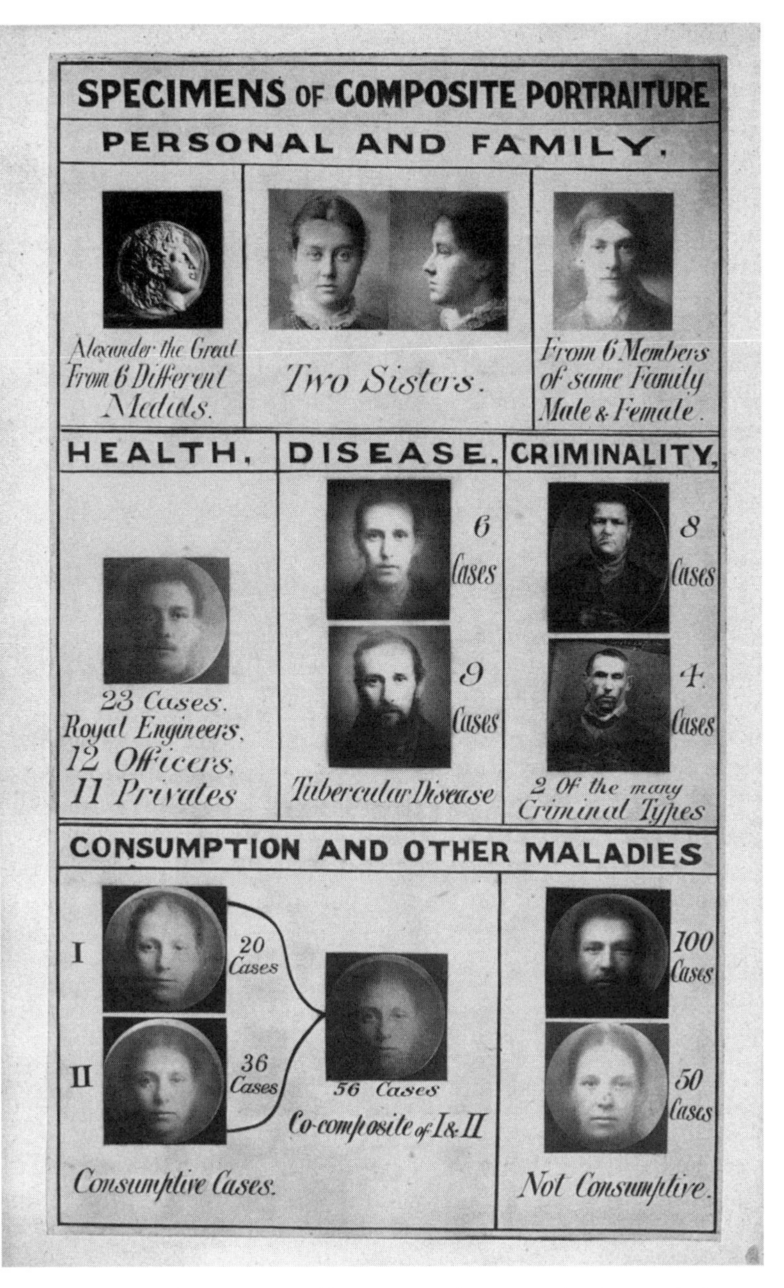

FIG. 4.6 Francis Galton, *Specimens of Composite Portraiture*, composite photographs, from Francis Galton, *Inquiries in Human Faculty and Its Development* (London: Macmillan, 1883), frontispiece.

be improved."[142] The Colorado composites similarly offer a eugenicist fantasy of the regenerated, white, middle-class American of 2001. Critics of immigration were frequently invoking the trope of the "future American race" to promote restrictive laws, posing a choice between a utopia of physiologically and mentally healthy Americans and a dystopia of degenerate and defective mongrels.[143] Some even envisaged the nation's downfall through the dilution of Anglo-Saxon blood, as in John Ames Mitchell's postapocalyptic novel, *The Last American* (1889).[144] Fears of ethnic and racial intermixing surfaced in utopian novels too. In his sequel to *Looking Backward*, Edward Bellamy clarified that his utopians of 2000 had left racial relations intact so as not to "offend" white southerners, socialism being "perfectly consistent" with "race separation." A Bellamy-inspired novel by an American diplomat, titled *Looking Forward*, went further, imagining that by 1999 blacks had been (willingly) deported to a Venezuelan reservation, thus rendering the United States a purified, white republic.[145]

Such racial visions were espoused by several time-vessel participants. In his message for the Colorado chest, Roosevelt, who began sounding the alarm of "race suicide" the following year, pictured his Rough Riders' "descendants" a "century hence." They would be "gentler and more refined" while retaining the same "whipcord fibre, . . . iron strength and courage."[146] He thus echoed Galton in positing a regiment of predominantly Anglo-Saxon soldiers as the seed of a reinvigorated race. Writing his time-vessel message at Louis Ehrich's house, Hamlin Garland—the regionalist novelist and advisor to Roosevelt on western issues—identified the rugged westerner in similar terms. "Out of these stern highlands," he wrote, "a new type of man is sure to come."[147] Ideas of racial "ascent" were central to Ehrich's own concept of posteritism. In his dedication speech (reprinted not coincidentally in Julius Sterling Morton's magazine *The Conservative*) he yearned for that "sense of deepest responsibility" that could "develop" out of the scientific theory of heredity, that notion that "we are a summing-up of all our ancestors" and "that we shall live again in our descendants." This responsibility to posterity was felt, moreover, by the "best and noblest of the race." Addressing the vessels' recipients, he wished that "your generation [may] give heed to the physical breeding of the race, controlling and restricting the right of reproduction in order that Society may be relieved from the burden of the physically unfit and the mentally and morally defective!"[148] In these messages, contributors posited the body itself as a repository of "organic memory" that would be transmitted down the generations through inheritance.[149]

Fantasies of racial purity, however, commingled in these vessels with hopes for racial equality. In the same breath as speaking of eradicating undesired types, Ehrich—again echoing Mosher—wished "all prejudice of sex, of race, of color, of religion, and of class, [would] ceas[e] to exist!"[150] Eugenics, while clearly linked to pernicious racial policies from Jim Crow to the Holocaust, was thus compatible with racial egalitarianism. And although few members of oppressed minority groups contributed to the vessels, some white contributors did attempt to represent them. The widow of a US Indian agent in Colorado paid tribute to the neighboring Ute Indians and in particular their chief, Ouray.[151] A Harvard student similarly memorialized his fellow student, "a very bright negro"—the son of Blanche Bruce, the first black senator to serve a full term—as a great debater and "fine speaker." For all its paternalism, the tribute at least implies support for increasing black enrolments at Harvard and for their integration into its clubs and social life. Thirty-five years after first gaining admittance, they remained a tiny minority and continued to experience discrimination and exclusion, while at other leading universities they were excluded altogether.[152]

A more outspoken critic of racial inequality, however, was David Augustus Straker, the only identifiable black contributor to a centurial vessel. A Barbadian immigrant, Straker was a civil rights lawyer, founding member of the Afro-American League, and (as circuit court commissioner) Michigan's highest-elected black official until after the Second World War. His message, written over a decade before the Great Migration, asserted that Detroit's blacks—then numbering only 4,111, or 1 percent of its population—had made great strides over the past century. They had obtained new "civil, and social privilege[s]," gaining access to public schools and public accommodations, savings accounts, and homeownership. Far from waiting for white Americans to confer those rights, they had "brought [them] about" themselves, led by such activists as William Lambert and Ben DeBaptiste.[153] Straker omitted his own, crucial role ten years earlier, when, as the first black attorney to appear before the Michigan Supreme Court, he successfully argued for the unconstitutionality of the "separate but equal" doctrine in *Ferguson v. Gies*, a landmark case for civil rights in Michigan and a precedent for *Brown v. Board of Education*.[154] Yet despite these advances, Straker's message was forthright about continuing injustices: the persistence of racial discrimination and the lack of full equality of opportunity. The "future progress of the race" was not inevitable but contingent on removal of these obstructions. Through his counterhistory of blacks' past efforts, he anticipated a decisive challenge to the racial status quo.

He thus concluded with a concrete utopian wish: "May one hundred years hence find us in the full enjoyment of those rights and privileges which prejudice now deny us," and may that "demon prejudice" be driven out of "the beautiful city of the straits." In employing the vessel to call forth a millennium of racial integration, equality, and harmony, Straker intimated his disapproval of Booker T. Washington's accommodationist policy of building black institutions. His visionary future was a tacit rejection of Washington's more immediate and modest goals.[155]

By 1900, as violence and segregation spread across the South and accommodationism grew in popularity, some viewed Straker as "a politician whose time had passed."[156] Yet he was not alone in articulating a future of full racial equality. In a series of essays for a Boston newspaper in 1900, the biracial novelist and activist Charles Chestnutt intervened in debates over the future of the American race. Implicitly opposing eugenicists, racists, and nativists but also defenders of immigration who envisaged the coming American as a "harmonious fusion" of European traits only, Chestnutt insisted that the "future American ethnic type" must and would be "formed of a mingling" of *all* its races, "white, black, and Indian." Within "three generations"—that is, by 2000—this process of "amalgamation" would have erased the very existence of, and distinction between, races—"all would be tarred with the same stick." It would have overcome the "temporary" obstacles of racial prejudice, segregation, and marriage laws, and far from bringing about sterility or weakening, as some feared, it was a "necessary condition of harmonious social progress." Chestnutt's postponement of this postracial utopia until the millennium was arguably a concession to his white readers. By emphasizing that the process would take place "slowly and obscurely," he hoped to alleviate their fears of radical change. He also renounced (or underestimated) black people's desire to preserve their racial identity and history.[157] Nevertheless, his serialized article was a powerful response to entrenched racial projections of the future American. A contributor to the Colorado vessel echoed him in her vision of "the making of that great American—the product of *all* races which must be the culmination of all that is great in nature—the perfect man."[158]

Hopes and Impediments: Debating the New Woman

Further anxieties surfaced in the centurial vessels around the threat posed to the nation by women who challenged traditional gender roles. In 1895, a transatlantic debate had erupted over the so-called New Woman.[159] For traditionalists in the mainstream press, and in the novels and plays that ensued, she was an implicitly young, (upper) middle-class, white, nonimmigrant figure who renounced domestic duties, resisted marriage, and flouted social conventions, not least in her "mannish" dress, and thus augured a dystopian future of sexual anarchy, eroded families, and national decline. She was also encroaching on the masculine territory of the corporate office and college campus. Enlargement of her intellect, warned psychologists like G. Stanley Hall and politicians like Roosevelt, would compromise her reproductive and mammary functions, resulting in lower birth rates, sickly offspring, and ultimately the "race suicide" of the Anglo-Saxon elite.[160]

Despite historians' inferences about "frontier egalitarianism," these fears arose even in the Colorado vessel.[161] While extolling the young American woman as "bright, clever, quick-witted, self-possessed, independent . . . a charming companion and friend, and a good wife," Dr. Caldwell noted that "childbearing is not to her taste." He remarked how some, plied with cocktails at the country club, "gamble, tell risqué stories and flirt with men who are not their husbands, and dance Sunday nights," which "leads sooner or later to scandal and immorality." And he disapproved of new fashions that bared not just their arms and necks, but also a portion of their breasts and backs—a trend that, if allowed to continue, would culminate in future generations dressing "like the Venus de Milo."[162] The New Woman was more closely identified, though, with the "shirtwaist," or buttoned blouse. For office work or bicycle rides, its clean lines, modeled on men's shirts, were more appropriate than the frills and bows of the Victorian era, and it thus became a symbol of feminine independence; indeed, one female Coloradan expressed a hope that by 2001, women will no longer be "slaves of custom" and will wear clothes that are "healthful for the body."[163] But although an illustration of the shirtwaist was included in the vessel as one of the "fashions of the day" (**fig. 4.7**), it was roundly condemned in an essay on "Colorado Springs from a Woman's Standpoint." The author expressed her "arden[t] hope" that such "masculine . . . attire" would prove a passing fad. Certainly, "no mother . . . ought to indulge in one!"[164]

Guardians of gender convention often invoked posterity to reattach women to an essentialized function of motherhood. The two Detroiters commissioned to write on the topic posited that women's chief contribution to American life has been—and would continue to be—as mothers. Over the past century, one concluded, mothers were the "greatest inspiration in all the good results," and in the century to come their

FIG. 4.7 Anonymous, untitled illustration with handwritten caption, "Shirt waist," n.d. A fashion illustration submitted by Mrs. Henry McAllister Jr. [Phoebe McAllister] to accompany her essay on "Fashions of To-Day." Folder 53, Colorado Springs Century Chest Collection, 1901, Ms 0349, Colorado College Tutt Library, Special Collections.

"added strength will bring to pass a higher civilization."[165] The other, a mother of nine, concurred that "upon the mothers of our day . . . rests the fate of millions, in the mysterious gloom of futurity." She thus denounced women's growing public activities as undermining their domestic duties and capacities. "Distract[ed]" by "various cults—fads—and philosophies," they overextend themselves to the point of nervous exhaustion. "The boasted liberty of our great century, . . . [which has] professed to emancipate woman and place her on an equal footing with man, has really made of her a slave to its conflicting theories."[166]

The feminists who defended and reclaimed the New Woman as the embodiment of their millennial hopes for social and political equality were not represented in the centurial vessels. Yet other contributors—male and female—did suggest a more complex picture of gender relations.[167] Many of them invoked posterity not to return women to domestic and maternal roles but to signify transformations and struggles that remained ongoing and thus dependent on current and future generations.

In particular, the vessels deposited at Harvard, Colorado College, and Mount Holyoke illuminate women's status in higher education. Recent historians have argued that women, far from invading college campuses, made limited progress there. Thirty years after entering the first major institutions, they remained a small minority: just 19 percent of bachelor degrees and 6 percent of doctorates in 1900.[168] By then, a conservative backlash was under way, with some institutions restricting the numbers of female students through quotas and other measures.[169] College enrollment, moreover, was not filtering down through the class structure. While the first generation was typically from middling backgrounds, the second (graduating 1890–1920) were largely from the upper middle class, which embraced college as a socially acceptable and morally safe environment.[170] Institutions that did admit them did not treat them as equals. They were largely excluded from fields other than domestic science and from student government, campus newspapers, and other activities and societies—restrictions that drove many to the then proliferating all-women colleges. On all these campuses, women's freedom of movement and interaction were severely constrained.[171]

Several diaries in the Harvard chest do confirm the sense of women as second-class citizens there. For more than two decades, a growing number of women had been attending classes through the Harvard Annex, which by 1894 had evolved into Radcliffe College but had not yet achieved full integration into Harvard because of President Charles Eliot's persistent concerns about the mingling of the sexes. Professors referred in the vessel to having to repeat their lectures in separate sessions for the female students.[172] There was also resistance to allowing women full access to the library, an issue that had just come to a head that month with a controversial letter in the student newspaper. The leading opponent of coeducation, Professor Wendell complained in his message about how a recent faculty meeting was largely spent "discussing Radcliffe, which seems to me—who am prejudiced—constantly trying to steal marches, and . . . [become] one with Harvard" (a step not taken until 1999).[173] Some female contributors themselves renounced demands for equal status, indicating their internalization of these prejudices. Lucy Allen Paton, a graduate student, agreed with her friend that the protestors need to remember "this is not a co-educational institution" and that "one of the best lessons that a woman can learn is how to adapt herself intelligently to circumstances." She even confessed doubts about the "attempts of women"—herself included—"to win true scholarship."[174] Although western and midwestern institutions have been deemed more progressive regarding gender, Colorado College students also complained in their vessel about the strict regulations on interaction, such as the prohibition against men escorting women to and from their dormitories.[175]

Women struggled to gain acceptance at colleges not only as students but also as employees. Paton, one of Radcliffe's first PhD recipients, published widely in her field but failed to secure a faculty appointment.[176] Another contributor, the astronomer and pioneering female scientist Williamina Fleming, expressed that gender discrimination particularly poignantly. Originally hired as research assistant in the college observatory by default—the director, exasperated with his male assistants, allegedly declared that his maid could do a better job, and Fleming happened to be his maid—she had gone on to make important discoveries of star clusters.[177] Yet as she disclosed in her diary, she received only a token promotion—Curator of Astronomical Photographs—and had to continue performing monotonous computation while subordinating her own research or publishing it under her director's name.[178] Two decades earlier, such injustices had prompted another contributor to the vessel, Alice Freeman Palmer—former president of Wellesley College—to help establish the Association of Collegiate Alumnae, forerunner of the American Association of University Women.[179]

While corroborating recent arguments about gender constraints in American higher education, however, the vessels also complicate that picture. Mount Holyoke students deposited copious testimony to the rich homosocial culture they had forged through activities and societies from dramatics to the mandolin club.[180] A college catalog further testified that physics and chemistry received as much emphasis as the humanities.[181] Meanwhile, Colorado College's female students were challenging men's dominance of campus. Ella Graber reported that their glee club was finally attracting as much interest as the men's and that women were dispelling the notion that debating and public speaking "belonged entirely to men." Although not yet on "equal footing," they shared "a very helpful and pleasant comradeship" with their counterparts. They also found "many little ways" to circumvent the restrictions on their social interaction. They would go downtown without a chaperone or take the streetcar to the western outskirts for picnics and campfires, sometimes with men—"and no one is the worse for the imprudence."[182] The assumption that male students uniformly resisted gender equality, moreover, is belied by expressions of sympathy in the Harvard diaries. One man rejected the assumption that "coeducation would never find a foothold at Harvard." He wished the college would "wake up to the fact" of intellectual parity and wagered that his diary's recipient would "as likely be a woman as a man," thus invoking a future body to concretize his utopian hope for social change.[183] That women participated in the Harvard and Colorado chests in itself constitutes a powerful claim to inclusion. Still marginalized in the classroom and on the sports field, they could speak out through the vessel.

For all their advances on campus, however, college women would encounter even more entrenched discrimination in the job market. Indeed, their embrace of the time vessel implies awareness that they would probably not enter the historical record through professional achievements. "Every name on our roll may be unknown to fame in your time," wrote one Mount Holyoke student; "Possibly it is because of doubt as to the lasting brilliancy of our fame that we leave you this bequest."[184] Another article in that vessel described how the class of 1900, despite investing so much in their education, would graduate to find "no lucrative opportunity for the exercise of" their latent powers besides teaching, bookkeeping, or typing. Women's right to gainful employment remained unrecognized "even after generations of self-supporting and man-supporting women." To make matters worse, these graduates would suffer "a host of unsympathetic, I-told-you-so relatives."[185] Incoming female students increasingly consented to those constraints, renouncing any claims to men's jobs and viewing college as a finishing school. Graber wrote that the "average girl" at Colorado College does "not want to be a 'new woman' and to be equal with men in the outside world—far from it. She only wants to get the best training to make her a good and useful woman wherever she may be." Graber even extended these limited goals into the future. She expected college women to "enter public life more than we" but hoped their "life purposes" would remain the same: "to be womanly women in our own place—home."[186]

Those contributors who did secure jobs revealed their difficulties in the workplace. Fleming reported her ongoing struggles to raise her salary at the Harvard observatory from $1,500 per annum to a level commensurate with that of her male colleagues ($2,500). Every time she broached the subject with the director, he replied that she received "an excellent salary as women's salaries stand." Although Fleming had been deserted by her husband and had to support herself and her child, as a woman she was assumed not to be the principal breadwinner. Does the director "ever think that I have a home to keep and a family to take care of [too]?. . . . I suppose a woman has no claim to such comforts. And this is considered an enlightened age!" Moreover, like many other working women, she was not exempted from domestic labor. "My home life is necessarily different from that of other officers of the Universities since all housekeeping cares rest on me, in addition to . . . their expenses." The unfair compensation and long hours left her despondent, exhausted, and "on the verge of breaking down." Although "tempted" to resign, she continued working—a decision that may have contributed to her death within ten years.[187]

If opportunities in the workplace remained limited, the rise of the club and charity movements by 1900 offered middle-class women an escape from the confines and chores of domesticity. Understudied by historians, clubs and charities constituted an alternative public sphere—albeit one segregated along the lines of race, class, and religion—in which even women who were not college educated could engage with the social, political, intellectual, and cultural issues of the day.[188] "There is a large element, composed of bright, intelligent, philanthropic, public spirited, high minded and cultivated women" in Colorado Springs, effused one female contributor, and they "do much toward moulding public opinion and giving character to the town."[189] The Detroit and Colorado vessels contained several essays on women's charitable work through churches, synagogues, temperance organizations, and fraternal societies as well as increasingly through secular, coordinating bodies such as the office of the Associated Charities.[190] Further essays documented the blossoming of women's clubs—similarly coordinated under a new "general federation"—and their role in the "development and enlargement of woman['s] sphere."[191] Some were literary and historical reading groups, which—in one member's view—taught "women . . . to find real enjoyment exclusive of 'the man.'"[192] Others, such as preservationist societies and civic improvement leagues, empowered women to intervene in public affairs or collect data in support of legislation.[193] Through the Colorado State Federation of Women's Clubs (CSFWC), time-vessel contributor Virginia Donaghe McClurg—a crucial but overlooked figure in the history of preservation—had appealed in 1900 to the "new woman of the West" to support a female-administered state park at Mesa Verde that would preserve the cliff dwellings and the Ute Indian community there.[194] The president of the Detroit Federation further clarified that such clubs were not exclusively for middle-class women; they were forming "auxiliaries for working girls," which imposed reduced "obligations."[195]

But despite the growth of the movement, which peaked at two million in 1900, clubwomen still encountered disapproval and derision. "Sometimes in these days we hear the question 'Should a wife and mother be a club woman[?],'" reported one Coloradan. She responded in the affirmative and hoped that by 2001 "the benefit of women's clubs will have been realized and duly appreciated"—a hope founded on her counterhistory of that movement.[196] Another Coloradan suspected that perhaps through progress in gender relations, they will have become obsolete, leaving no "Club Woman of 2001" to receive her letter, and that whoever does read it will be "smiling indulgently as we do upon the samplers and quilts of our great-grandmothers."[197] The message's failure to reach or be taken seriously by its addressee would here mark a deeper success. Rather than attempting to delineate that utopia of gender equality, the author negatively glimpses it—and renders it concrete—through the bodily absence of the Club Woman.

Women's participation in clubs and charities was contingent, however, on their exemption from domestic labor—their ability, that is, to hire servants. The latter were increasingly scarce because of the "disinclination of even recently imported immigrant girls to do general housework" for less than six dollars per week. According to an essay on housekeeping for the Colorado chest, the "servant question" was especially dire there, requiring "many a delicately nurtured lady" to become uncomfortably "familiar" with housework. But the same author was confident that "mechanical contrivances will continue to be invented to lessen drudgery." [198] By 2001, predicted one Coloradan woman, mass production will make clothes cheap enough to be disposable, freeing women from the chores of cleaning and mending.[199]

Finally, despite assumptions that the suffrage movement reached its nadir in this period (reemerging around 1910), female contributors' most strident hopes were for political equality.[200] The rise of the New Woman provided fertile ground for the ensuing growth of the suffrage movement. In 1893, Colorado had become the first state to enfranchise women by popular vote (and, in 1894, the first to elect female representatives to a state legislature), thanks to suffragettes' mobilization of a diverse coalition, including populists.[201] Female contributors to the vessel there spoke proudly of their participation in those efforts and of how their subsequent voting inspired their daughters.[202] One reported how, four years earlier, she was among the first female delegates elected to a national political convention.[203] Not resting on their laurels, they expressed concern that they "may have not fully appreciated our opportunities, . . . [or] fully lived up to our privileges." Rather than presenting gender equality as assured, they retained an open sense of the future that entailed an ongoing defense of those rights by the coming generations.[204] They might have been pleased to learn that a four-year-old named Berthe Arnold, whose photograph was deposited in the vessel, went on to become a leading suffragette.[205]

Although the National American Woman Suffrage Association (NAWSA) remained weakened by divisions between moderates and radicals and by the passing of the "old guard," the achievements in Colorado inspired women across the nation.[206] In a time-vessel essay on the issue, a Detroit suffragette declared it "one of the largest questions" now confronting the "civilized world," because enfranchised women would go on to solve other problems. She prophesied that through reeducation of the male electorate, all women would gain suffrage.[207] The dedicatory speaker for the Mount Holyoke vessel also hinted at these fervent hopes. Implying that suffrage would be granted earlier rather than later in the twentieth century, she believed that "the reforms seen by us only in the hazy future of the world" would be "records of history" to the class of 2000: "You will be happy in looking backward to see the well-defined road between our day and yours."[208] If the route to that postsuffrage utopia remained vague in her own time, through the embodied figure of her future reader she could not only glimpse its possibility but also imagine it as having emerged out of the hopes and actions of her own time.

"Did It Come?" Anticipations of Social Revolution

Of all the anxieties aired through the centurial vessels, the deepest centered on class. Rapid industrialization had cleaved vast rifts in American urban society. Kansas City, Kansas, was now populated by impoverished Irish, Slavic, and Croatian workers who lived in the shadow of its burgeoning packing houses, a world away from the mansions of Hyde Park across the state line.[209] Similar contrasts were crystallizing between the ore-reduction works and slums of Colorado City and the hotels and parks of its twin, Colorado Springs, the country's wealthiest city (by per capita income) in the century's first decade.[210] Detroit experienced even greater social and spatial differentiation. The local, ethnically owned workshops that had enabled social mobility were rapidly supplanted during the 1890s by larger metalwork and railroad factories owned by wealthy Anglo-Saxon Protestants. As those factories spread outward from the city center and lots for single-family homes filled, tenement slums followed in their wake. Thus, even before the rise of the automobile, Detroit was already evolving from an ethnic checkerboard to a class-stratified city.[211]

As products of the social elite, each of the centurial vessels reflected—and reinforced—these growing class divisions. To contribute to Kansas City's, one had to receive an invitation to the Century Ball, pay at least ten dollars ($280 today) for a ticket or as much as $350 ($10,000) for a box, and wear a dress suit or historical costume (aristocratic figures only; no "plebeian" costumes allowed).[212] Class biases also imbued the other vessels. The two contributors in Detroit and Colorado commissioned to summarize their city's "social life" described the afternoon tea parties and opera soirees that punctuated the bourgeois calendar. Neither thought to document the social life of the lower classes or even acknowledge that omission.[213]

Those who did look beyond their class did not necessarily admit to social inequalities. Colorado College's student body, wrote one of its privileged students, has "no system of caste. . . . Everyman . . . , be he poor, rich, black, or white, . . . has an equal show to make the most of himself."[214] The stove manufacturer in Detroit viewed large-scale industry as an unmitigated blessing for Detroiters, who are now "happier, more prosperous and contented" than ever.[215] If social inequalities are emerging, wrote newspaper magnate James Scripps, this is not undesirable or unnatural: a "wide chasm" between the wealthy and the "wrecks" is the by-product of a republic's maturation and stems from necessary distinctions of intellect, morality, and talent, perpetuated across generations through heredity. We are now "laying the foundations," he believed, for "great and influential families," the Rothschilds of the future. Scripps hoped the idea of posterity would inspire in "every American citizen" not a sense of social duty but a striving for the highest "honor"—to be remembered as the founder of a dynasty.[216] Furthermore, in memorializing charities, the vessels implicitly endorsed those organizations' limited response to poverty, which they viewed as a moral failing rather than a structural phenomenon. Dispensing only temporary relief, making it conditional upon work, and excluding the "worthless pauper," those charities arguably functioned as instruments of social control over the ethnic ghettoes through the inculcation of Protestant middle-class values.[217]

Others were more critical. An early prospector's widow lamented how the new mines of Cripple Creek had sharpened economic differences by "giving out their golden treasure to some, withholding from others."[218] Already a town with a top-heavy social structure and high cost of living, concurred a fellow resident, Colorado Springs was now marginalizing its lower classes: "land is so dear as to preclude many from owning their own homes, except upon far outlying district[s]." Meanwhile, "lavish display and expense" were becoming more prevalent among the "wealthier class," or what another contributor, alluding to Thorstein Veblen's 1899 treatise, called "the aristocratic, leisure class."[219] Such critiques sometimes took the form of nostalgia for the supposed equality and solidarity of the town's early days. "There was hardly more than one social grade," one old-timer reminisced, "and all were in it"—"no squalor . . . nor poverty," no labor conflicts nor crime, just "a happy and contented and prosperous community."[220] Detroit's schools superintendent similarly feared that "class distinctions" would intensify and "ultimately rend society" if private schools were to lure away all wealthy students, leaving public schools for the "poor, ignored, and unlettered."[221] Harvard's clubs, according to Frothingham, were also perpetuating inequalities, as only "richer men can afford to belong."[222] So, too, were the dormitories. The head janitor's message revealed how his men rose as early as four in the morning to ensure that students would wake to a warm fireplace, polished shoes, and a clean toilet.[223]

We tend to downplay the threat posed by class inequities to the status quo during this period. Citing the ethnic fragmentation of the working class, the conservatism of the union movement, the ideological power of the American dream, and the weakness of socialism, historians depict the capitalist order as largely unthreatened, at least until the 1930s.[224] Yet many time-vessel contributors feared an imminent outbreak of revolution or class warfare, especially in Colorado's mining belt, frontline in the struggle between labor and capital. Seven years earlier, a violent battle erupted just twenty miles from Colorado Springs at Cripple Creek, where mine owners raised a private army to suppress a protracted strike by the Western Federation of Miners (WFM) only to be defeated by the state's populist governor, who controversially called out the state militia in support of the strikers, the only such instance in

US history.[225] Strengthened by that victory, the WFM organized further strikes, including one at Telluride just three months before the vessel was sealed. Unlike the American Federation of Labor (AFL), the WFM was fully committed to the overthrow of the capitalist wage system.[226] Such conflicts on their doorstep would have particularly disturbed Colorado Springs businessmen, who were attempting to lure eastern tourists and health seekers by promoting the region as an unblemished landscape, a task that necessitated expunging all traces of labor.[227]

The vessel allowed concerned Coloradans to vent their darkest apprehensions while also adopting a detached, postapocalyptic perspective. An Episcopalian reverend wondered whether the luxurious Antlers Hotel would still be standing in 2001, "or did some terrible mob of strikers sack it—killing every one within reach[?] Some of our contemporaries predict for us as great and bloody a revolution as the French one of 1793. Did it come? And were . . . the great palaces built by our millionaires . . . also destroyed?"[228] This apparently imminent upheaval led some to question prevailing notions of undeviating progress. "Before this danger is overcome," brooded William Platt, "we may have to go through a period of tentative socialism . . . [or] anarchy." Such bourgeois prophets of doom did not hesitate to assign responsibility for the impending chaos. As a Republican newspaper editor, Platt predictably preblamed the Democrats—that "party of discontent, of socialism, of anarchistic tendencies, of wild finance and crazy economics." But he saw labor unions as "the greatest danger [to] the Republic," as they "indirectly preach disloyalty and treason."[229] Dr. Caldwell agreed that "the aggressive spirit of trade-unionism" now provoking "our" already overpaid "house servants" to organize "will have to be met and subdued before another century rolls around."[230]

The social unrest gripping Colorado largely bypassed Detroit. Even as inequality deepened, the labor movement could not overcome ethnic divisions and forge class solidarity there. Of the country's major cities, Detroit was the least unionized: only 6.2 percent of its workers, down from 20 percent in 1886. The few strikes that did occur around 1901 lacked union support.[231] Yet the specter of class apocalypse still haunted its time vessel. The board of trade's president envisioned "strife between capital and labor that may shake the fabric of our republic to its very foundation, if not change our entire system of government." Already, labor organizations had become so strong as to induce "uneasiness" among businessmen. The nation's fate "may depend very much" on the character of those unions' future leaders.[232]

Although conceived and overseen by the elite, the vessels were not univocal in their characterization of unions. As collaborative memory projects, they included the voices of working people and union leaders. Even at Kansas City's Century Ball, an advocate of the News Boys' Union, whose 1899 strike had forced some concessions from Hearst and Pulitzer, slipped union materials into the vessel.[233] Union members in Detroit and Colorado submitted more thorough testimony, emphasizing what they had accomplished—albeit within the moderate, exclusivist, craft-based model of the AFL that had supplanted the Knights of Labor by the 1890s. Malcolm McLeod, of the Detroit Railway Employees' Association, described how unions had enabled the city's workers to raise their wages, limit hours, educate themselves, and "take their place in society which has been denied them."[234] Such efforts, remarked David Boyd, statistician for the AFL-chartered Council of Trades and Labor Unions, tend to go unmemorialized.

> It is strange that a movement that has . . . done so much to bring better conditions to the toiling masses . . . should have so little literature, especially of an historical nature. The labor movement lives and grows stronger though its champions die and are forgotten, for labor organizations, like republics, are ungrateful and erect few, if any, monuments to its heroes and martyrs.

He therefore deposited his historical essay on local unions' struggles and successes—an outline of a larger "labor history of Detroit for the benefit of the agitator of the future."[235] Anticipations of an equitable society were thus grounded in counterhistory.

Charles Collais, of Colorado Springs Local No. 515, the United Brotherhood of Carpenters and Joiners, and former president of the county's Trades Assembly, expressed similar pride in unions. Against capitalists' allegations of sedition, he insisted they were merely a response to the organization of massive trusts and to attempts to decrease wages and extend hours. A subscriber to the AFL's moderate approach, he admitted to "serious mistakes . . . by . . . leaders with wrong ideas, and hasty dispositions"—a clear reference to the WFM. Collais recalled serving on a citizens' committee that entered the miners' armed camp to arbitrate the Cripple Creek strike. Nevertheless, it irked him that "many things are layed [*sic*] at the door of labor organizations." Unions, he maintained, kept out of politics and locally supported civic improvement, education, temperance, and the YMCA. "In our city . . . labor and capital are at peace," something that "cannot be said all over our broad land."[236]

The AFL is remembered for renouncing visions of a classless, cooperative commonwealth in favor of incremental, immediate gains.[237] Yet these AFL contributors articulated surprisingly utopian dreams of the millennium. Collais dreamed of the "common people" rising through the power of "organization" and the Christian "spirit of brotherhood" until they "stand as a whole mighty Phalanx all over this great land of ours." Boyd hoped that "the hosts of organized labor will be triumphant in their efforts to secure to the laborer the full product of his labor." And McLeod's "fondest hope [was] that . . . when this letter is read organized labor will have accomplished that end for which it was intended[:] the liberty and exaltation of the people who toil."[238] Rather than cast an abstract utopia, these writers conceived an embodied link to their future readers: a kind of labor solidarity across time. The metalworker who made the Colorado vessel hoped that his pride in closing it would be matched by that of the man who would "unrivett" it, while Boyd expressed his "desire to greet the wage workers of the 21st century if there be any."[239] With those last four words, Boyd acknowledged the contradiction between transhistorical class solidarity and hopes that wage labor and "class society will be wiped out of existence."[240] Whereas club women embraced their future nonexistence, Collais resolved the contradiction by postponing the classless utopia still further, calling on the workingmen of 2001 to "carr[y] on the struggle in your various branches of trades."[241] This postponement, along with the claim to be acting for future generations, allowed union leaders to disavow the immediate self-interest of higher wages and limited hours. They could also recuperate failures and reverses into this narrative of eventual glory. "[Once] the seeds of unionism are . . . sown among the workers of any trade," wrote Boyd, employing that recurring metaphor for posterity, "the effort is not wasted, for though the union may go down, yet will this seed on some future occasion spring into life and bear fruit."[242]

These workers did not explicitly appropriate Ehrich's term *posteritism*, but labor advocates elsewhere did. His neologism, along with his vessel idea, was cited approvingly in *The Public*, a mouthpiece for Henry George's single tax movement that was edited by a leading labor advocate.[243] The American Association for Labor Legislation's president Henry Farnam elaborated posteritism as an alternative rationale for labor protection laws—"distinct" from yet "not necessarily antagonistic" to the more common, socialist rationale. Instead of arguing that legislation is needed to elevate one class (workers) at the expense of another (capitalists), posteritism appeals to the care of society as a whole; instead of reckoning with "contemporary relations," it considers "the element of time"; instead of challenging "injustice," it challenges our "shortsightedness." Child labor laws, workingmen's insurance, and similar measures relieve the "burden" on women and children and thus ensure that the "succeeding generation is brought up under more wholesome conditions"—an argument that regrettably reduced those women to child bearers and their children to future workers while also arguably legitimating the exploitation of current workers. Farnam explicitly tied labor legislation to a growing chain of actions for posterity, such as public health policies, the provision of playgrounds, conservation of nature, and eugenics—all exemplars of an American tradition he traced back to the founders' efforts to "secure the blessings of liberty to . . . our posterity." Although "not new," this tradition is "so little conscious of itself" that it only recently acquired a name. "It is capable of so many applications, that it almost implies a revolution in our social ideals." Farnam's theory of posteritism was well received by progressives and labor advocates, while the British economist and opponent of antitrust laws David H. MacGregor declared a "profound aversion" to this kind of "ethics of posterity [which] would pile up a golden age for the last generations."[244]

Hopes for labor rights and social equality were not confined to union leaders and intellectuals. The centurial vessels stimulated a surprising range of contributors to endorse the cause and project alternatives to capitalist exploitation. The retired dressmaker who instigated the Rockland vessel decried how "the system [of manufacturing] has so changed that the working man is as near a slave as he can be and not be property."[245] Noxon also confessed "a strong leaning toward socialism in its most radical form, but the world is not ready for that yet. I hope it may be in your day"—a hope she shared with her husband, a hydroelectric entrepreneur who vowed to "always fight that condition which creates a 'millionaire at one end of the line and numerous tramps at the other.'"[246] A doctor's wife similarly wondered whether the millennial generation will have attained that "happy and just economical condition" of absolute equality, and she welcomed the new Servants' Union, the provision of comfortable houses for workers, and the agitation for shorter hours as steps toward that goal.[247] Socialist dreams reached even Harvard's hallowed halls. One student predicted that by the time the chest was opened the "centralization of capital" will have provoked "a clash between the classes and the masses," leading to a state takeover of monopolies. Far from fearing it, he hoped to "live to see . . . [this prophecy] realized!"[248]

Of these time-vessel critics of capitalism, the most vociferous were Protestant clergymen. Historians have characterized the Social Gospel movement, the religious wing of the progressive movement, as hostile to socialism, thus obscuring the Christian Socialist tradition in the United States. Yet a significant number of ministers, including four in Colorado Springs, enunciated explicitly socialist views in the vessels.[249] August Kieffer, former rector of the upscale Grace Episcopal Church, disclosed how, eight years earlier, lay leaders had declared "war" on him for welcoming the lower classes into the congregation, forcing him to resign and leave the state. Revisiting Colorado at the time of the vessel's sealing, he seized the opportunity to address a larger congregation, namely posterity. He criticized Christianity for having become such a bastion of social "indifference" and "selfishness" and having so "alienat[ed] . . . the masses" as to raise doubts that it will still exist "when this Chest is opened." He himself retained faith in Christian Socialism's ability to restore the church to its original mission of establishing "an earthly kingdom of Social Righteousness," that is, a socialist state. As a result of "awful" political corruption and a growing belief in the necessity of "state and municipal ownership," socialism is no longer "a scare-word—for people are beginning to understand it better!" One finds "great discontent" even "among . . . the wealthy."[250]

Kieffer's views were shared by his successor, who feared the workingman was "turn[ing] his back" on Christianity because it was "timid in attacking the problem broadly and fundamentally." Confined to a "debased" kind of charity, it has failed to "discover the true foundations of Social Justice."[251] Christian Socialists often distanced themselves from the more secular and militant forces for change. A Presbyterian pastor in Colorado warned in his message that the "disciple of socialism" must avoid both demagoguery and mechanistic schemes that substitute class agency for "personal responsibility." Nevertheless, he was certain that "political systems must suffer revolution, more or less radical, to meet the rational demands of an age moving into clearer light."[252] Such was the fate of all things, believed the Harvard preacher and time-vessel contributor Paul Frothingham, a Unitarian who considered himself "about three-fifths a Socialist" and who lectured to the local section of the Socialist Labor Party. In his time-vessel diary, Lane summarized Frothingham's sermon in the college chapel thus: "Mastodons, slavery, dueling etc. all have their place for a time and in due course give way and disappear. Capital + labor replaces feudalism + will be itself replaced by something better."[253]

The centurial vessels also contained implicit anticipations of a postcapitalist, classless utopia. Of all the books that the head of the Colorado Springs Public Library could have contributed, she chose Bellamy's *Looking Backward*. The future could thus learn how the past had envisioned it and could compare the utopian hopes with its present realities.[254] Bellamy's influence in Colorado was evident in another message that perceived "glimmerings of a society founded upon an Exchange of Service,—a service of love rather than of money," an allusion to the cooperative movement then spreading across America through the Bellamyite clubs.[255] The reference to "glimmerings," like Lane's vague "something better" beyond capitalism, evoke Bloch's notion that utopia, always exceeding our capacity to represent it, can only be grasped negatively.[256]

The same vessel's inclusion of a stock certificate for a Cripple Creek mining company (par value one dollar), although likely an expression of commercial pride—or perhaps a monetary gift, one that might have appreciated in value had the company survived—also hints at awareness of money's impermanence (**fig. 4.8**). Just as photographs in the vessel transfigure new technological devices into future-historical specimens, so does the archival preservation of a financial document imply its eventual obsolescence. In *Equality*, Bellamy himself employed a time-vessel device to radically debase the monetary and speculative instruments of capitalism. Along with his fiancée's letters, West's safe transports a stash of gold and "several drawers full of securities" (property titles, corporate stocks, utilities shares, government bonds)—all

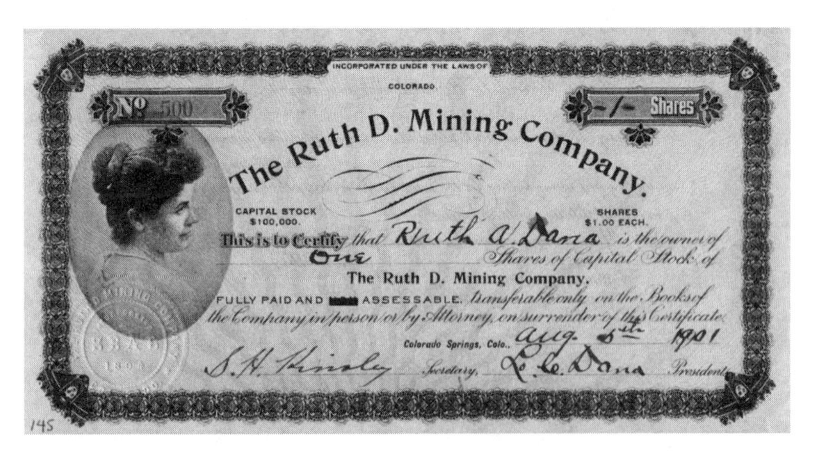
INCORPORATED UNDER THE LAWS OF
COLORADO
No 500
-1- Shares
The Ruth D. Mining Company.
CAPITAL STOCK
$100,000.
SHARES
$1.00 EACH.
This is to Certify that Ruth A. Dana is the owner of
One Shares of Capital Stock of
The Ruth D. Mining Company,
FULLY PAID AND ASSESSABLE. Transferable only on the Books of the Company in person or by Attorney, on surrender of this Certificate.
Colorado Springs, Colo., Aug. 5th 1901
Secretary,
President
145

FIG. 4.8 Stock certificate for the Ruth D. Mining Company (one share), submitted to the Colorado Springs Century Chest. Folder 145, Colorado Springs Century Chest Collection, 1901, Ms 0349, Colorado College Tutt Library, Special Collections.

now worthless in a socialist society. "I suppose we might as well make a bonfire of them," he surmises, "for they seem to have no more value now than a collection of heathen fetiches after the former worshipers have embraced Christianity." But Dr. Leete replies that they have acquired a new value, a historical one. They are testimony to the inequities of capitalism, "evidences that their possessors were the masters of various groups of men, women, and children." As the people had, during the "great Revolution," collected up such certificates from all over the country and gleefully incinerated them in an enormous bonfire on the site of the New York Stock Exchange, West's surviving certificates are "quite scarce" and should be donated to the history museum.[257] It is this kind of radical transvaluation that the Colorado stock certificate undergoes when imagined as a relic.

Even Louis Ehrich himself, a wealthy owner of real estate and banking stock (despite his losses in the Panic of 1893), vehemently attacked the capitalist order. Unlike most businessmen, Ehrich refused fealty to the Republican Party. Five years earlier, he had become a leading member of the Gold Democrats—a short-lived third party that rejected both William McKinley's protectionism and William Jennings Bryan's inflationism.[258] He had also published an article denouncing the use of troops against the Pullman strikers of 1894 and instead advocating "stock sharing" as the best solution to workers' discontent, which he attributed to their reduction in the corporate-industrial system "to the level of a tool," lacking any personal stake.[259] He reiterated this idea in his dedication of the chest, praying that by 2001 "every capitalist . . . [will be giving] to society a full equivalent for benefits received . . . and . . . every laborer . . . shar[ing] in the results and profits of his own labor!" In his sealed message, he added that the "primary cause" of the present troubles is the national abandonment of the ideals of equal rights and popular sovereignty for the sake of "material gain." Invoking a Judaic theology of atonement and purification and hope for a messianic age, Ehrich declared that only "a great national calamity can . . . purify our temper and lead us humbly back to the ark of our national covenant. Unless we atone for our present sin against the holy spirit of Liberty, and again re-consecrate this nation to our original ideal of the sacred equality of man, . . . we will follow the downward path of the nations that have disappeared." Reserved for the closing lines of his message, these hopes for equality—and fears of destructive revolution—were what inspired Ehrich's time vessel. By embodying and inculcating posteritism, it was to redirect capitalists from their immediate, economic interests to a higher, sacred cause.

If Ehrich's message was a medley of contradictory visions of the present and the future, so too were the centurial chests themselves. Far from making future historians' task easier by endowing them with a coherent representation of the world of 1901, each presented a host of conflicting assessments. But in addition, and more importantly, each of them constitutes a rare archival gift: a heterotemporal assemblage of the multiple futures that vied for dominance during the Progressive Era.

FIVE

"A Living History of the Times": The Modern Historic Records Association, 1911–1914

On December 9, 1911, several men entered an ornate brownstone overlooking Manhattan's Gramercy Park. Once the residence of Democratic governor and presidential candidate Samuel Tilden, it now housed the National Arts Club (NAC), founded thirteen years earlier to foster the fine arts in America. At 8:15 p.m., they adjourned to the wood-paneled gallery to listen to experts discuss how the present might be archived for posterity. They also listened to Thomas Edison on phonograph, apparently the only recording of his voice at that point. The highlight, however, was probably the presentation by architect Arthur Dillon and sculptor Carl Heber. They inserted items into a glass jar, sealed it with a glass stopper, then placed it into a terracotta pipe, which they filled with concrete and capped with a copper disc bearing an inscription and list of the contents. The elaborate ceremony "was followed with great interest by those present," reported the *New York Times*.[1]

This vessel was one of several in the years preceding America's entry into the Great War. In 1905, a masonic lodge in Cedar Rapids, Iowa, established a docket in its vault for records addressed to the future historian and pledged to top it up annually. There were "century boxes" launched by the towns of Bowdoinham, Maine, in 1912;

Westbrook, Maine, in 1914; and Ashburnham, Massachusetts, in 1915. The most ambitious of these prewar vessels was Oklahoma City's century chest, a six-foot copper sarcophagus entombed in a church's basement in 1913. Smaller but more ornate was the bronze casket sealed in 1914 by the Lower Wall Street Businessmen's Association.[2]

As some of their names indicate, these other vessels largely followed the centurial formula. They were local efforts conceived to mark some historical anniversary or to bind some community in a collective pledge to its posterity. Besides printed material, they contained personal messages (typically sealed) to their future counterparts and—in the Oklahoma chest—photographic surveys of the built environment, phonograph recordings of songs and speeches, and material artifacts such as shoes, a Kodak camera, and a telephone. Indeed, the Oklahomans explicitly acknowledged their debts to Colorado Springs' century chest.[3]

By contrast, the Gramercy Park presentation, though sparsely attended, hinted at new avenues for—and barriers to—communicating across time. To begin with, it was not confined to an anniversary celebration but was part of an ongoing project. Those visitors were witnessing the inaugural meeting of the Modern Historic Records Association (MHRA). Incorporated a month earlier, its goal (as indicated on invitations printed in indelible ink on sheepskin vellum, copies of which were deposited in the glass jar) was to "preserve in imperishable form the records of History" (**fig. 5.1**).[4] While the MHRA shared other time-vessel architects' concerns regarding the inadequacies of traditional institutions of memory—inattention to the memory of the present, neglect of ordinary experience, and failure to address the vulnerability of modern materials, especially paper-based ones—it offered more ambitious solutions. Rather than an isolated deposit, Dillon and Heber's mixed-media assemblage was a prototype for multiple vessels. And those vessels were just one component of a broader program of schemes and proposals. The aim was nothing less than to bequeath "the completest possible records of the present," which it would not merely gather but also actively produce.[5] It expanded its scope from the local to the national, seeking to encompass the rural as well as the urban and the Native American as well as the European American. Prompted by the internationalism and pacifism of several of its members, it resolved even to go beyond US borders. It was also temporally ambitious; besides launching hundred-year vessels, it targeted more distant recipients. And, perhaps influenced by the emerging genre of postapocalyptic fiction, it considered that those recipients might no longer be American citizens.

FIG. 5.1 Cover of an untitled pamphlet detailing the Modern Historic Records Association's aims (New York: MHRA, 1912). Miscellaneous Personal Name Files, box 73, miscellaneous file, Manuscripts and Archives Division, New York Public Library.

The MHRA's originality resided above all in the degree to which it embraced the new media technologies of photography, phonography, and cinematography. The enhanced capacity to register images, sounds, or movement—or, with Edison's kinetophone, all three together—was what prompted its founding. It conceived of those media as crucial tools for preserving traces of everyday life, from regional dialects and folk songs to indigenous rituals. It also intended to use the camera and the phonograph (including certain new prototypes and formats) to transmit books and documents, thus anticipating librarians' adoption of microfilm by more than a decade. This receptivity to new technologies linked it to contemporaneous European efforts, similarly inspired by utopian and internationalist visions, to construct vast multimedia archives.

While the MHRA exceeded its predecessors in embracing modern media, it did so without fetishizing or overinvesting in them. Some scholars characterize early advocates of the archival potential of modern media as naive or outright deceptive about their durability.[6] Yet the MHRA appears to have been cognizant of this problem and sought ways to merge newer technologies of inscription with ancient media such as terracotta to produce hybrids combining the advantages of each. Further key objectives were to raise public awareness about the fragility of modern, pulp paper and to pressure newspapers and government departments to produce archival copies on rag paper. This concern that modern civilization's reliance on what Harold Innis called light, space-binding media was condemning it to ultimate oblivion led the MHRA to more extreme forms of entombment involving remote locations, duplicates, high-tech vaults, and, above all, concrete.[7]

The MHRA's adoption of audiovisual media, however, was not without negative consequences. Its quest to capture indigenous culture aligned it with the pernicious myth of the Vanishing Indian. Indeed, Edward Curtis, the photographer most synonymous with that myth, was a contributor (although his contribution hints at resistance as well as complicity). The MHRA's embrace of new media as ethnographic tools reinforced the further notion that Indians were vanishing partly because of their inability to perpetuate their own culture through such technologies. The association also contributed to a weakening of ethical obligations to posterity insofar as it redefined the present's duty as a merely archival one of producing (or salvaging) records of what was disappearing.

This seductive belief in the power of the camera and the phonograph to produce authoritative and objective records generated contradictions for the broader project as a whole. It induced the MHRA to imagine itself less as a collection of messages for posterity and potential sources for social and local history than as an archive of historic events. This self-conception implied the fallacy that certain events could be identified *in the present* as "historic" and thus worthy of audiovisual recording. Such decisions, moreover, were to be made by the association's board, thus signaling a trend toward top-down control of time vessels. The focus on recording "events" ultimately lured the MHRA—despite its avowed investment in the "new history" and the everyday—back to the realm of "great man" history and the memorialization of the illustrious. "Thousands of years hence," wrote the journalist who attended the sealing at the NAC, when the explorer may delve in the ruins of New York, then perhaps buried under the accumulations of time, he will come across strange-appearing cylinders of terra cotta and concrete, and in them, in thick [g]lass encasements, he will find the record of our period. . . . The singers of the present age will sing, our actors talk their lines, and our great men speak again.[8]

Toward an Archaeology of the Present

The MHRA was the pet project of Alexander Konta, a wealthy Wall Street banker and civic leader. Born in Budapest in 1862, he emigrated to the United States in 1887, settling in St. Louis before moving to New York in 1901. As he ascended in the banking world, he accumulated memberships of, and honorary appointments to, the city's leading clubs, cultural institutions, and scientific and historic associations.[9] Those connections proved invaluable. An active member of the NAC, serving on its board of governors and its arts committee, he established its Gramercy brownstone as the MHRA's official address. He also lured prominent figures to his governing board, including sitting president William Howard Taft and explorer Robert Peary, just back from the North Pole—or so he controversially claimed. At the inaugural meeting, the former was elected honorary president and the latter one of its vice presidents. He subsequently persuaded Edison and New York mayor William Jay Gaynor to participate and J. P. Morgan to provide funding.[10]

The idea for the MHRA arose years earlier. An 1887 visit to the Egyptian museum of antiquities at Bulaq sparked Konta's epiphany just as the pyramids had inspired Cogswell. As Konta later recounted, the eminent German Egyptologist Heinrich Karl Brugsch "was taking me over the museum . . . and regretting the scantiness of our knowledge of the remote past of the country and its people, owing to the lack of records. The thought came to me then: 'What is this advanced civilization of ours doing for the enlightenment of the historians of the far distant future? Do we think of them? Do we prepare records for them much more than did these Egyptians?'"[11]

Konta's mission to rescue the present from oblivion may have been further stimulated by the prominent British historian and liberal reformist Frederic Harrison. In "A Pompeii for the Twenty-Ninth Century," published in the distinguished journal *Nineteenth Century* in 1890 and reprinted in his essay collection of 1908, Harrison proposed a "National Safe" for the preservation of "a careful selection of those products of to-day which we think will be most useful and instructive to our distant descendants." The scheme stemmed from his perception of a disjuncture between Victorians' preoccupation with the fate of other civilizations and their disregard for preserving their own. "We live in an age of archaeological research," he began; "all corners of the planet have been ransacked to yield up their buried memorials of distant times." Yet this "archaeological zeal" has not prompted efforts to ensure that London's ruins furnish equally revealing sources. Painfully aware of how much of the past had been lost—"the Alexandrian Museum; . . . the statues of Praxiteles and Scopas burnt to make mortar; Greek dramas and Roman institutes erased to write over them patristic homilies; temples destroyed . . . , mediaeval cathedrals gutted"—he feared an even greater archaeological erasure of contemporary civilization given its relentless technological obsolescence and capitalist development and its increasing enthusiasm for architectural restoration. "Let us, once in a [while], take to looking forwards," he implored, "and, with all our archaeological experience and all the resources of science, deliberately prepare a Pompeii . . . for the . . . twenty-ninth century."[12] Although no one in Britain appears to have heeded the call, the MHRA's secretary W. T. Larned did acknowledge its similarity to the "enterprise" conceived in Harrison's "eloquen[t]" essay. The latter affirmed this transatlantic connection, offering the MHRA his "warm approval" and a sealed prophecy for one of its vessels.[13]

Konta's project was thus a response to a perceived blind spot within the modern memorial process. Preoccupied with the colonial and revolutionary past, museums, historical societies, and preservation societies were neglecting the present.[14] "With all our modern science and with all our historic consciousness," he told the *Times*, "we ourselves are not doing much better than our ancestors did"—and "in some respects" are doing worse. The MHRA was claimed to be the first "association formed to systematize the gathering and preservation of *contemporary* records for posterity."[15]

Popular Fiction and Archival Obligations to the Deep Future

This growing impulse to preserve a broader picture of the present was reinforced, at the other end of the cultural spectrum, by popular dystopian fiction. The utopian authors who pioneered the futuristic novel in the late nineteenth century had remained confident that the cultural products of modern civilization would endure—a confidence echoed in the absence of great books and arts from early time vessels. Edward

Bellamy's Julian West was relieved to discover on waking in 2000 that because of the largely peaceful transition from capitalism to collectivism, his host's library still boasted all the great authors—from Shakespeare and Milton to Hawthorne and Irving—and that the lapse of time "had aged [those books] as little as it had myself."[16] Doubts about bibliographic durability were largely confined to the minor genre of apocalyptic fiction, to Ambrose Bierce's and Mark Twain's satires, and to radical critics such as Henry George.[17]

Scenarios of catastrophic cultural loss became more prevalent and mainstream, however, with the flowering of postapocalyptic science fiction toward 1900. A pioneer of this genre, H. G. Wells, projected the protagonist of *The Time Machine* (1895) into a future (AD 802,701) in which no historical records have survived nor the knowledge of reading and writing. A late-Victorian "Palace of Green Porcelain," containing a library and a museum of natural history and science, still stands, but its books have decomposed into "brown and charred rags, . . . warped boards and cracked metallic clasps"—"every semblance of print had left them."[18] More immediate cultural loss occurs in Wells's subsequent invasion novels, such as *The War in the Air* (1908). The historian who narrates the aerial bombardment of New York, which precipitated the collapse of capitalism and eventual dispersal and regression of the urban population, can draw only on a few "surviving fragments of literature" and "scraps of political oratory."[19] With the discovery of how to isolate radium two years later, Wells's predictions grew even darker. *The World Set Free* (1914) envisions a nuclear war that destroys cities around the world, reducing their "museums, cathedrals, palaces, libraries, galleries of masterpieces, and a vast accumulation of human achievement" to "charred remains."[20]

Yet scientific romances (as they were then called) also singled out texts and relics that somehow eluded the depredations of time and man. Even the bleakest of apocalyptic narratives imagined such remnants as precious links across time, sometimes even as catalysts for the eventual transition to utopia. In Richard Jefferies's *After London, or Wild England* (1885)—an early prototype of postapocalyptic science fiction, with its vision of a nation inexplicably reduced to rural feudalism and its capital to half-submerged, uninhabitable ruins—the young protagonist Felix possesses a handful of parchments and books, which he keeps in a sturdy oak chest. Their pages are stained and their covers rotted, but they crucially enable this future ruler to glean the "knowledge which was the common (and therefore unvalued) possession of all when they were printed."[21]

The authors following Jefferies differed over the size of such repositories. Some imagined an entire library, miraculously preserved. After the collapse of urban-capitalist civilization in Ignatius Donnelly's *Caesar's Column* (1890), a collection of books and inventions in the mountains of Uganda safeguards humanistic and scientific knowledge for "when the world is ready to receive it."[22] In M. P. Shiel's *The Purple Cloud* (1901), the apparent lone survivor of volcanic fallout, who subsequently burns London and other cities to the ground, retrieves "a good hundred-weight of Ordnance-maps" and "three topographical books" from the "dim sacred galleries" of the British Museum, sidestepping the corpses of librarians and the complex cataloging system.[23] Similarly, the hero of Van Tassel Sutphen's *The Doomsman* (1906), exploring a Manhattan now occupied by the descendants of criminals, discovers in midtown "one of the magnificent libraries upon which the ancient municipality had prided itself." From the rows of dusty shelves, a guidebook and an encyclopedia stand out as most useful.[24]

Other novels, however, envisioned more limited links between pre- and postapocalyptic times. In *The Scarlet Plague* (1912), Jack London questioned his earlier faith in bibliographic and linguistic continuity expressed in *The Iron Heel* (1908). Sixty years after the pandemic destroyed civilization, the narrator—once an English professor at Berkeley and now one of about forty people living in or around San Francisco—clings to his small collection of books. Yet as the last literate man, he has to deposit them in a "dry cave on Telegraph Hill" with a pictorial "key" to the alphabet so that "some day men will read again" (**fig 5.2**).[25] In some cases, no written texts survive, only images or material artifacts. The Persian archaeologists who rediscover "Nhû-Yok" in AD 2951 in John Ames Mitchell's *The Last American* (1889) retrieve some coins from a house (one bearing the head of "Dennis, the last of the Hy-burnyan dictators"), a figurine of a Native American from a store, and some metal printing blocks (featuring magazine illustrations) from "the basement of a high building, all laid carefully away upon iron shelves." Befitting its visual orientation, "Mehrikan" civilization appears to have left behind no words besides the legends on those coins, the half-erased inscriptions on a hotel and the stock exchange, and the hieroglyphs on an Egyptian obelisk.[26] As media for the long-range transmission of writing, those artifacts and structures show up the inadequacy of paper.

Nevertheless, each of these authors intimated how such texts, images, and artifacts would have limited historiographical value because they were accidental, unintended sources. The materials recovered by

[have stored many books. In
wisdom. Also,
ıave placed
he alpha-
ɔne who
re-writ-
› know
e day
d again;
no accident
my
will

"In the Cave I have stored many Books"

FIG. 5.2 Gordon Grant, illustration from Jack London, *The Scarlet Plague* (1912; New York: Macmillan, 1915), 175. In hope that a future age will rediscover literacy, the former English professor deposits his books in a cave in San Francisco. He subsequently admits that this was a futile act; whether the books "remain or perish, all their old truths will be discovered, their old lies lived and handed down" (179).

the Persians thus lack any explanatory text, which leads to misinterpretations, such as that the Egyptians had invaded Mehrika.[27] Similarly, in Jefferies's *After London*, most of the surviving books are extremely ancient volumes that had lingered in neglected museums and that therefore reveal nothing about the culture just preceding the Great Change. Contemporary works, by contrast, had been abandoned to the elements in "decaying houses"—a self-reflexive meditation, perhaps, on his own novel's fate.[28] Postapocalyptic fictions thus implied a need for deliberate efforts to speak to posterity, via intended sources. Harrison's call for a National Safe five years later may have been a response to Jefferies. We should "leave fewer things to chance," the historian urged. "It is the duty of an age to be self-conscious, and to reflect how its acts and its thoughts will appear in the eyes of a distant posterity."[29]

One scientific romance that did present the recovery of an actual, rather than accidental, time vessel was George Allan England's "The Last New Yorkers." Serialized in the pulp magazine *Cavalier and Scrap Book* in 1911—the year MHRA was incorporated—and retitled *Darkness and Dawn* when issued as a book, it opens with engineer Allan and his stenographer Beatrice waking up in the Metropolitan Life Tower a thousand years after some mysterious catastrophe had devastated the earth. They survive at first by recovering various unintended deposits in the ruins of midtown stores. A locked cedar chest in a fur store furnishes them with warm clothes; a security vault in a jewelers offers a gold clasp to fasten them; a sealed bottle has preserved toothbrushes; and a glass showcase in a hardware store contained scissors, knives, and guns, which resemble "Egyptian relics . . . [Allan used to see] in museums."[30] These help them triumph over the Horde, the monstrous offspring of nonwhites and apes who had also survived the catastrophe. Yet their efforts to understand how civilization collapsed and how it might be rebuilt are stymied by the lack of textual remains. In the fortieth episode, however, they learn that in 1957 a band of white survivors had left two intentional deposits in the Colorado Rockies and in the Hudson Valley documenting their history since the cataclysm of 1920. Allan and Beatrice's transcontinental search for one of those deposits is motivated by their larger racial quest to reeducate the Folk, a tribe of white barbarians they found living in a deep abyss, and thereby construct a white socialist utopia on capitalism's ruins. "We need their history," Allan explains, "even the little of it that the records must contain, for surely there must be names and events in them of great value in our work of trying to bring these people to the surface and recivilize them."[31]

Whether England's time capsule inspired Konta's MHRA (or vice versa), both were symptoms of an increased sense of the present's ephemerality and at the same time of the possibility—indeed, *responsibility* of "civilized" men—of preserving it. Harrison's insistence that such a task was not a frivolous exercise but an ethical obligation was echoed by a number of scholars. In "Our Duty to the Future" (1916), Colorado chemistry professor C. E. Vail lamented that "in our present age of hurry and worry, hustle and bustle, strain and stress," such needs are overlooked. Given our scientific knowledge, he urged, "there should be a concerted effort and aim of living civilized men to preserve a representative portion of the present for the future." Even "our present-day museums" are failing to do this, focusing their energies instead on the relics of past civilizations.[32]

Such suggestions were also symptoms of a lengthening of temporal consciousness. Just as nineteenth-century geologists, paleontologists, biologists, and prehistoric archaeologists had extended timescales into distant pasts—what Stephen Jay Gould called the discovery of deep time—so were science fiction authors now extending them into the future.[33] Readers were projected not merely a hundred years into the future (as in utopian novels), but a thousand or more. This extension was paralleled by the time spans proposed for vessels. For his National Safe, Harrison targeted the twenty-ninth century (or, when he reprinted his essay in 1908, the thirtieth) and even entertained the possibility of another ten centuries.[34] Vail, too, called for longer time frames. Initially he upheld the Egyptian pyramids as the benchmark but subsequently asked, "Why . . . stop at . . . five thousand years as the limit of our responsibility? What of us and our descendants ten—twenty—fifty—one hundred thousand years from now?"[35]

To target such deep futures was, of course, to confront a time after the downfall of the United States. Whereas earlier vessels addressed their compatriots, Konta imagined postnational others as his recipients. He warned against "assum[ing] . . . that our civilization will continue to develop without a break," as historical evidence shows that "civilizations reach a certain point, then decline, and even disappear, leaving the human race to start anew at the bottom." The records that the MHRA proposed to make and preserve "would tell a new civilization still to be born far more than the records of the past have told us of to-day."[36]

More specifically, they would inform the yet to be born historian. Earlier vessels had more often addressed elected officials, biological descendants, or broader groups, and the historians they did address were amateur and local. The MHRA, by contrast, targeted the professional, "scientific" historian. Its self-declared purpose was "to be of vital service and inspiration to historians, educators and others" studying "our own time."[37] This was a reflection of the incorporators' own commitment to history. Few historians were involved with the earlier vessels—some declined even to contribute to Harvard's chest—but several signed up with the MHRA, including William Milligan Sloane, a history professor at Columbia whose books ranged from seventh-century Arabia to nineteenth-century France and who was president of the American Historical Association in 1911; Joseph Edgar Chamberlin, the forgotten pioneer of counterfactual history with *The Ifs of History* (1907); and Josephus Larned, a similarly forgotten pioneer of what is now called big history with his *Seventy Centuries of the Life of Mankind* (1905).[38] The board included prominent collectors of historical documents too. Dr. Smith Ely Jelliffe, an early Freudian, collected antiquarian books and ancient texts relating to psychiatry, while George Plimpton, an educational publisher (and grandfather of the eponymous journalist), collected educational materials from antiquity onward.[39] Also on the board were two founders of historic preservation: Edward Hagaman Hall, executive secretary of the American Scenic and Historic Preservation Society, and George Frederick Kunz, president of the same body as well as mineralogist and chief gem expert for Tiffany.[40] With the inclusion of such figures, the MHRA delineated its duty to posterity as an archival one to future historians. Such a role rested on the assumption that even if Western civilization collapsed, a scholarly curiosity about the past (and a commitment to studying it empirically and "scientifically") would endure. It thus overlooked modern historical consciousness's own historical specificity—a common blind spot of time capsules.

An Alternative Archive

If the MHRA was an archive, it was one that eschewed—and implicitly critiqued—the prevailing model. Since its emergence in the aftermath of the French Revolution, the historical archive had been conceived as an essentially passive repository to which governments and institutions consigned records whose original value had expired so that they might be preserved for some vague posterity. In the process, those records acquired a "secondary" value as public records.[41] Concerned about the relegation of that secondary value to an afterthought, the MHRA would function as a pressure group to push deliberative and administrative bodies to keep proper records in the first place. In early 1912, it lobbied governors to investigate recordkeeping practices in their states, enforce existing laws, and advocate new ones if needed.[42] It also petitioned Congress for a national archives building in Washington. Although the American Historical Association (under its former president J. Franklin Jameson) led this campaign with a strong resolution at its 1912 annual meeting, the MHRA may have played a prior role in influencing President Taft—its own honorary president—to embrace the cause that year.[43]

The MHRA's most significant departure from the traditional, passive notion of the archive lay in its conception of the historical document. Not content to preserve existing records, it would produce its own purpose-made ones. One initiative was to document the Republicans' and Democrats' upcoming conventions.[44] Another was to dispatch four blank parchment pages to public figures "earnestly request[ing]" them to inscribe and return them to the MHRA.[45] The resulting volume, "Message of the Records," ranged from an account by the captain of the RMS *Carpathia* of his rescue of survivors from the *Titanic* twelve days earlier to a defense of Progressivism by Theodore Roosevelt and a lament by Taft about the Senate's recent rejection of his international arbitration treaties.[46] Through the systematic production of further such testimony, the MHRA would eclipse not only conventional archives but also cornerstones and earlier time vessels.[47] Harrison had criticized cornerstones—with their indiscriminate collections of coins, documents, and newspapers—as representing "a futile and trivial mode of providing for the historic research of ages to come." His National Safe would elevate that local, "aldermanic" fad into a "complete, systematic, and scientific mode," performed on a "truly national scale."[48] Konta's archive would resist cornerstones' boosterish tendencies, too, by admitting unfavorable as well as favorable evidence so that "future generations may know the exact measure of our wisdom and our ignorance, our achievements and our failures."[49]

To ensure "the completest possible records of the present," the MHRA adopted an expansive definition of history.[50] In addition to political conventions, diplomatic negotiations, and other items of event history,[51] it wished to take into account the deeper realms of culture, economics, and everyday life and thus allied with the New History introduced by Frederick Jackson Turner and James Harvey Robinson. This "modern" school of history, Konta explained, "attaches more importance to the evolution of the daily life of the common people, the development of their civilization and culture, their economic and social status, than to the doings of their leaders in peace and war, to what has been so aptly described as 'past politics.'"[52] As we shall see, Konta's definition of history also encompassed American Indian life, then relegated to ethnology departments and natural history museums and overlooked even by the New Historians.

Believing historians' "scope" would "continue to expand," Konta developed a wide-ranging archival agenda.[53] Guided by his cultural interests, he approached prominent figures in the arts, securing parchments from H. G. Wells, playwrights George Bernard Shaw and Maurice

Maeterlinck, actor Arthur Wing Pinero, and artist Elihu Vedder, while Richard Strauss contributed a bar of music. Yet he also intended to encompass culture with a small *c*, that of the ordinary. This stance again suggests the influence of Harrison, a prominent critic of Matthew Arnold's elitist discourse on culture.[54] While conceding the National Safe should include "the best works, memorials, and specimens of our own age," Harrison believed there would be more "interest a thousand years hence in the ordinary books of information which are very likely to perish. Our curious . . . [ruin explorer] of 2890 would no doubt much prefer a Whitaker's *Almanack* or a Bradshaw's *Railway Guide* of 1908 to all the works of . . . Robert Browning." He also nominated *Encyclopaedia Britannica*, the London *Times*, the British Museum catalog, and city directories. Recognizing capitalist urbanization's threat to the natural and built environment, he suggested Ordnance Survey maps along with "plans of great cities and great public buildings." Rapid obsolescence of technologies and fashion, too, would necessitate material artifacts such as a model locomotive or a doll displaying Victorian clothing.[55] Such a program may have influenced Konta to announce his organization's embrace of "the daily life of the common people, or their economic arrangements and social ways."[56]

Konta's archive was also to be geographically inclusive, recording "our daily life in town *and* country." To avoid the metropolitan or eastern bias implied by its New York address, it would be structured as a federation. Municipal chapters would be founded in cities and towns across the country and would report, in turn, to their state association. This decentralized structure was intended also as a means to extract records of "wider importance." The most significant records would be relocated from the municipal to the state association, and ultimately—if of "national importance"—to the headquarters to be built in Washington. The chapters could also exchange records among themselves through a kind of interlibrary loan. With this ambitious national scope, the MHRA exceeded the local orientation of the centurial vessels—and state historical societies—while retaining grassroots participation. And unlike the national archives established in various countries (including eventually the United States), this was to be a private body.[57]

Nor did the MHRA confine itself to the nation. A majority of the parchments it collected were from non-Americans, including not only cultural figures but also King Alfonso XIII of Spain and former Mexican president Francisco Madero, and it planned to add King George V, Kaiser Wilhelm II, Tsar Nicholas II, Sultan Mehmed V, and the Emperor

Meiji, among others. To address the lack of records from France and Italy, the MHRA treasurer traveled to Europe in April 1912 bearing a letter of introduction from the State Department.[58] The aim was thus to record not merely American civilization but "our age."[59] This planetary scope was also evident in the MHRA's calls for international laws to protect archives from military bombardment and for international collaboration between libraries.[60] Similarly, its projected collection of American Indian records would encompass the entire hemisphere, from the Inuit of the Arctic to the Tehuelche of Patagonia—an ambitious project that Konta argued would require the founding of and "co-operation" between "similar associations in different parts of the world" as well as between governments.[61]

These agendas echoed not only Taft's hopes for arbitration but also Konta's own internationalism. A leader of the Hungarian American community, Konta must have feared war between his native and adopted countries. In January 1913, he issued a plea in the *New York Times* for the Balkan states, already at war with the Ottoman Empire and threatening war with Austria-Hungary, to "remember that they have serious international duties now as well as national rights" and that "a premature attempt to stand alone . . . will but lead to . . . further wars and alarums of war on a gigantic scale."[62] Other MHRA officials were also active in promoting peace and internationalism. The previous year, Kunz proposed (with Konta's endorsement) the erection of a Museum of the Peaceful Arts in New York's Riverside Park by 1914, both to celebrate the century of "peace between English speaking peoples" since the Treaty of Ghent and to counter museums that celebrated war and empire. Although never realized, substantial funds were raised and plans drawn up for twenty buildings, including one dedicated to "Historic Records."[63]

The MHRA's internationalism and archival experimentalism aligned it with a number of contemporaneous projects. In 1895, the Belgian pacifists and visionary pioneers of information science Paul Otlet and Henri La Fontaine began to inventorize all the facts in the world in a Universal Bibliographic Repertory, a searchable catalog of index cards eventually numbering nearly 16 million.[64] A decade later in Paris, the Jewish banker and internationalist Albert Kahn launched an equally ambitious project: a photographic and cinematographic Archives of the Planet. As Paula Amad has shown, these (and other) efforts to produce a systematic and exhaustive inventory of the present were the product of both utopian dreams for world peace and technological faith in the universal language of photographic media. They also derived from a belief in the infinite possibilities and universal applicability of the decimal system of classification invented by librarian Melvil Dewey—yet another board member of the MHRA.[65]

A Hall of Fame for the Future

For all its efforts to enlarge the scope of what was to be preserved for future historians, the MHRA could not quite renounce the core function of Mosher's and Deihm's vessels, namely, memorializing the illustrious. Only noteworthy individuals were permitted to contribute to its "Message of the Records."[66] Besides luminaries of the arts and heads of state, famous scientists, inventors, and financiers accepted the invitation to commit to parchment a "thought or achievement for which they would like to be remembered for all eternity."[67] As the MHRA released their messages to the newspapers (and to its paying members), it allowed them to burnish their public image or assert a claim in the present. Peary's contribution, a transcription of the message he supposedly left at the North Pole in 1909, was thus part of his larger campaign (coordinated by journalist and MHRA president Herbert Bridgman) to discredit his rival Frederick Cook's claim to have got there a year earlier.[68] The MHRA planned to immortalize their faces, too. Leading portrait photographer Pirie MacDonald pledged six portraits a year, starting with one of poet Edmund Stedman, while an ethnological sculptor from the American Museum of Natural History offered his new technique for taking a plaster-cast impression, "quickly and without pain," of a living face.[69]

In seeking to preserve the famous on parchment, silver nitrate, or plaster, the MHRA attests to a resurgent interest in erecting pantheons to society's "greatest" individuals. Temples of fame, dedicated initially to ancients and subsequently to contemporaries, had been imagined since the early Renaissance and satirized by Geoffrey Chaucer and, later, Alexander Pope.[70] Westminster Abbey may be considered the earliest physical hall of fame; originally a burial place for monarchs, by the fourteenth century it added the bodies of (or memorials to) statesmen, later admitting poets (beginning, ironically, with Chaucer), composers, scientists, and others. The rise of the nation-state spawned similar shrines across Europe. In 1791, the French Assembly converted a prerevolutionary church into the Panthéon, while in 1807 Ludwig of Bavaria conceived perhaps the first purpose-built temple of fame: the Walhalla memorial to soldiers, philosophers, artists, scientists, and other "heroes" of the embryonic German nation.[71] Yet the concept took some time to cross the Atlantic. Although Pierre L'Enfant included in his 1790 plan for Washington, DC, a nondenominational national church to house monuments to the republic's founders and subsequent heroes, the idea was thwarted by republican misgivings about deifying leaders, secularist concerns about a state-sanctioned place of worship, and sectional and partisan disagreements over whom to honor.[72] The closest approximation to a temple of fame was either Statuary Hall, established in the US Capitol in 1864 but with the limitation of two statues per state, or short-lived, commercial versions, such as Charles Willson Peale's American Pantheon or Mathew Brady's gallery. If Statuary Hall was too restrictive, the latter were arguably too inclusive.

By 1900, however, a growing mania for monuments, frustration with Statuary Hall, and sectional reconciliation allowed the national pantheon idea to take root.[73] That year, architect Stanford White constructed a Hall of Fame—the first to use that name—on New York University's Bronx campus in the form of a semicircular colonnade containing niches for the statuary busts of 102 Americans. To distance itself from its European precursors, it democratized the process. Instead of empowering royal, executive, or legislative bodies to make arbitrary selections, it allowed the general population to nominate individuals (deceased for at least a decade), and a board of electors from across the country would convene every five years to vote on them.[74] Even grander edifices were proposed—but never realized—for Washington, DC. In 1902, the McMillan Commission for the redevelopment of the National Mall suggested adding a "Pantheon" to the "illustrious men of the nation," while in 1915 sculptor Frederic Ruckstull envisaged a sumptuous "Palace of Fame."[75] Pantheons also emerged in the form of academies—the National Institute of Arts and Letters in 1898, the Academy of Arts and Letters in 1904—with membership restricted to those deemed to have contributed most to their field. Meanwhile, an outpouring of biographical books offered their own version: *American Immortals* (1901), *Men of America* (1908), *Distinguished Successful Americans of Our Day* (1912), and, of course, *Who's Who in America* (1899–). These went beyond the mug books of the Gilded Age, which had been limited to individual cities.

Although it included non-Americans, the MHRA was a by-product of this cultural preoccupation with enumerating and enshrining the nation's greats. It borrowed the term *American Pantheon* for its proposed headquarters in Washington and had an institutional connection, through its director Kunz, to the Bronx Hall of Fame.[76] Konta even suggested building the headquarters of the MHRA's New York chapter as an annex to that monument.[77] The MHRA echoed the latter also in drawing up a protocol for deciding whom to memorialize. The records of living men would not be deposited in its central archives "until perhaps fifty years after death, and then only by a majority vote of the State Associations, whose directors should be men eminent in their respective communities."[78] And the public was to nominate "living men and women whose reputations are likely to endure."[79] The two institutions also shared significant social biases. The Hall of Fame initially refused to consider foreign-born (naturalized) citizens until forced to change its rule in 1915 (very few were ever admitted), and it excluded women from its initial list and blacks until 1945. So, too, with the MHRA: secretary Larned's initial shortlist of 119 immortals included just three women (scientist Marie Curie and actresses Sarah Bernhardt and Eleonora Duse), one Jew (literary critic George Brandes), and no blacks, while its portraitist MacDonald branded himself a "Photographer of Men," refusing even to photograph his female relatives.[80] The list also favored leaders, scientists, and representatives of the "high" arts while barring sportsmen, popular musicians, and other celebrities. Thus, where Deihm's Century Safe was perceived by some as ushering in "cheap fame" by immortalizing anyone who paid a fee, Konta's project sought to shore up an older, more restrictive notion of fame. The history of renown evidently does not follow a linear narrative of gradual inclusion.

"All Possible Mechanical Means"

Ultimately, the gathering of parchment messages from famous individuals was, as Larned himself revealed, "only a minor undertaking of our association"—perhaps a publicity stunt to promote their broader archival agenda.[81] That agenda included not just redefining history to include the cultural and the everyday but also transmitting it through new media technologies. In embracing "all possible mechanical means," including those "yet [to] be made," Konta further challenged contemporaneous institutions of memory.[82]

This was to be achieved by working closely with Edison. For the MHRA's second meeting in November 1912, the inventor submitted a phonograph machine and some of his new "indestructible" records to be played between speeches, promising to donate several more to the MHRA to be preserved for posterity.[83] The following year, they collaborated on a project to produce historical records using his brand-new kinetophone, a sound-film system consisting of a cylinder phonograph connected via belt and pulleys to a projecting kinetoscope.[84] For their first assignment, they invited Mayor Gaynor and his police, fire, and street commissioners to Edison's studio in the Bronx to explain to posterity—in just three minutes each—the "workings of the city government."[85] Two weeks later, Edison appeared at city hall to formally present that recording to the MHRA.[86] In response to the MHRA's request for further recordings, Edison pledged to deposit his first "dozen" kinetophone films and "thereafter any . . . deemed of importance to posterity."[87]

Such an initiative was ahead of its time. Film was still largely considered a medium of entertainment. And once movies have "served their immediate commercial purpose," observed Konta, they are "not likely . . . preserved."[88] Indeed, although paeans to the medium's archival potential date back to its early years, historians of film preservation in the United States typically begin with Senator James Phelan and Hollywood reformer Will Hays's efforts in the 1920s and Iris Barry's at MoMA in the 1930s.[89] Libraries and archives—and the historians who used them—were particularly slow to embrace film. Although the Library of Congress had allowed paper prints of films to be deposited from 1894 onward, it had done so for purposes of copyright, not conservation. Logistical obstacles (particularly film's flammability) and the cultural conservatism of its head librarians (before Archibald MacLeish) delayed the establishment of a Motion Picture Division until the 1940s.[90] The studios were hardly interested in preservation either, preferring to destroy reels to prevent piracy. Consequently, only about 10 percent of films made in the 1910s are estimated to have survived.[91] Yet in assuming that film preservation began only in the 1920s, we overlook earlier efforts (albeit short-lived ones), such as the MHRA's.

To some extent, however, Edison's involvement enticed the MHRA back to memorializing the elite. The Gaynor film moved one reporter to extol that medium as a guarantor of "permanent fame." In its capacity to capture "personality," he wrote, film surpasses the "pyramids, obelisks, [and] huge mausoleums" employed by past rulers. Edison subsequently produced a kinetophone recording of Andrew Carnegie for the MHRA in which Carnegie read from his *Gospel of Wealth* the same passage he contributed on handwritten parchment.[92]

This impulse to store away famous voices can be traced back to the early years of the phonograph. Edison himself, despite promoting it as a dictation device, envisaged "every town and hamlet in the country" assembling on national holidays to hear past leaders' greatest speeches.[93] Such predictions were accompanied by a recurring lament that earlier leaders, from Alexander the Great to Abraham Lincoln, had died before the phonograph's advent, their voices lost to history.[94] There were also attempts to preserve great singers for posterity, most notably the hundred-year time capsule containing twenty-four discs with arias by Enrico Caruso and Nellie Melba, among others, which an American gramophone executive sealed in a locked basement vault of the Paris Opera House in 1907, thereby inspiring Gaston Leroux's prologue to his 1910 novel, *The Phantom of the Opera*.[95] The phonograph promised these prominent figures a kind of indemnity not just against mortality, but more specifically against the fickleness of celebrity culture and audience tastes.

The MHRA was, in part, an outgrowth of this culture of technological mummification. While phonographic messages to posterity were typically isolated efforts, the MHRA set out to organize an entire archive of them, beginning with Edison's own voice. It thus followed in the tentative footsteps of E. W. Scripture, a leading speech scientist at Yale, who had recently begun collecting recordings of "great" Americans as selected by "statesmen, college presidents, writers, etc." To avoid controversy, Scripture kept the selections secret, releasing transcripts only of two messages from Kaiser Wilhelm II that he had recorded and deposited in three major libraries.[96] The MHRA went a step further than such single-medium archives, adding Edison's "Kinetophone portraiture" and MacDonald's photographic portraits to its phonographic program. In its multimedia approach, the MHRA was fulfilling Harrison's call for a national time vessel that would store both visual and auditory traces of "our foremost statesmen, poets, thinkers, and men of mark."[97]

The MHRA was not content, however, to create a kind of virtual wax museum. It would harness the new media to its larger archival goal of "transmitting to posterity a vivid and comprehensive record of the life and civilization of the day."[98] This would encompass "important events," such as the upcoming political conventions, which the MHRA requested to record on disc and celluloid. Both promised to be historic, as Roosevelt was challenging Taft and raising fears of third-term Caesarism, while the Democrat candidates remained in deadlock (*Moving Pictures News* wondered whether the presence of cameras would influence their "behavior").[99] The outbreak of war two years later provided the MHRA with even more momentous "history" to preserve through films, phonograph records, written documents, and artifacts such as uniforms and weapons for the edification of "children of one hundred or one thousand years hence," but also for exhibiting in a "great museum" to that conflict "as soon as [it] . . . is over" (somewhat contradicting its earlier-proposed museum to the "peaceful arts").[100] Konta also embraced these media to record anthropological forms of culture, in particular those believed to be vanishing, such as regional dialects and indigenous languages, artifacts, and rituals.[101] This broad program echoed the Vienna Phonogrammarchiv's idea to "bottle up" all the sounds of the fin de siècle: not just statesmen's messages but also "specimens" of the world's languages, "contemporary music, dramas, sermons, lectures . . . [and] even . . . conversation of the various classes of society."[102]

Like its European counterparts, the MHRA lauded the new media's capacity to convey such phenomena directly rather than through words. "At present, we get our information [about the past] at second and third hand from books," explained its secretary William George Jordan, "but the future generations will be able to see and hear things of great historic importance exactly as they occurred and from the lips of [those involved]."[103] They would learn, for instance (added Konta), that "Bismarck had a thin, piping voice in ludicrous contrast to his enormous bulk."[104] This immediacy would yield a historiography not only more vivid (the term Konta used was a *living record*), but also more objective.[105] The "uncertainty" and "constant revision" that plagues historians, which is due to their reliance on (often contradictory) written sources, will be "completely eliminated" by audiovisual ones.[106] The MHRA cited, among others, the example of the Gettysburg Address, of which "over a hundred different versions" have been published. "Had it been possible to secure a phonogram record of the address when delivered there could be no dispute as to the actual words."[107] The association's "watchword" was thus that "light and sound waves never lie."[108] This myth of transparency not only obscured the biases in audiovisual texts (presidential candidates were already conscripting film as a propagandistic tool by 1912) but also ironically implied a rejection of historiography altogether. With such a "perfect reproduction of present day life," future historians would have nothing left to do. The kinetoscope and phonograph, proclaimed the *St. Louis Post-Dispatch*, will be the "Macaulay and . . . Gibbon of twentieth-century events."[109]

But the MHRA's faith in the archival potential of new media was not based entirely on their indexicality. Another rationale was their supposedly greater durability than wood-pulp paper documents (although, as we will see, the MHRA was far from oblivious to issues of media degradation). The new phonograph discs, especially Edison's "indestructible" condensite ones, appeared to circumvent the fragility of wax cylinders.[110] Konta similarly advocated the archival use of glass photographic plates as a solution to the long-standing problem of prints that faded or yellowed over time.[111] There was even a plan to preserve motion pictures on glass—presumably using a little-known camera, invented in 1912, that used a moving lens to expose five hundred and seventy-six tiny frames on a single, nonflammable plate.[112]

A third rationale was the greater compressibility offered by phonographic discs and celluloid film. Concerns about "economy of space" prompted Konta also to welcome newer materials, such as Oxford India paper (whose thinness could drastically reduce the bulk and weight of books), and to anticipate an "indestructible and even thinner material" that would ultimately replace paper altogether.[113] Taking advantage of similar advances in microphotography, Konta proposed to transfer books and other records onto glass plates, thus anticipating the microfilming projects that libraries initiated in the late 1920s.[114] Inventors had explored photography's miniaturizing possibilities since its birth, but their microphotographs remained the stuff of novelty shops and carrier pigeons. What Konta sought to fulfill was Sir John Herschel's 1853 dream of using the camera to produce "concentrated microscopic editions of works of reference . . . and manuscripts, etc."[115]

Yet another feature that rendered these technologies ideal for archival uses, in the MHRA's view, was reproducibility. With the aid of arc-lit photographic enlargers (introduced in the early twentieth century), future recipients could not only view but also duplicate images and documents preserved on the glass plates. Konta further hoped to employ electrotyping—a process involving copper and electricity—to generate copies of phonograph recordings.[116] This logic of reproducibility—indicated in the MHRA's inaugural meeting with the transference of that wax-cylinder recording of Edison's voice onto a newer disc—negated the traditional archival preference for the original.[117] Facsimiles also found their way into contemporaneous time vessels, such as the bronze chest sealed on Wall Street in 1914, which commemorated the 140th anniversary of the first written appeal for a "union" of the colonies by depositing not the original (recently rediscovered) letter but a printed copy.[118] This trend prompted some reservations among time-vessel advocates such as Vail, who wondered whether posterity would not prefer to receive "the originals of many historic documents, photographs, and the like." Copies "are not always desirable, nor can these things always be copied to the best advantage."[119] Nevertheless, the MHRA believed that technological reproduction averted a worse fate: eradication of the historical record. By replacing fragile paper documents with media that can be "easily stored and . . . countlessly duplicated," wrote an MHRA incorporator, "there will be no lost history in the future. . . . There will be complete pictured and phonographed records of everything worth keeping."[120]

The multimedia records produced for the MHRA, however, were not exclusively addressed to the future. Konta hoped that the undertaking would influence contemporaneous discourse and that "emotional politicians" would become more careful about what they said.[121] Several of their records, moreover, were released to the public rather than kept sealed. Three days after the ceremony in city hall, Gaynor's kinetophone message to posterity was screened at four vaudeville theaters in Manhattan. His cooperation may have been motivated by a desire to communicate with the electorate directly rather than through newspapers. In particular, the progressive-reformist mayor sought to advocate for Workmen's Compensation bills and to challenge certain "corrupt newspapers" that were pressuring him to dismiss some of his commissioners. Yet by ostensibly addressing posterity, Gaynor could disavow any intent to gain political advantage or popularity in the present.[122]

The rhetoric of posterity also served those behind the camera (or phonograph). While Edison envisaged the kinetophone as a provider mainly of theatrical entertainments, his collaboration with the MHRA allowed him to claim a higher civic, pedagogical, and historical purpose. Enlisting film to document "history" would appeal to a broader, middle-class audience by countering its association with the commercial and the ephemeral.[123] Such rhetoric might have served also as a smoke screen masking the kinetophone's technical limitations: its short running time (under six minutes), unreliable synchronization, and poor sound.[124] Accordingly, movie and phonographic trade journals warmly welcomed the MHRA for the aura of legitimacy it would cast on their media. *Edison Phonograph Monthly* proclaimed that "the formation of this society marks a new era in the world's history," while *Moving Picture News* applauded its plan to record and film the upcoming conventions.[125] It helped that the latter's founding editor, Alfred Saunders, a leading advocate of the "educational uplift and refinement of the moving picture," was on the MHRA board.[126] Gaynor, too, was an advocate for cinema; shortly before his kinetophone appearance, he vetoed an ordinance to regulate and censor movie theaters, claiming they were "a solace and an education" to the poor rather than "schools of immorality."[127]

"A Living Record of the Indian"

Another who stood to gain through association with the MHRA was Edward Curtis. In 1906, the ethnographic photographer had embarked on a vast undertaking (also funded in part by J. P. Morgan) to document—in twenty volumes of words and images and another twenty portfolios of larger photogravures—the "history, life and manners, ceremony, legends and mythology" of all Indian tribes in North America still living in a "primitive condition" west of the Mississippi.[128] Seven years later, perhaps to attract new subscribers to this costly project in the aftermath of Morgan's death, Curtis offered the MHRA one of his photographs, *The Oath—Apsaroke* (**fig. 5.3**). Originally taken in (or before) 1908, and published in 1909 as the frontispiece to the fourth volume of the *North American Indian*, on the Apsáalooke (Crow Nation) and the Hidatsa of the Northern Plains, it depicts three Apsáalooke men, one of whom is consecrating an oath of truth. According to the description in his book, the ritual involves "thrusting an arrow through a piece of meat, placing it upon a red-painted buffalo-skull, and then raising it toward the sun" and "touch[ing] the meat to his mouth."[129] Before submitting the print to the MHRA's "Message of the Records" for long-term preservation but also immediate reproduction and distribution to its members, Curtis added an inscription alluding to the scope and importance of his larger project.

Curtis may also have offered his photograph in recognition of the MHRA's own pledge (in its prospectus) to make a "systematic" record of all aspects of American Indian life. Indeed, he appended an endorsement to his inscription: "The complete scientific study of [the American Indian] seems a natural and imperative duty of the MHRA."[130] Having included Canadian tribes in his project, Curtis would have approved of Konta's intention to extend this ethnographic work beyond American borders.[131] The MHRA's objective of a multimedia, "living record of the Indian" would also have gratified Curtis, who had recently won acclaim for his incorporation of motion pictures, lantern slides, and music into an "Indian Picture Opera" and was currently at work on his pioneering film *In the Land of the Headhunters*.[132] Already, the MHRA's board included journalist and amateur anthropologist Charles Fletcher Lummis, an early adopter of the phonograph as an ethnomusicological tool who gathered nine hundred "perishing songs" from American Indian and Mexican American communities between 1904 and 1912 and founded the Southwest Museum to house the cylinders

The Modern Historic Records Association

THE OATH – APSAROKE

The American Indian affords science its best opportunity to study primitive man as a living human document of supreme importance. His life history in its evolution of thousands of years, is a unique revelation of beginnings, early stages and development in language, customs, religion and institutions. The complete scientific study of this subject seems a natural and imperative duty of the Modern Historic Records Association. The rapid changes make quick action essential. It can not be deferred if it is to be done at all. For this work there is no tomorrow.

Edward S Curtis

FIG. 5.3 Edward S. Curtis, *The Oath—Apsaroke* (copyrighted 1908), printed, inscribed, and submitted to the Modern Historic Records Association's "Message of the Records" in April 1913. MssCol 2025, Manuscripts and Archives Division, New York Public Library.

and accompanying photographs.[133] Konta emphasized the urgency of such efforts. "The American Indian is approaching," he told the *Times*,

> if indeed he has not already reached, the last stage of his native existence. Either he is gradually disappearing or he is adopting the white man's civilization and adapting himself to it. In both cases the records of his own civilization will be lost unless something is done and done speedily.

Curtis concluded his MHRA message with a similar warning: "For this work there is no tomorrow."[134]

In pledging to record American Indians' culture before they "disappeared," Konta and Curtis were not necessarily criticizing government policy or advocating for Indian rights. As numerous historians have shown, the pervasive discourse of the Vanishing Indian, although often expressed in sentimental tones of regret or in the soft focus of pictorialism, located the indigenous in a nostalgic past, denying them any future. It also presented their predicament as inevitable and natural, the result of fixed Darwinian laws rather than of specific historical crimes against them. All that could be done was salvage their artifacts and cultural practices—a duty to posterity that, again, was merely archival.[135] Curtis did acknowledge in the general introduction to *The North American Indian* that the "treatment" of the Indians was "in many cases . . . worse than criminal" even as he disclaimed that his purpose was the "rehearsal of these wrongs."[136] In the Apsáalooke volume, he described how they had been decimated not only by smallpox outbreaks and wars against neighboring tribes but also by the incursions of white settlers and the slaughter of the buffalo and most recently by the "cession" of thousands of acres of their reservation.[137] None of this, however, is included in the photograph Curtis deposited with the MHRA. By eliding the problems faced by the Apsáalooke, *The Oath—Apsaroke* conveys to the future historian the impression that these last Indians of 1913 were still living a pristine and untroubled existence, practicing their traditional customs without interference from the government or, for that matter, ethnologists and photographers.

Salvage ethnography was viewed, moreover, as an opportunity to study the hazy origins of the white race. In his inscription beneath *The Oath*, Curtis described the American Indian as a kind of time capsule that perfectly preserved the story of civilization's emergence. "[He is] a living human document . . . [whose] life history, in its evolution of thousands of years, is a unique revelation of beginnings, early stages, and development in languages, customs, manners and institutions."

This entailed the depiction of Native Americans as though they were still living in a pure, prehistoric (or precontact) state, through careful posing of subjects, the introduction of props, and the removal of any trace of the European or modern from the scene—all of which belied the rhetoric of the record as direct transcript of reality.[138] It also entailed the cultural appropriation of art objects, the plundering of burial sites, and sometimes the offering of money in exchange for their performance for the camera or phonograph.[139] The MHRA was thus embracing ethnographic practices that arguably legitimated and accelerated the destruction of indigenous cultures. It is telling that two of its seven vice presidents—Frederick Grant (son of Ulysses) and David Brainard—had served as military commanders in the Indian Wars, while a third—Peary—had notoriously delivered six Inuits to the American Museum of Natural History in 1897 to serve as "living specimens."[140]

There was also a racial subtext to the faith, expressed by both Konta and Curtis, in the preservative powers of audiovisual media. American ethnologists had long attributed the disappearance of indigenous languages to Indians' alleged failure to develop an alphabet. Indigenous arts and institutions, too, were "dissolving" because of that fixed state of orality, the late-nineteenth-century ethnologist John Wesley Powell implied. Conversely, the written word's capacity to store sophisticated ideas was viewed as a key factor in the white race's evolution and superiority. The culmination of this inscriptive capacity was the modern audiovisual triumvirate of photography, phonography, and cinematography.[141] New media, then, were not simply innocent tools with which to preserve Indian culture for posterity but were considered proof of the evolutionary process apparently condemning it to extinction. Stories of Indian resistance to such technologies served to reinforce this idea. "The tribes of the Southwest," declared Kunz at the MHRA's inaugural meeting, retained a "superstition" toward the phonograph that "was almost impregnable"—yet which Lummis, with his "long, intimate and wide acquaintance" with them, has been able to "overcom[e]."[142]

The very idea of a time vessel appeared to confirm that higher stage of cultural development. Indians were often characterized as indifferent to the future in accordance with the myth of the happy savage. For archaeologist and ethnologist Alice Palmer Henderson, the tribes of the Yukon "have very little idea of the future, and, to tell the truth, very little interest in it. They follow the instruction of the poet: 'Worry not over the future, present is all thou hast, / For the future will soon be present, and the present will soon be past.'"[143] The self-styled expert on Plains Indians Richard Irving Dodge had attributed this disinterest in the "possibilities of the future" to their supposed lack of a proper calendar or of any "word synonymous with our word year."[144] Only whites, this implied, could undertake the task of photographically and phonographically preserving Indian life for posterity.

From this perspective, *The Oath* appears to deny a future for the Apsáalooke, thus construing posterity in racially exclusive terms. Unlike messages written by MHRA members to their own lineal descendants, this photograph does not reinforce a racial bond across time. Its assumed recipients are not Indians but rather whites, for whom Indians will have become the stuff of legend, their rituals and practices obscured by the mists of time. Already the Apsáalooke's numbers were dwindling, from a population of "approximately nine thousand" in 1825 to "less than eighteen hundred" by 1908, according to Curtis. Yet, photographs such as this would serve, in his words, as "transcriptions for future generations that they might behold the Indian as nearly lifelike as possible."[145] The bison meat raised to the heavens here stands for his photographic offerings to posterity.

A closer examination of Curtis and other MHRA associates, however, complicates this impression. Curtis's project was arguably guided by a desire to challenge negative perceptions of the indigenous other not only through verbal accounts but also through portraits (especially close-up ones) that humanized the Indian and by documenting rituals that the government had banned.[146] Konta's "Message of the Records" also included critiques of the racial ideologies of the period. H. G. Wells's handwritten parchment denounced how "the politically ascendant peoples of the present phase are understood to be the superior races" while those "not at present prospering . . . are represented as the inferior races, unfit to associate with the former on terms of equality, unfit to intermarry with them on any terms, unfit for any decisive voice in human affairs."[147] Lummis went even further. His encounter with the Pueblo Indians of New Mexico led him to reject salvage ethnography and its guiding myth of the Vanishing Indian and to assert indigenous cultures' capacity to adapt and endure. He was also a fierce critic of the US Bureau of Indian Affairs and its post–Dawes Act policy of forced acculturation and land confiscation. In 1899, he launched a campaign against the removal of Indian children to boarding schools and against an educational program designed to turn them into a "class of constitutional peons," and in 1902, he organized a league to protect Indians' legal rights in California. He was the "forerunner by half a generation" of the American Indian Defense Association of 1923, according to its founder John Collier.[148]

The assumption that Indians were simply duped or bribed into posing for Curtis's camera or singing for Lummis's phonograph also seems reductive. Many shared Curtis's assumption that catastrophic cultural loss was inevitable and imminent and wished to bequeath a visual memorial. There is also evidence that some tribes (especially the Apsáalooke) were aware of his project and invited the "Shadow Catcher" to come and photograph them.[149] The Apsáalooke initiated their own efforts at salvage ethnography during this period through autobiographies, such as that of the warrior Two Leggings, and the photographs taken by the Cree Indian but adopted Apsáalooke Richard Throssel, who preserved their ancient ceremonies and current living conditions in more than a thousand images between 1904 and 1911.[150] Indians may have brought these concerns and motives to their sessions with Curtis. There are suggestions that they may have had a say in the roles they played, the postures they assumed, and the costumes they wore. They may even have subverted Curtis's mission by performing their rituals wrongly.[151] Curtis's contribution to the MHRA thus registers an encounter between two peoples—a moment of cross-cultural contact and negotiation that was mediated by a lens and preserved by an archive.

Moreover, if Curtis had wanted to affirm his belief in the notion of the Vanishing Indian, a more obvious choice of photograph to submit to the MHRA would have been one that mournfully depicted Indians riding away from the viewer into a blurry darkness, such as his celebrated *The Vanishing Race—Navajo* (1904).[152] In contrast, *The Oath*'s depiction of two rifles contradicts the essential rule of salvage ethnography, that of presenting a precontact vision of the Indians; indeed, one finds no further photographs featuring firearms in that fourth volume and very few in the other volumes.[153] The rifles' inclusion may have been on the insistence of Goes Ahead and Pretty Tail, another sign perhaps of how Curtis's project was subject to cultural negotiation. The photograph thus presents a more defiant image of survival into an uncertain but open future.

Confronting the Shelf Life of Modern Media

In promoting photographs, phonograph records, and films as sources for future history, however, the MHRA and its affiliates did not skirt the question of the durability of those media nor succumb to a naive faith in their powers of preservation that some scholars have ascribed to this period.[154] Although it backed Edison's "everlasting" condensite and never acknowledged celluloid's flammability—despite the evidence of contemporary fires at theaters and film exchanges—the MHRA did consider the temporal challenges posed by modern media.[155]

From the outset, one of the MHRA's principal objectives was to raise awareness about the inadequacy of modern, industrially produced paper as an archival medium. The sulfite process of manufacturing woodpulp paper had been introduced in the United States in 1884 and was widely used to publish newspapers and other ephemera included in the centurial vessels. It was clear by 1910 that this cheaper paper disintegrated more quickly than its predecessors—as little as fifty years in the MHRA's estimate—and that modern records are thus "writ on water."[156] One of the MHRA's solutions was to revert (as time vessels had done since 1876) to older, low-acidity writing surfaces, such as pure cotton (or rag) paper. It launched a campaign for newspapers to print library editions of each issue on rag with special ink—an idea that appealed to the *New York Times*, which aspired to become the "newspaper of record" with the launch of its index in 1913.[157] For its own documents (including the Message of the Records), the MHRA reverted to even older writing media, adopting parchment or handmade sheepskin vellum imported from England, "identical" (it claimed) to that used in the fourteenth century.[158] The phonograph and kinetophone records were also to be accompanied by parchment documents that offered a transcript of the speech or an authenticating signature of the speaker.[159] This turn to parchment as an archival supplement suggests doubt about the stability and self-sufficiency of modern media and a belief in the need to bridge new and old.

As readers learned from "The Last New Yorkers," however, not even parchment was immune to decay. When, after seventy-five episodes, Allan and Beatrice finally come upon the time vessel in the vault of a cathedral overlooking the Hudson River at Storm King, they are disappointed. A thousand years of dry rot have destroyed the wooden boxes, exposing their parchment contents. Those historical documents are now so "crinkled" and brittle they cannot be "moved . . . or read . . . or handled . . . in any way." With a conservationist's tools and chemicals—"proper appliances, glass plates, transparent adhesives"—and a "year or two at our disposal," the two explorers might have recovered some information. But they lack such luxuries; and in any case, a moment later, a vibration reduces them to a "mass of rubbish, powder, dust."[160]

Fortunately the vault contained a second box: a time vessel that a certain Philavox Society had soldered up in 1918, two years before the catastrophe. Along the lines of the Paris Opera House vessel of 1907, this coffin-size leaden casket contained "eighty or a hundred" recordings made and deposited in the Metropolitan Opera House, to be opened in 2000. In the 1950s, a band of white survivors had rescued it from Manhattan's ruins and stowed it in the cathedral alongside their parchment documents. Prying it open with his hunting knife (**fig. 5.4**), Allan and Beatrice discover "instrumental as well as vocal masterpieces" by Rossini, Schumann, Brahms, and others. These were preserved not on conventional shellac nor on "everlasting" plastic such as condensite—materials that would have succumbed to extremes of heat and cold—but on steel, using a process Allan remembers was invented in 1917 (metal records were in fact invented in 1921). With the provision of a record player "swathed in oiled canvas" (a device that also accompanied the phonograph records in the Oklahoma time vessel of 1913 but not those of the Colorado Springs century chest —perhaps indicating a growing awareness of the risk of technological obsolescence or regression), Allan and Beatrice presumably may use these records to recivilize the Folk. Meanwhile, a recording of an Episcopalian wedding ceremony allows them to consecrate and consummate their own relationship after seventy-five episodes of strained abstention.[161]

Thus, while England doubted modern shellac, celluloid, or paper—or even medieval-style parchment—would survive a thousand years, he maintained faith in an even older medium of inscription: metal. The steel phonograph records were among a number of metallic texts in the story. In an earlier episode, Allan and Beatrice discovered a complete set of encyclopedias printed on "thin sheets of nickel" using Edison's "electrolysis process."[162] Their delight at finding such a rich source of practical information recalls Frederic Harrison's belief that future archaeologists would prefer encyclopedias to literary texts. And Edison had indeed predicted in 1911 that nickel would replace paper, enabling books to "contain 40,000 pages[,] . . . weigh only a pound[,] . . . [cost only] a dollar and a quarter," and last for millennia.[163] The metal used for printing also proved durable. The ruins of a newspaper press near Buffalo contain "lead types" that were still "recognizable," allowing Allan, crucially, to determine their location.[164] Even the Folk have embraced metal. When their last surviving book from preapocalyptic times (less usefully, a copy of *Pilgrim's Progress*) begins to crumble, and efforts to piece it together with seaweed and sharkskin fail, they resort to transcribing it onto gold plates.[165]

The MHRA shared England's confidence in *heavy media*. At the inaugural meeting, Kunz lectured on the history and longevity of writing

FIG. 5.4 "Giving the torch to Beatrice, Allan set to work. In his powerful hands the hunting-knife laid back the metal of the mysterious leaden chest." Illustration by P. J. Monahan from George Allan England, *Darkness and Dawn* (Boston: Small, Maynard, 1914), frontispiece.

surfaces from antiquity to the present. While reiterating modern paper's inferiority to medieval parchment, he considered metal the most durable. A record of Roman military service inscribed in the fourth century on "thin sheet-lead," he observed, has remained legible despite having being buried for around 1,500 years. Of all metals, he recommended the MHRA use silver, which develops a protective surface of black sulfide when it tarnishes, unlike the rust that penetrates and corrodes iron.[166] Meanwhile, the MHRA used metal to disseminate historical records in the present. In 1915, it conceived—and funded—the idea of decorating the entrance to New York's Williamsburg Bridge with sixty-five plaques, each displaying the Declaration of Independence in a different language, a scheme typical of the Americanization efforts implemented by progressives, especially during the Great War.[167]

Yet, the most durable medium, Kunz maintained, is one of the oldest: the clay tablet. His talk was accompanied by exhibits of Babylonian tablets from J. P. Morgan's collection (to which he was adviser), whose inscriptions remain "as distinct and legible" today as they were five thousand years ago, and can withstand even flood and fire (fire in fact preserves them). However, Kunz did not conclude his talk by lamenting the superiority of ancient over modern media. Instead, he proposed—and performed—a procedure for marrying terracotta to the new linotype machine, thus creating highly accurate and reproducible "modern tablets" (**fig. 5.5**).[168] Further vitrified tablets were presented at the subsequent meeting, courtesy of the Atlantic Terra Cotta Company, with Lincoln's Gettysburg Address (one wonders which version) among the first documents to be selected.[169] Hybridizations of other media, such as photographs on parchment, were also envisaged, while Harrison suggested "photographs on stone" or "inscriptions cut upon lava and cased with glass."[170] In fact, hybrids already existed as by-products of reproduction processes. Kunz, for instance, recommended preserving the metal "stereotype" plates used to copy wax (or shellac) records.[171] A reader of *Science*, responding to Vail's article, similarly proposed acquiring unwanted electrotype plates from printers. Rescued from the melting furnace, a set of electrotypes for the *Encyclopaedia Britannica* "would provide for future generations a considerable knowledge of almost everything of importance pertaining to this era"—a scenario already imagined by England.[172]

Ultimately, though, time-vessel advocates (including the MHRA) questioned the wisdom of relying on any single medium, whether ancient, modern, or hybrid. "No one of the materials known to man" can withstand all of the "destructive agents" that might assail them, observed Vail. Neither can we predict whether the agent will be natural (volcano, floods, wildfires, etc.), man-made (war, despoliation, arson), or an assortment. Thus, whereas clay tablets may be "quite resistant to the weather, and to a certain degree of heat," they remain vulnerable to "severe weather conditions or intense heat"—and, presumably, earthquakes. In the absence of any "ideally resistant material," he therefore suggested "a combination of materials."[173] Not only would this permit a hedging of one's bets; it would also, as Harrison suggested, yield a valuable research sample to "future scientists," enabling them to determine which is the most durable: a certain "metal, . . . porcelain, or a compo-

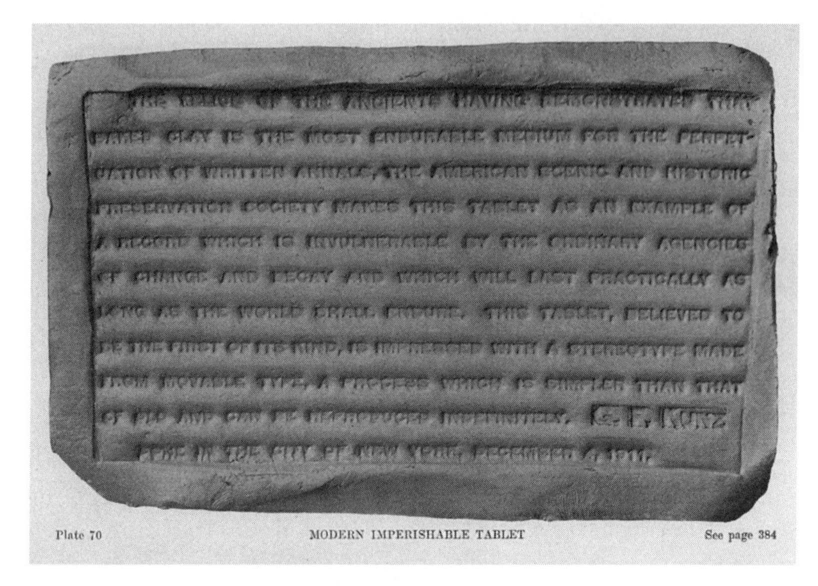

FIG 5.5 At the inaugural meeting of the Modern Historic Records Association, mineralogist George Kunz presented his "modern imperishable tablet"—a hybrid of the ancient terracotta tablet and the modern linotype machine—as a medium for transmitting records across long time spans. From George Kunz, "The Imperishable Records of the Ancients, Compared with Methods in Use up to the Present Time," *Documents of the Assembly of the State of New York* 32, no. 59 (1912): plate 70.

sition [i.e., hybrid]."[174] Such rationales informed the MHRA's similarly diversified approach to preservation. Resisting blind faith in new media and antimodernist nostalgia for old media, the MHRA sought syntheses that compensated for the limitations of each.

Concrete Fantasies

The MHRA betrayed doubts about the durability of modern media also in its deployment of time vessels. Although its prototype vessel of 1911 contained only vellum invitations, Kunz believed the method could be adapted to store "photographs, phonograph records and moving-picture records."[175] Two years later, the MHRA did reportedly launch two larger time capsules addressed to New Yorkers of 2013. These contained "documents, tablets, and photographs relating to the life and affairs of the people of 1913" along with any personal photographs and messages that members wished to submit in vellum envelopes supplied by the association.[176] President Taft was among those who submitted a sealed envelope, prompting much speculation as to its contents.[177]

Awareness of those materials' perishability and aspirations to eventually communicate across longer time spans prompted the MHRA to refashion new kinds of vessels. Wood, which had been used for the Harvard chest of 1900 and proved inadequate in "The Last New Yorkers," was rejected. The prototype vessel instead combined the material properties of terracotta, glass, and concrete with a copper plaque—thus constituting another hybrid—while the 1913 vessels consisted of "indestructible" steel chests lined with lead and labeled with an inscription on a copper plate.[178] Both the shape of the prototype and the idea of an engraved description may have been inspired by the cylinder seals that Babylonians employed to stamp their clay documents and to seal containers and that they deposited in tombs and temple walls. As Kunz described in his paper, these "strange cylindrical seals," a collection of which had recently been donated by Morgan to the Metropolitan Museum, were stamped with "images of the gods" to render the documents "sacred and inviolable," leading "many" to believe they had "a talismanic, possibly also a phallic, significance."[179] The sculptor Carl August Heber, despite his Beaux-Arts training, stopped short of applying such ornamentation to the MHRA's modern cylinder. Perhaps he was heeding Harrison's warning not to offend the "artistic, economical, political, or religious susceptibilities" of subsequent generations.[180] Embellishing one's vessel—as Deihm did with presidential portraits and as the Lower Wall Street Businessmen's Association (LWSBA) did with the faux rope handles, paw-shaped feet, and crowning finial of its bronze chest—might render it vulnerable to some future iconoclastic movement (**fig. 5.6**).

Whereas earlier depositors entrusted their vessels to some existing library, university, or government building—a convention continued by the LWSBA, who consigned theirs to the New-York Historical Society—the MHRA proposed its own purpose-built structure. A rendering of its American Pantheon, drafted by architect (and cocreator of the prototype cylinder) Arthur Dillon, was unveiled at the inaugural meeting (**fig. 5.7**). It would house the records in vaults modeled on those of banks but with "some modifications" to address the needs of "permanent storage." Those vaults would be "hermetically sealed" and "the air in them exhausted and replaced by some dry inert gas . . . to obviate all chances of decay."[181]

Dillon's first consideration was location. "It is seldom an architect

FIG. 5.6 A bronze time vessel, with decorative finial and lion's paw feet and measuring 17½″ × 25″ × 15½″, made by the renowned New York foundry, Jno. Williams Inc. The chest was launched by the Lower Wall Street Business Men's Association in 1914 and entrusted to the New-York Historical Society. Object number: INV.9152, NYHS. Photograph by Glenn Castellano.

FIG. 5.7 "The building that was to outlast the pyramids as a treasure house for future archaeologists. It had one weakness; it was never built." Proposed headquarters for the Modern Historic Records Association, designed by architect Arthur Dillon. Boyden Sparkes, "Civilization, He Fears, Is Slipping," *New York Tribune*, November 13, 1921, 7.

looks ahead more than a hundred years, but [this] building . . . should be nearly everlasting so the site is important."[182] Harrison's article had warned against depositing the National Safe in London, "or indeed anywhere near the haunts of man," where suburban or infrastructural development might endanger it. Instead, he suggested a remote place like Salisbury Plain, even as part of Stonehenge, "thus link[ing] the centuries A.D. to those B.C."[183] Dillon acknowledged the further locational hazards of "flood . . . excessive meteorological extremes, . . . fire and earthquake," perhaps alluding to San Francisco's fate in 1906 or Galveston's in 1900. Concerns such as these may have led England's postapocalyptic survivors to deposit their vessels in the Hudson Highlands and the Colorado Rockies.[184]

A second consideration was the building's design. To withstand flood, fire, and earthquake, Dillon suggested it be "massive in construction and low rather than high" and made with "concrete, reinforced by steel imbedded in it." The choice of concrete for a civic edifice was ahead of its time. Since its rediscovery in Europe and the United States in the 1820s, concrete had been considered suitable mainly for utilitarian structures such as foundations, retaining walls, sea defenses, and military fortifications—and, by 1900, grain elevators and factories. Even the introduction of reinforced concrete in the 1870s failed initially to dispel most architects' suspicions of it as a base material, a "low-grade substance."[185] Dillon himself conceded that it "is not desirable esthetically. It assumes an unpleasant color."[186] In the ensuing years, modernist architects came to embrace and promote concrete as a "modern" material that promised bold cantilevers, a new monolithism, and freedom from ornament and tradition—an association that has persisted, on and off, ever since. But its earliest architectural advocates, such as Auguste Perret, considered it compatible with classical traditions. This approach is evident in Dillon's Beaux-Arts design for the MHRA and also in his earlier proposal that Ernest Flagg's Romanesque plan for the New York Public Library be built in concrete. Dillon thus chose concrete not for its new aesthetic possibilities but for its supposed resilience. It was the "obvious" choice, he declared, because "theoretically it will last forever."[187]

Walter Benjamin famously explored how the glass and iron architecture of nineteenth-century expositions and arcades aroused the temporal imagination.[188] Arguably, though, concrete—with its putative durability—was, by the early twentieth century, the preeminent material through which to imagine and speak to the distant future. Harrison called for "the strongest known cement" for his National Safe, while Vail drew on his chemistry expertise to confirm that reinforced concrete is the most resistant of building materials to earthquake and fire. Oklahoma City's century chest of 1913, enclosed in double cement walls and buried in a church basement under a foot of concrete, was a product of such faith.[189]

The material attained its apotheosis in "The Last New Yorkers." The cathedral on the Hudson was a "perfect" place to deposit the phonograph vessel because it had been built (around 1916) entirely from concrete, its crypt forming a "solid monolith" more impervious to the elements than the "Great Pyramid of Ghizeh." This fictional structure—which anticipated the exposed concrete cathedrals actually constructed after the Great War—is "practically intact," Allan notes, "after ten centuries of absolute abandonment." This theme of concrete's imperishability—and the transience of other materials, especially stone, brick, and wood—recurs throughout the novel. From his first view from the Metropolitan Tower, it is apparent that only Manhattan's concrete structures have survived intact; steel structures are in "rotten shape," while all others have crumbled. They send out a futile distress signal from a wireless station, "fortunately constructed of concrete." And they find the nickel encyclopedias in a millionaire's "still intact" concrete summerhouse by the New Jersey Palisades. As a former building engineer, Allan realizes the "lesson to be learned," namely that concrete must be used "in our rebuilding of the world."[190] Not only has concrete preserved certain artifacts of capitalist society; it will also cement the socialist utopia to come.

While Dillon's plans for the MHRA drew on such fantasies of concrete's endurance, however, he resisted any assumption that modernity had resolved the problem of architectural temporality, just as Konta and Kunz questioned the supposed superiority of modern media. For one thing, concrete could be considered ancient as well as modern. Some years earlier, Dillon had translated an article by Auguste Choisy that described how the Romans had constructed walls in concrete faced with brick. Contemporaneous articles in trade journals such as *Cement Age* similarly invoked ancient Roman bridges and buildings as proof of concrete's durability.[191] As a combination of an ancient material with a modern invention (twisted steel rods), reinforced concrete was itself a hybrid medium analogous to those the MHRA was proposing. Dillon acknowledged, moreover, that reinforced concrete had not yet attained perfect immunity to oxidization and moisture; "it will in time be worn away by the constantly recurring attacks of frost." He therefore suggested the MHRA building would need some kind of exterior cladding.

Such innovations in the design of time vessels, and of structures to house them, were a response not just to modern media's perceived fragility but also to other threats modernity appeared to pose to archival memory. Both Dillon and Konta cited the catastrophic fire that engulfed the New York State Capitol in Albany six months earlier, destroying eight hundred thousand books and manuscripts in its state library. That disaster, attributed to faulty wiring, lent support to the MHRA's campaign to fireproof libraries and to use the "photographic plate" to copy and thus "perpetuat[e]" precious records.[192] The MHRA's time vessels and concrete designs were also a means to raise awareness of how libraries and archives around the world were vulnerable to modern, industrialized warfare. In language reminiscent of Wells's *War in the Air*, Konta envisioned those repositories of "the precious blood of master spirits" succumbing to bombardment by invading armies unless international laws were passed to protect them.[193] With the recent upsurge of anarchism and labor militancy (signaled by the dynamiting of the Los Angeles Times Building in 1910), a specter of internal unrest also loomed over those buildings. In our quest for "permanent storage," warned Dillon, "we must guard against mob violence, riot, and revolution" too.[194]

These hazards, however, did not drive the MHRA to the conclusion reached by Harrison and by later time-capsule advocates, namely, that no aboveground structure, however well built, could be rendered invulnerable. "Fire, war, insurrection, greed, taste, caprice, and necessity," Harrison had warned, will "have it down in the end. The Tower of Babel, Babylon itself, the Colosseum, and the Temple of Ephesus, have all gone the way of all brick and stone. Besides, a building would cost much money. It would provoke the communists, the contractors, and the art societies to destroy it, or convert it." Instead, he suggested digging an underground chamber or appropriating a disused mine and subdividing it into galleries for each century. "When finished and filled, the museum would be solemnly closed up with twenty or thirty feet of cement. . . . There would be neither doors, keys, nor locks. Nothing but a gang of navvies [laborers], working for weeks under a staff of engineers, could ever open it again. It would need no guarding, no insurance, and no outlay." A granite portal, inscribed with donors' names, would be the "sole architectural feature."[195] Rejecting this proposal to entomb the records of civilization, the MHRA insisted on its building remaining in constant use as a museum, reference library, and national headquarters, to ensure the sealed vaults would not be forgotten. Its archive, as an externalization of memory (or hypomnesis), would thus remain tied to the ongoing work of human memory (or anamnesis).[196]

While retaining faith in aboveground architecture, the MHRA was wary of putting all eggs in one basket. "Unique works of art, or . . . historic records," Kunz recognized,

> should never be grouped in a single building, or even in a single city. Whatever be the safeguards provided, however well protected the building may be from ordinary accidents and chances, there is always and everywhere the risk of some unforeseen accident, of some catastrophe against which no adequate provision can be made. . . . They really belong to the world . . . [and] should, therefore, be widely distributed.[197]

This was a further reason for founding municipal chapters and for issuing facsimiles of its parchment records to "the guardianship of our members, and to the great libraries, colleges, and societies throughout the world."[198] It also prompted them to issue their 1913 time vessel in duplicate. Contributors were invited to submit two copies of every message, one for a time vessel to be consigned to the New York Public Library until their headquarters was complete, and the other for a vault near the Pyramid of Cheops in Egypt.[199] In adopting such an insurance policy, the MHRA again echoed England's story. The depositors of that time vessel in the cathedral on the Hudson had intended it as backup for the one in the Rockies. And to ensure neither was forgotten, they had disseminated small golden cylinders containing a parchment roll documenting their existence and location.[200]

Archival Overreach

For all its concerns about paper-based records, the MHRA failed to anticipate its own association's impermanence. After the initial flurry of planning and publicity, it lost its way, and the press lost its interest. The 1913 vessels turned out to be among its last undertakings. There is no evidence of further experiments with terracotta or collaborations with Edison or Curtis (the latter fading from public view himself).[201] Nor did Dillon's concrete "treasure house" get off the drawing board despite claims they were still raising funds for it in 1914.[202] After the war, a *New York Tribune* reporter called to find out what happened to it. "Why, that building never materialized," Dillon responded. "It was thrown into the discard. It would take an archaeologist to find the plans so deep are they buried in my office records."[203] The cylindrical vessel, linotype tablet, and other records thus never found their permanent home in a grand headquarters but were condemned to a permanent limbo in the vaults of the New York Public Library.[204]

The demise of the MHRA might be attributed in part to Konta's personal difficulties. In early 1914, his estranged wife Annie sued him for $101,000.[205] He was also hampered by the nation's entry into the Great War, which diverted funds and interests away from cultural projects while also prompting questions about the loyalties of Austro-Hungarian-born citizens. Konta publicly urged Hungarian Americans not to support their former country, led a delegation that met with President Woodrow Wilson to affirm their allegiance, and founded the American-Hungarian Loyalty League—all of which earned him the praise of Committee on Public Information chairman George Creel.[206] However, in September 1918, he was called before the Senate to respond to allegations that, three years earlier, he had advised a German propagandist on how to purchase an American newspaper. While admitting to having been pro-German then (like others with ties to St. Louis breweries), Konta claimed that the sinking of the Lusitania had changed his mind. Nevertheless, questions about his loyalties persisted. According to Creel, "vicious factional elements threw every possible obstacle in his path."[207]

Yet Konta survived these allegations to serve after the war as consul general to the Republic of San Marino and on the state parole board of New York among other things, and he could have resumed the MHRA or at least relinquished its leadership to others. The latter more likely collapsed from within, victim of its unorthodox and experimental approach to history and archiving. Its radical espousal of new media technologies rendered it vulnerable to the high failure rate in that field. As its initial novelty wore off, the kinetophone struggled to establish itself commercially, partly because of the technical limitations cited above but also because of Edison's decision to present it in the old vaudeville theaters rather than the emerging movie theaters. After a fire broke out in his West Orange laboratories in December 1914 and consumed all the original kinetophone masters, he abandoned the apparatus, shelving his quest for sound film.[208] The MHRA's embrace of microphotography was similarly premature, as dedicated, high-speed microfilm cameras and user-friendly reading machines were not available until the 1920s.[209] Its dream of phonographic and film recordings of public meetings, legislative sessions, and trials was also impractical, according to the *Tribune*.[210] Even its proposal of requiring newspapers to print library editions on rag met with resistance; the *New York Times* did not follow through until 1927.

The span and range of activities envisioned for the MHRA may also have been its undoing. Like other utopian archives of the period, the aspiration to encompass everything, from the municipal to the transnational, ran up against the trend toward state and national archives. The MHRA's resolve to create its own records also fell foul of the emerging codification of the professional archivist as passive and noninterventionist.[211] The recording of Indian tribal cultures across the Americas by itself would have been a vast undertaking. Its additional role of serving as a pressure group lobbying for the passage of national and international laws or for the modernization of archival practices overextended it still further. Such an archive would have been prohibitively expensive, exceeding even Konta's fundraising skills.

Of all its departures from the traditional archive, the MHRA's aspiration to record contemporaneous historical events may have most fatally undermined it. Not only did its emphasis on events, unlike the centurial vessels' emphasis on sources, inevitably draw it back to "great men history," but it was also clearly a logical fallacy to identify "historical events" as they were occurring. As E. H. Carr influentially argued in 1961, only subsequent historians can retrospectively distinguish historic events from the more general category of past events. The philosopher Arthur Danto amplified the point four years later by positing an "Ideal Chronicler" who is endowed with "the gift of instantaneous transcription [of] everything that happens . . . as it happens, the *way* it happens," yet who lacks "knowledge of the future" and thus cannot know "the whole truth" about any contemporary event.[212] The MHRA, then, may have noted the outbreak of the Italo-Turkish War in 1911–1912 but could not have realized its significance in revealing the weakness of the Ottoman army and thereby encouraging nationalist movements in the Balkans, thus contributing to the outbreak of the Great War. There were events that would have appeared even more minor at the time—such as the Balkan League's formation in 1912 or the announcement of Franz Ferdinand's decision to visit Sarajevo in June 1914—yet which proved to be historic. There are also events that the MHRA could not have recognized at all, such as those that took place behind closed doors: the formation in 1911 of the Black Hand, the secret military society responsible for Franz's eventual assassination, or the Anglo-French naval agreement after the Second Moroccan Crisis of 1911, which was not even revealed to the British Cabinet until August 1914. The MHRA thus overreached in presuming its vessels could record contemporaneous historic *events* rather than merely convey potential *sources* such as letters to the future, reports, or confessions. This presumption was perhaps a symptom of the cultural effect of new media technologies, which by "narrowing the lapse of time between the actual events themselves and their concrete record," Lewis Mumford subsequently argued, made the present seem already "historic." The logical culmination of this ascendency of "life as recorded" over "life as lived" were "fictitious performances"—Gaynor's comes to mind—"staged [for the motion picture camera], before or after the real event, in order to leave an 'accurate' record for posterity."[213]

Although this critique of our ability to historicize the present is associated with Carr, Danto, and subsequent figures, it was leveled against the MHRA at the time. "Almost all that we do or say" appears to be "of importance" to us, wrote one newspaper critic in 1913, but only "relatively few things" will be judged important by the future historian. "Who can say now that future American generations will care a button about [Woodrow Wilson's] papers or his speech? The people who inhabit this country during the twenty-first century may look back upon Dr. Wilson as a very great figure or as a figure of only transient importance and no one who lives today can say which it will be." From this perspective, the MHRA's challenge to the traditional conception of the archive was misguided.

> You cannot select now the historical material which will be of value in the opinion of those who are to live a hundred years later. What you can do is to save all the material that would probably go toward making up that later judgment, and take your chances on having stored away the essential things. The historians of that older age will then make their own selections and draw their own conclusions.[214]

The MHRA's attempt to historicize the present was not only philosophically suspect but also liable to generate controversy. This was particularly evident when its secretary Larned announced the shortlist of "historic" individuals whose testimony would be preserved. The literary journal *Dial* not only questioned the "Occidental" bias of that shortlist ("Japan, China, India, and all the rest of the Far East, are ignored") and odd choices (such as including journalist Henry Watterson but not politician and diplomat Whitelaw Reid), but also the very notion that one could adjudicate the greatness of contemporaries, "so often does the verdict of posterity reverse the most carefully considered judgment of this sort."[215] The list even generated disagreements within the MHRA. During one heated meeting, Senator Henry Cabot Lodge was included and then "scratched," while William Jennings Bryan was grudgingly added, as was Mark Twain despite the objection that he was "a poor specimen of a man of letters." The controversies left Larned despondent. "I tried to think up a lot of [names of prominent Americans] but nobody would help me. All I got was criticism."[216]

Despite its premature demise, the MHRA should not be considered a total failure. It argued for film archives, for microphotographic preservation of documents, for rag editions of newspapers, for media conservation, for international agreements to protect archives from bombardment, for a National Archives, and for inclusion of American Indian records in such archives rather than in ethnological and natural history museums—all of which eventually came to pass. Indeed, in authorizing the construction of a National Archives building, the Public Buildings Act of 1913 may have taken the wind out of Konta's sails; and it also affirmed that such an archive should be a public body. Furthermore, the MHRA expanded the time vessel beyond its centurial formula by conceiving new kinds of containers, fail-safe measures, and time spans.

Ultimately, the MHRA's most important contribution may have been a symbolic and unintended one. Future generations may question its archival selections, one critic conceded, but they "ought" to view the "effort in their behalf" as

> significant proof of the altruistic spirit now so prevalent in the common thought. They ought to wonder how so busy a people as we are could let our provident care run forward for a hundred or two hundred years. . . . They are likely to concede that we meant well. . . . Even the worthless things that it stores away will serve as a bridge of unselfish thought between the men of today and those who live later.[217]

SIX

Mausoleums of Civilization: Techno-Corporate Appropriations of the Time Vessel, 1925–1940

Seven months after Westinghouse Electric and Manufacturing Company buried its "Time Capsule" in Flushing Meadows Park, the 1939 New York World's Fair officially opened (**fig. 0.1**). The capsule was one of the most popular exhibits, reportedly visited by "thousands daily." Like mourners at an open-casket wake, large crowds lined up to pay their respects to the metal cylinder lying at rest in the "Immortal Well," awaiting its final sealing-in (**fig. 0.2**). Inside the Westinghouse Pavilion, visitors viewed a full-size replica, cut away to reveal its innards, alongside duplicates of the selected objects.[1] For those yet (or unable) to visit, this effort to communicate across the ages was simultaneously broadcast across the continent. It was featured in newsreels, magazines, and radio broadcasts about the fair and in Westinghouse's own promotional film, *The Middleton Family at the New York World's Fair.* "Boy," exclaimed young Bud Middleton from Indiana to the company man giving them a guided tour, "they weren't fooling when they made that capsule!"

The publicity surrounding the Westinghouse Time Capsule—together with the catchiness of the name, the recognizability of its gleaming, missile-like design, the impressive depth of its burial (fifty feet) and time span (five thou-

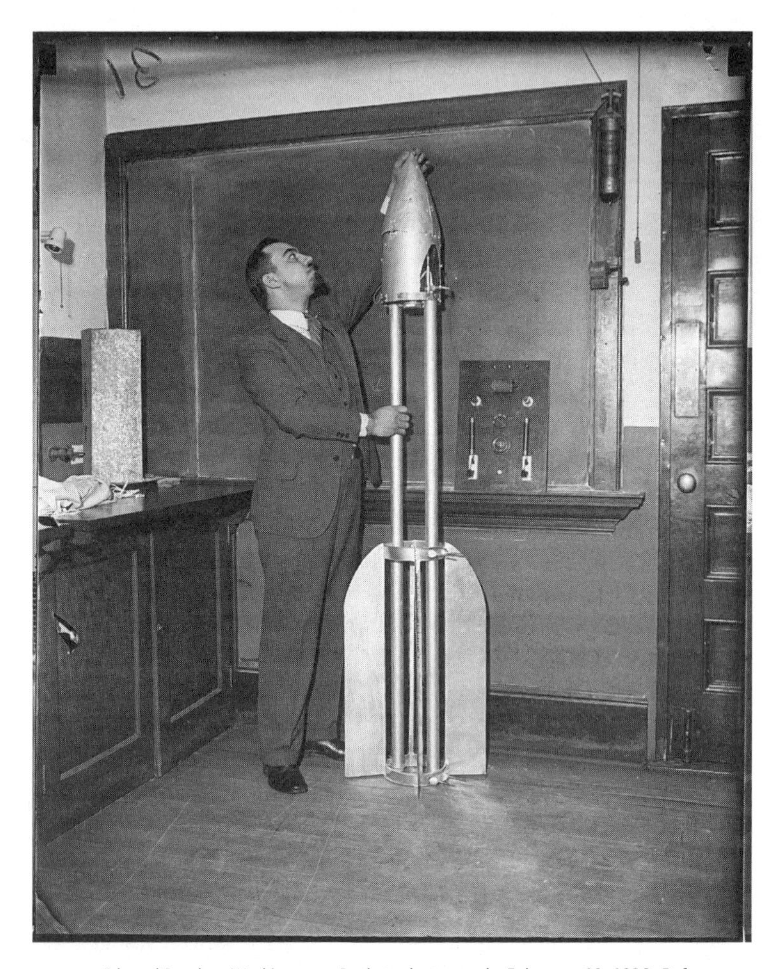

FIG. 6.1 *Edward Pendray Working on a Rocket*, photograph, February 19, 1932. Before conceiving and coordinating the Westinghouse Time Capsule, Pendray cofounded and led the American Interplanetary Society. Bettmann Collection, Getty Images.

sand years), and the dramatic spectacle of the sealing ceremony—has contributed to its misidentification as the inaugural and definitive effort to deposit a microcosm of the present for the future. Westinghouse vice president David Youngholm fostered this impression at the sealing ceremony by declaring it "unique in history," as did Westinghouse's public relations specialist G. Edward Pendray (**fig. 6.1**), who claimed to have dreamed it up on vacation.[2] Yet far from representing a new practice, the Westinghouse capsule marked the corporate and technocratic appropriation of a civic tradition that had thrived for more than six decades.

Westinghouse's publicity machine, in particular, stole the limelight from three other interwar projects. In 1925, after a brief downturn in time-vessel deposits, the once famous populist pamphleteer William ("Coin") Harvey began constructing on his estate of Monte Ne, Arkansas, a 130-foot concrete obelisk he called "the Pyramid," with three chambers for books and other objects (**fig. 6.2**).[3] In 1928, Duren

FIG. 6.2 W. H. Harvey's Pyramid as imagined on the cover of his pamphlet *The Pyramid Booklet: With Ten Illustrations* (Monte Ne, AR.: Pyramid Association, [1928?]).

J. H. Ward, a university lecturer, evolutionist, and Unitarian minister, purchased vaults in two Denver mausoleums and began filling them with thousands of cellophane-wrapped books and documents, which he christened Records to Future Ages. And in 1935, Thornwell Jacobs, Presbyterian minister and president of Atlanta's Oglethorpe University, announced his plans for a "Crypt of Civilization," a vast stockpile of media and artifacts sealed behind thick steel doors. In fact, Pendray's anecdote about his vacation epiphany is contradicted by evidence he had poached the idea from Jacobs, much to the latter's irritation. In his previous job as science editor of the *Literary Digest*, Pendray had penned one of the first articles about the crypt.[4]

Although Ward, Harvey, and Jacobs had divergent political and religious views, lived far apart, and (like Pendray) claimed sole credit for their idea—the projects shared certain features that distinguished them from earlier vessels.[5] One was their targeting of deep futures. Vessels with long-range time spans had been envisioned earlier—in written proposals such as Harrison's and Vail's, in fiction by Twain and England, and by Konta's association—but now they were implemented. Jacobs selected AD 8113 for the opening of his crypt, a date devised by adding to 1936 the number of years since 4241 BC, supposedly the first recorded date. (Pendray borrowed this idea of "standing at history's midpoint," albeit dating recorded history more conservatively as 5,000 years old, giving a target of AD 6939.)[6] Ward settled on AD 4000, with advance viewings every two hundred years, but he also wrote a series of "Letters to Future Ages" (inspired by Petrarch's) that specified a range of opening dates from 1999 to the year 10,000 and even 100,000.[7] Meanwhile, in the Ozarks, Harvey was thinking in a deeper, geomorphic time frame. He opted to construct the Pyramid in a valley overlooked by mountains that, according to his geological readings, were once as high as the Rockies and were continuing to erode. Their debris would eventually fill in that valley, but the Pyramid would be high enough to protrude, so that postapocalyptic explorers would spot the metal plaque near the top bearing the inscription, "When this can be read, go below and find a record of and the cause of the death of a former Civilization."[8] Harvey thus conceived a kind of geological time lock, one effectively set for around two million years' time. Such expanded time spans precluded the kinds of direct, embodied engagement with posterity fostered by the century chests.

These three vessels also betray a heightened fear of civilizational collapse, which had been only a minor theme of their predecessors. Working in the depths of the Great Depression, Harvey, Jacobs, and Ward believed, albeit for differing reasons, that this collapse was imminent. That fear is evident not only in their writings but also in the range and type of materials they set out to include; in their strategies for making them intelligible to those living in a postapocalyptic world; and in the construction materials and locations for their vaults. Westinghouse's choices, by contrast, implicitly denied such fears.

Moreover, whereas the deadline of a world's fair required Pendray to complete his capsule within five weeks of announcing it, the synoptic ambitions of these other projects of the 1930s necessitated longer gestations. Ward's Records for Future Ages took twelve years to compile, or a total of four decades from its initial conception in Iowa City around 1900 to its belated completion in March 1940. By 1933, Ward claimed to be dedicating ten hours a day and all his savings to it.[9] Jacobs also failed to deliver on time, sealing his crypt two months later after five years of active work on it and at least a further eight mulling over the idea, while the Pyramid—with its overambitious design—consumed Harvey's energies and finances for sixteen years until his death in 1936, by which time it was an aborted construction site.[10] As long-term commitments, these projects represented a further break with those earlier vessels that were executed within months as celebrations of some kind of anniversary—centennial, centurial, or otherwise.

At the same time, these three projects subtly anticipated Westinghouse's corporate co-optation of the time vessel. Their response to the threat of political, economic, and moral collapse was to reaffirm the redemptive promise of scientific and technological expertise. In part, this was pragmatic; the need to fund their vast projects impelled them to greater collaboration with corporations and scientists. But the rapprochement with the corporate-industrial system was largely ideologically driven by their fundamentalist faith in the scientific method. Their vessels—like Westinghouse's—not only showcased the latest industrial products but also acted themselves as demonstrations of new media technologies, new techniques of media preservation, and new systems of communication. They further exemplified the cult of corporate management in their organization. Rather than multivocal and collaborative products of a community, they were tightly controlled from above by various technical experts, all male. Instead of personal letters and confessions that could convey an embodied, affective, and obligational relationship to posterity, they were to transmit objective, historical "data" addressed to another expert, the "scientific" historian (even Ward's "letters to future ages" were essentially impersonal essays). More insidiously, each of them used their vessel to promote eugenics.

Evident in earlier vessels, that "science" was now articulated in more extreme ways. Together with Westinghouse, they thus promoted a vision of the ultimate apotheosis of science, technology, and technocracy. This abstract techno-utopia played a compensatory role during the Depression, foreclosing alternative futures.

This technocratic message appears at first sight to be contradicted by the democratic, almost populist, rhetoric of the 1930s vessels, especially the Oglethorpe crypt and the Westinghouse capsule. Yet their claims to represent ordinary life and popular culture in microcosm concealed a new nationalistic agenda. Their predecessors had certainly been conceived as microcosms, but of *smaller* worlds—a city, a college, a congregation, a social class, or a political elite—or, in Konta's case, of the *larger* domain of the supranational. Drawing on a holistic theory of "culture" from contemporary anthropology and on a chauvinist discourse of "civilization," the new vessels now hewed to a national framework. Like other projects of the period, from polls and social surveys to the founding of American studies, they advanced a consensus image of "American civilization." This agenda inclined the vessels to deny the nation's internal conflicts and to exclude the voices of women, ethnic minorities, and working people.

A "Glorious" Future: Corporate Visions and Public Relations

Westinghouse claimed its capsule's contents spanned the breadth of "modern civilization," but an informed and observant fairgoer might have detected a narrower theme. Many of the electric devices and synthetic materials—plastics such as Micarta, textiles such as glass fabric, and alloys such as Hipernik and Hipersil—were Westinghouse products. Further articles that did not bear its name had a Westinghouse connection. Some were made by subsidiaries such as Bryant Electric Company, which furnished the Mickey Mouse drinking cup. Others, like the Remington-Rand close shaver, contained Westinghouse components or materials. The capsule also included photographs of its factories, addresses and reports testifying to its beneficence to the public and to its employees, a copy of its stockholders' quarterly, and a biography of George Westinghouse himself.[11]

Even the container functioned as product placement. Cupaloy was a copper alloy Westinghouse had recently developed in its Pittsburgh laboratories and hoped would be widely adopted in industry. A five-thousand-year time capsule made out of it would promote its supposed strength and resistance to corrosion stemming from its incorporation of chromium and silver—the secret formula for which was included as a gift to posterity.[12] Youngholm boasted, against a more cautious plant manager's advice, that it could last fifteen thousand years "if need be."[13] With its striking, double-headed bullet shape (whose resemblance to an unexploded shell might in fact deter future archaeologists), the capsule was an ideal trademark to reproduce on promotional brochures. Three-dimensional replicas also appeared; one was exhibited in a model home in Virginia featuring Westinghouse electric fixtures. There were even plans, later scrapped, to sell Cupaloy strips as souvenirs to fairgoers.[14]

The self-referentiality of Westinghouse's capsule was no accident. Pendray conceived it from the outset as a publicity stunt. Appointed in 1936 as special assistant to president Frank Merrick, he was responsible for the company's advertising and public relations, including its pavilion at the fair.[15] In recruiting Pendray, Westinghouse aligned itself with a larger trend of the 1930s. Public relations—hitherto denigrated as superfluous or spurious—was now incorporated into the managerial structure of large companies.[16] Previously content merely to present themselves as legitimate, corporations now encouraged people to actively "love" them.[17] Westinghouse, in particular, embraced public relations to compete with its larger rival, General Electric. Despite decent sales figures, it was concerned about a reputation for being "stodgy" rather than "forward-looking," hence its appointment of the visionary Pendray.[18]

The public relations "craze" of the 1930s was also stimulated by businessmen's larger fears—induced by the growth of unions and an upsurge in strikes and violence—that the entire capitalist system remained in the balance.[19] One way in which Westinghouse hoped to save it was by promoting the benefits of mass consumption at the fair. The capsule represented a publicity opportunity for all kinds of companies, from cosmetics and fashion (Elizabeth Arden, Lilly Daché) to sporting and tobacco products (A. G. Spalding, Alfred Dunhill). So effective was it perceived to be as a commercial showcase that Pendray was flooded with requests from manufacturers wishing to be included, most of which had to be declined.[20] This promotional rationale led him to disregard the earlier principle of confidentiality and reveal the capsule's contents to the contemporaneous public by exhibiting the replicas and itemizing them in printed materials. It was also predicated on a denial of the unequal spending power of Americans, 17 percent of whom remained unemployed in 1939. "In our day," Westinghouse proclaimed in the pamphlet it published to memorialize its repository,

The Book of Record of the Time Capsule of Cupaloy, "there is hardly a man so poor he cannot afford [to purchase a motor car]"—a flagrant denial of the economic realities of that year.[21]

Westinghouse's vessel was hardly the first to showcase commercial products. Earlier ones had relied on donations, ranging from photograph and autograph albums to the receptacle itself. Commercial organizations themselves were beginning to embrace what had remained a civic practice; in 1930, the New York department store Bloomingdale's sealed a vessel for 2130 containing products such as a cocktail shaker and radio along with predictions contributed by Henry Ford, Calvin Coolidge, and Babe Ruth, among others.[22] Jacobs's Crypt of Civilization had taken this a step further. As the president of a small college during the Depression, he decided from the outset to draw on his corporate connections. His ledger book reveals that as many as four thousand companies and individuals contributed materials totaling about fifty thousand dollars. In return, they received a certificate they could use for promotional purposes, while the CEOs of the American Can Company (who provided the canisters) and American Rolling Mill (who provided the stainless steel door) were invited as dignitaries to the dedication ceremony.[23] But whereas Jacobs appeared to embrace corporate publicity as a means to complete his vessel, for Westinghouse it seemed to be an end in itself.

Westinghouse also sought to assuage economic fears by celebrating the achievements of science and technology. Along with electrical innovations and man-made materials, the capsule contained articles on recent discoveries in fields from mathematics to medicine. As described in *The Book of Record,* such advances had already brought about a new era of comfort and security by overcoming nature's vicissitudes. The "forces of the earth and skies" had been "harnessed," electricity "tamed," diseases "suppressed." It even proclaimed the abolition of hunger, pointing to the capsule's "samples of the products of our farms, where machinery has turned scarcity into abundance; where research has produced plants never seen in nature; where science now is able to produce plants even without soil." The capsule itself was a further instance of this subduing of nature. Just as new means of food preservation have "arrested the processes of decay," so would new techniques of archival preservation inhibit the decay of paper and other material. The five-thousand-year time span, although yet to be fulfilled, was presented as yet another scientific "achievement" of the age.[24]

In promoting mass consumption and technological innovation as palliatives to the ills of the Depression and the gathering clouds of war, the Westinghouse capsule chimed with the larger program of the New York World's Fair. Although progressives on the planning committee such as Lewis Mumford had originally pushed for a fair that would enlighten the public about social reform and planning, they were overruled by the businessmen, who envisioned a "consumer's fair" that would showcase an "American Way of Living."[25] This was to be done by giving corporations greater control over their exhibition spaces. Confined at earlier fairs to shared buildings such as Philadelphia's Main Building and Machinery Hall, they were now invited to construct their own. This allowed them to go beyond the display of individual products to craft a total, integrated environment that would project—on multiple levels, from the streamlined exterior facades to the interactive exhibits—the corporation's larger "image" to an awestruck public.[26] The proliferation of corporate buildings gave a greater prominence to science and technology, which now permeated the entire fair rather than being relegated to a single, dedicated building. The time vessel thus appeared in a very different exhibitionary context in 1939–1940 than it had in 1876. Whereas the Philadelphia organizers had placed Deihm's and Mosher's exhibits in a cultural setting—the Art Gallery and Photographic Hall, respectively—the New York fair subsumed the time capsule within a larger corporate-industrial complex.

The main strategy for reassuring the public was to show how corporations were mobilizing technology toward a utopian future. Fairs had always been occasions to proclaim the blessings of progress, but the ones staged in Chicago in 1933 and New York in 1939 were unprecedented in their portrayal of better living through machines.[27] The latter, billed "The World of Tomorrow," was dedicated entirely to such abstract, compensatory utopias. Walking through the grounds, visitors encountered multiple futures: from General Motors' Futurama, a large-scale model of the smoothly functioning national landscape that expressways would enable by 1960, to Henry Dreyfuss's Democracity, a diorama depicting the harmonious constellation of a skyscraper city and its satellite communities that regional planning would achieve by 2039. In evoking a much deeper future, Westinghouse had to acknowledge the "cycles of development, climax, and decay" that would probably operate during the intervening centuries.[28] Yet its prognostication was ultimately positive. By 6939—Westinghouse chairman A. W. Robertson predicted at the depositing of the capsule—the United States will have expanded into a "great community, including both Canada and Mexico, . . . society will be better ordered . . . the threat of perpetual war . . . long since . . . eliminated," while work will have become "the happy exercise of man's desire to create."[29] If southern regions have become too warm through climate change, he added, those "undesirable sections" would simply become "game reservations and great national parks."[30] In short, concurred *The Book of Record*, "the future will be glorious."[31]

This futurism was echoed in the period's science fiction. Hugo Gernsback's pulp magazines rose to prominence in the 1930s mainly through techno-utopian visions of flying cars, rockets, and automated homes and workplaces. This affinity led Gernsback to publicize the Westinghouse capsule both through a nonfiction feature and through the 1939 story "The Warning from the Past," in which a rust-proof metal "Time Capsule," buried underground by an unknown civilization fifteen thousand years earlier, provides information that enables the scientist-hero to repel an alien invasion (although, as we will see, Gernsback also published stories that were implicitly critical of Westinghouse's capsule). Pendray reciprocated by depositing an issue of *Amazing Stories* in the capsule.[32] The relationship in fact dated back several years. Under a pseudonym, Pendray had been a successful science fiction author, publishing in various Gernsback magazines.[33] As cofounder of the American Interplanetary Society in 1930, he was also an early advocate for the construction of actual rockets. Indeed, his prototypes may have inspired the projectile-like design of the capsule (**fig. 6.1**).

If the fair, the capsule, and the pulps envisioned the future as utopia, their utopias were effectively extrapolations of the techno-corporate order of the present. Instead of remaining open to contingency and alterity, they articulated closed futures. Just as Futurama presumed that the automobile would sustain its dominance and Democracity presumed that inner cities would continue to thrive even as suburbs grew, so did Westinghouse's *Book of Record* endorse the assumption that "material and social development" would ultimately prevail and "humanity . . . march onward." The capsule itself presumed the survival (or revival) of scientific techniques and instruments. As *The Book of Record* indicated, its recipients would require electronic and magnetic detectors to determine its exact location if twentieth-century maps with geodetic coordinates were no longer extant. After piercing the concrete and "Electroplast" seal, they would have to sink a caisson fifty feet down to keep out the soft mud and water—or else freeze the soil artificially—and then use a crane to lift the eight-hundred-pound capsule out (**fig. 6.3**). Next, they would need power tools to cut open the Cupaloy. And finally, they would have to obtain their own projector and record player to play the films and audio recordings, as the capsule was too small to include such apparatuses.[34] Pendray also retained a confidence in the capsule's semiotic self-sufficiency. Rather than connecting it to ongoing institutions or officials, as earlier vessels had been, he envisaged it as a solitary transmission, one that might become the only surviving trace of the twentieth century. This entailed a new conceit that such a missive could independently communicate complex ideas—

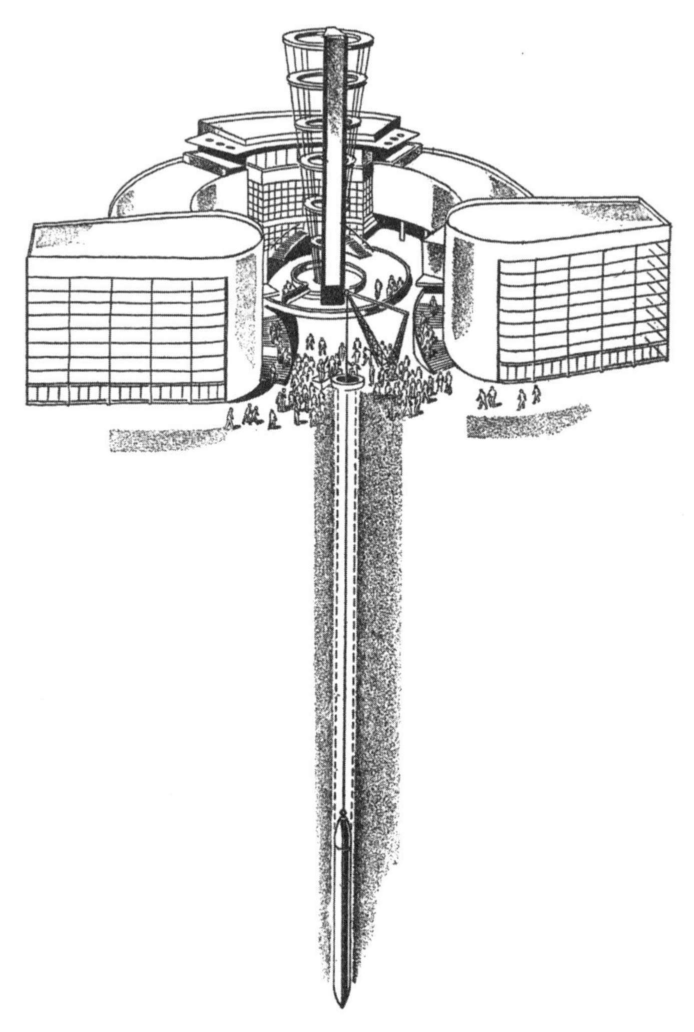

FIG. 6.3 Diagram of the Westinghouse Time Capsule. From *The Book of Record of the Time Capsule of Cupaloy* (New York: Westinghouse, 1938), 4.

and indeed, the very idea of a time capsule—across millennia, single-handedly overcoming all linguistic and cultural barriers.

Moreover, although Westinghouse presented its capsule as a noble gift to the future, its choice of location indicates a consideration only for its promotional value in the present. The company knowingly dug its Immortal Well in a former salt marsh that had been filled in with coal ash and garbage—hardly a stable environment for a long-range deposit. On top of that, the East Coast was sinking. Pendray reassured the public that the rate was "less than an inch a century," a misquoting of the US Coast and Geodetic Survey, which specified "less than a *foot* a century."[35] "There is no assurance," the magazine *Archaeology* opined after the fair, "that, five millennia hence, this part of Long Island, now twenty feet above sea level, will not lie beneath the Atlantic Ocean."[36]

Preparing for a Second Dark Age

Westinghouse's techno-utopianism departed from the futures projected by the other interwar vessels. Working for themselves rather than a corporation, Ward, Jacobs, and Harvey had free rein to express their deepest fears and misgivings. Undoubtedly, their own personal circumstances colored their outlook. All three were in old age and had serious health issues by the 1930s; Ward and Harvey were descending into blindness, and Jacobs and Harvey had almost died of illnesses. They had also suffered bereavements. Ward lost his wife of forty-two years midway through his scheme, while Harvey had earlier lost his son in a fire and had become estranged from his wife. Career disappointments and financial pressures may also have been factors. Never able to secure a professorship, Ward complained about his "life of struggle, hardship and small success."[37] Harvey wrote a further ten books but never repeated the triumph of his bimetallist treatise *Coin's Financial School* (1894), which had sold about a million copies and elevated him within the Democratic Party. After the crushing disappointment of the 1896 and 1900 elections, he retreated to the Ozarks to found a health resort, which soon fell on hard times. In 1932, he returned one final time to politics, founding the Liberty Party and running for president but winning just 53,425 votes.[38] The sense of having failed to produce a lasting legacy—exacerbated in Ward's case by the realization that "I shall leave no descendants" (or, indeed, graduate students) to perpetuate his memory—appears to have generated a morbid fear of death, then known as thanatophobia. Ward admitted that his mortality "has rested very heavily upon my mind for half a century" and that the "anxious strain" to exert a posthumous influence was a motivation for his vessel scheme.[39]

While venting personal anguish, however, they used their vessels to express larger forces they perceived to be threatening civilization. A fierce opponent of President Franklin Roosevelt, Jacobs considered the New Deal a particular threat. "The lower 51 percent," he explained to his readers in the year 8113, "is at present legally robbing our treasury led by a combination of Alcibiades and the Gracchi"—allusions to the Athenian aristocrat who became a notorious demagogue and to the Roman brothers who called for the redistribution of landed estates to plebeians and veterans. Federal programs were eroding not only "individual initiative" but also states' rights.[40] He also feared for the Anglo-Saxon race, predicting that, "if nothing is done about it, the United States will, in a few centuries, become a nation of quadroons ruled by an upper class of Jewish blood."[41] He detected a similar regression in morals and the arts, especially the popular arts.[42] But his greatest fear was war. A staunch isolationist, Jacobs argued that intervention in the Great War was a mistake encouraged by financial interests that were now threatening to ensnare America again. Premonitions that the next war would be even more catastrophic and "total" gripped interwar intellectuals on both sides of the Atlantic (Mumford included), contributing to a foreclosed sense of the future.[43] By the time he sealed the crypt on May 16, 1940, the day after the Wehrmacht pierced the Maginot Line, Jacobs had come to the conclusion—expressed in his message to the eighty-second century—that "the outlook is very dark . . . the old civilization seems to be perishing," and the crypt will ironically serve as its mausoleum. To convey those death throes, he included a recording of an "air raid over a European town."[44]

For Harvey, by contrast, the threat was largely economic. He had by now replaced his earlier arguments about the benefits of bimetallism with an apocalyptic theory that usury was leading civilization to its ruin. The perversion of money from "a medium of exchange" to a source of profit through interest, he warned in *Common Sense* (1920), has generated a host of disorders.

> The blood of civilization no longer flows unmolested thru the arteries and veins of the body politic. Public spirit is dying; opportunity monopolized; population congested in the cities; wastefulness and extravagance characterize the administration of public office; millions are suffering for the want of food; the absence of proper nourishment for a multitude of children bringing to us an anaemic race;

the people[,] ignorant as to whence comes their wrongs and suffering, are striking blindly, evidenced in race prejudice that is annually taking the lives of tens of thousands of people. If we, who are still in a position of semi-comfort, do not act successfully to correct the evil at work, civilization is doomed.[45]

Failure to abolish usury, in other words, would yield *"terrible"* results: *"further usurpation of power, the constitution overthrown . . . , a dictator enthroned . . . , and a reign of terror, anarchy and despotism!"* In the coming "Dark Age," moreover, there were will be no "undiscovered countries to which liberty-loving people [can] flee."[46] Harvey conceived the Pyramid, he explained in a promotional pamphlet, to alert contemporaries to these dangers and thereby avert ruin. Failing that, it would serve as civilization's cenotaph. With its copy of *Common Sense* (and his subsequent books), it would at least "inform and admonish [future civilizations] of the cause of the death of this one."[47]

As a historian of civilization (with a focus on the evolution of religion), Ward's pessimism may have derived from his readings of the intellectual prophets of doom who dominated the interwar years. His pamphlets cited, among others, H. G. Wells and the German philosopher of history Oswald Spengler, author of *Decline of the West*.[48] "Everyone who reads the stronger literature of the times knows the dangers of civilization," Ward wrote in 1934 to his former colleague at Iowa, political science professor Benjamin Shambaugh. "No words can ever describe the terrible calamity that a decline and fall of our present lofty stage would mean to the generations of the Future. . . . Slumping because of carelessness and indifference, into another longer stage of Dark Ages and Barbarism" would be an "eternal disgrace."[49] Given its advanced and complex nature, he expanded in a pamphlet, modern civilization's collapse would be a "far greater calamity than that of any ancient fall. . . . The total human loss would be beyond description."[50] Like Jacobs, Ward viewed the emergency in racial terms: "The White Race is facing its third Crisis," following those of the Fall of Rome and the Reformation.[51] And like Harvey, he offered his scheme as a last-ditch effort to rally a defense of civilization.

This cultural pessimism is evident not only in their rhetoric but also in the precautions they took for their vessels. Whereas Pendray settled for a low-lying, urban site, they chose higher, more remote locations. Ward proclaimed Denver, at the foot of the Rockies, "the safest place in America" to deposit records; Harvey considered the Pyramid immune from earthquake or volcano in the Ozarks; and Jacobs boasted about his crypt "resting on the granite bed rock of the Appalachian[s]."[52] Anticipation of social and political cataclysm led them also to eschew the practice of entrusting a metal box to a local college, library, or other institution. Instead, they opted for their own dedicated, full-sized vaults, thus fulfilling Konta's and Harrison's fantasy of entire, hermetically sealed and impregnable chambers full of records. Harvey's Pyramid was to have contained three vaults: a main room, sixteen feet square, supplemented by two smaller ones in the shaft in case future archaeologists inadvertently destroy the contents of the first with their dynamite. Surpassing the MHRA's faith in reinforced concrete, he called for eighteen thousand sacks of portland cement and eight-foot-thick concrete walls to render it indestructible.[53] Indeed, Harvey conceived the thickness of the Pyramid as a kind of civilizational time lock supplementing the geological one. "Before they can break into [it]," the new civilization, rising "slowly" from the "ashes" of ours, must have reached "a period when steel and dynamite have been discovered." They must also have developed an archaeological "appreciation of what they find [there]"—a rare acknowledgment of how such deposits were not universally intelligible but rather dependent on a historically variable sense of curiosity about the past.[54]

Ward also opted for multiple vaults embedded in concrete but economized by relying on two mausoleums already under construction in Denver by 1929. Fairmount Cemetery's neoclassical mausoleum and Crown Hill Cemetery's seven-story, neo-Gothic Tower of Memories (**fig. 6.4**)—both built in "Class A" fire- and earthquake-proof reinforced concrete, faced with "time-defying" granite—catered to a growing desire for aboveground burial so as to slow the corpse's decay. Through special "desiccating systems" of heating and ventilation, the deceased would be "far removed from the pagan precincts of damp earth burial." Ward purchased, at reduced rate, two family rooms, each containing several "air-tight crypts" behind a "huge bronze-grilled plate glass doo[r], held by double secret loc[k]" (**figs. 6.5, 6.6**). He thus gained ready-made time vessels in the only kinds of buildings that would not get torn down "for bigger buildings within 100 years."[55] Jacobs also saved money on his crypt by appropriating an old swimming pool in the basement of the campus' granite-hewn administration building, built to last "two to five thousand years" (**fig. 6.7**). Already waterproofed, the vault would be further safeguarded by pouring a concrete floor, lining the walls with vitreous porcelain enamel, and installing a stainless steel door that would be welded shut rather than fitted with hinges and locks (**fig. 6.8**).[56]

The vessels' inaccessibility and vast time spans necessitated measures to ensure they would not be completely forgotten. Science fiction

FIG. 6.4 The Tower of Memories, Crown Hill Cemetery, Denver, Colorado. Duren Ward considered this mausoleum ideal for preserving records as its crypts were constantly dehumidified and its concrete walls could purportedly withstand fire, hurricanes, and earthquakes. Photograph by author.

offered technological solutions: the ancient time capsule in "Warning from the Past" was designed to remain dormant until finally announcing its existence in a cornfield in central Illinois by automatically emitting smoke, a bright flashing light, and a radio-jamming signal.[57] But Jacobs and Ward opted instead to conscript living, generational memory. Evoking the golden cylinders in "The Last New Yorkers" and the facsimiles distributed by Konta, Jacobs issued "plaques" describing the crypt and its location (in multiple languages) to enduring institutions around the world, from a Masonic lodge in Washington, DC, to "strange monasteries hidden up in the shadows of Himalayan snows."

He also printed "several thousand" metal tickets for individual contributors, "to be handed down from father to son until the time for the opening of the Crypt, so that [its] memory . . . will be kept alive." The stainless steel plaque welded to the crypt's door, meanwhile, would alert subsequent generations of Oglethorpe students and staff to its presence.[58] Indeed, a member of the class of 1961 described the crypt as "an everyday fixture of our existence, its gleaming stainless steel door occupying one wall of the 'game room' where we danced after dinner to . . . [songs] on the juke box."[59]

Similarly, Ward hoped his vessels' location near the entrance of the mausoleums would keep them in the public eye. Their illumination by stain glass windows depicting mountain scenes, he added, "will stop the visitor, and the tablets with names and explanations will partly satisfy his curiosity" (**fig. 6.9**). His contract with the mausoleums, moreover, guaranteed his vessels "dignified maintenance forever."[60] Yet, for them to reach their long-term target, Ward realized (like Konta

FIG. 6.5 *A private room. One of the many designs of rooms in the "Tower of Memories."* At the discount rate of $5,000, Ward purchased a family room near the main entrance containing five vaults and several niches and storage boxes (*The Tower of Memories*, pamphlet [n.d.]). Illustration from *A Wonderful Tomb: Tower of Memories*, prospectus (Denver: Crown Hill Memorial Park, n.d.). Shambaugh Papers, Special Collections, University of Iowa Libraries.

but unlike Pendray) that a dedicated, self-perpetuating institution would need to watch over them. He thus founded The Far-Reaching Foundation toward Safer Civilization in 1931, under a managing board of three trustees—"young men of fine vigor, broad scholarship, and great interest in the welfare of humanity"—who were appointed for life, paid a small salary, and required to "devote" themselves to the

FIG. 6.6 Ward's time-vessel room in the Tower of Memories, Crown Hill Cemetery. His records lie ensconced behind an eight-foot bronze-grilled door with "double secret locks." Photograph by the author.

FIG. 6.7 View of the interior of the Crypt of Civilization before it was sealed for 6,173 years. Framed photograph [1940?]. Philip Weltner Library Archives, Oglethorpe University.

cause. He granted them "a reasonable discretion" to alter certain arrangements, as "conditions may greatly change in the course of time"; and, on their death, the vacancy was to be filled "without delay."[61] The foundation would also sponsor lectures every ten years, print and distribute free books every twenty, and explain the foundation's "origin and purpose" every forty.[62] These outreach activities would emanate from its yet to be built headquarters, an institute or "Temple to Civilization" (**fig. 6.10**).[63] The foundation, Ward told Shambaugh, "will stand a far better chance of security than do the state and municipal institutions, for they depend upon the maintenance of an exceedingly high degree of education and culture, and the continuity of political arrangements." It would also eclipse colleges, which "last two-hundred, three-hundred, or more years . . . [and] are limited to their city, state, or nation."[64]

Additional precautions betrayed even deeper concern for civilization. In denying that the crypt contains jewels or precious metals and imploring intervening generations not to open it, the plaque on its door conjures visions of a lawless future of looting and pillaging. One of Jacobs's collaborators warned that it should not contain anything

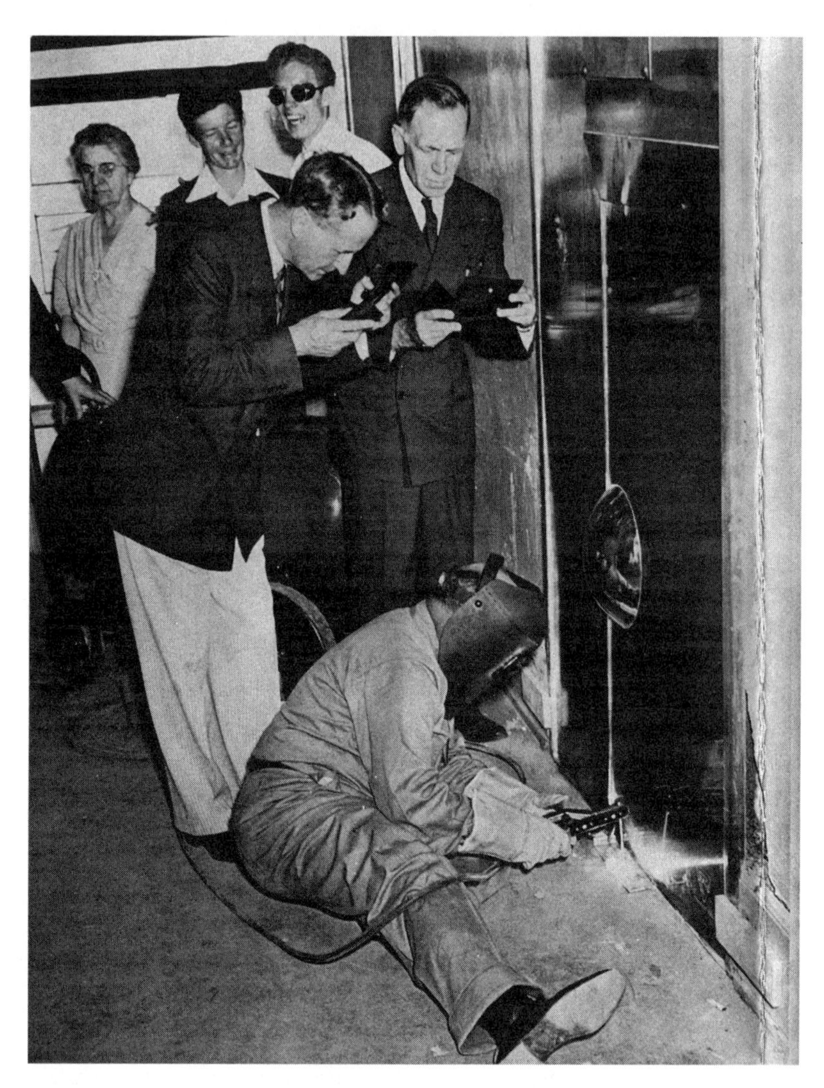

FIG. 6.8 Anonymous untitled photograph (May 25, 1940). A worker seals the Crypt of Civilization by welding the stainless steel, hingeless door shut. Oglethorpe University president Thornwell Jacobs (*left*) and his archivist Thomas Peters oversee the work through protective screens. Series 3, sub-Series 2 (Project—The Crypt of Civilization), box 4, folder 4, George Carlson Collection, 1913–1962, D. B. Dowd Modern Graphic History Library, Washington University in St. Louis.

FIG. 6.9 A. T. Von Lauter (artist), stained-glass window depicting Colorado's Snowmass Mountain in Duren Ward's time-vessel room in Fairmount Mausoleum, Denver (1932). Ward, who supplied the words, believed that this window, visible through the bronze-grill door, would "stir up [the visitor's] interest" (letter to Shambaughs, April 5, 1934, 3, Shambaugh Papers). Photograph by the author.

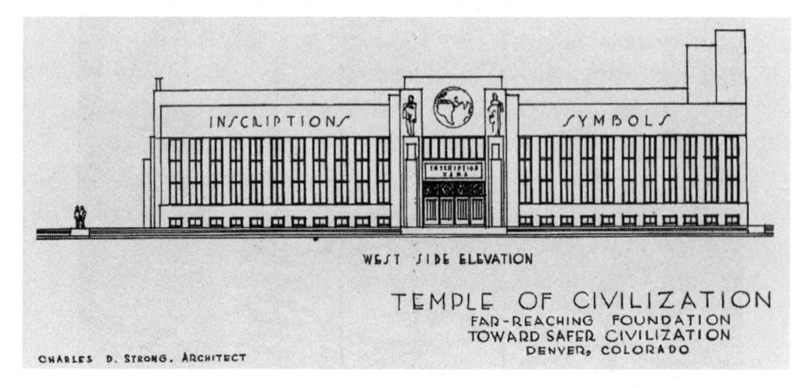

FIG. 6.10 *Temple of Civilization*, designed by the architect Charles D. Strong for the Far-Reaching Foundation toward Safer Civilization. These elevations were projected as stereopticon slides at the dedication of Ward's "Records to Future Ages." Illustration from Ward, *Temple to Civilization*, undated pamphlet. State Historical Society of Iowa, Iowa City.

of "practical value" either, or else it "would not survive the first period of anarchy immediately following the effects of the New Deal"—and that even decorative objects "would have value to an illiterate horde" if made of metal and glass.[65] Another obstacle was technological regression. Instead of assuming—as Louis Ehrich did in 1901—that future recipients would have their own record players, Jacobs provided a Victor machine. In case they had also lost all knowledge of electricity, he opted not for a motor-driven Victor (already popular by the 1920s) but for a hand-cranked model. He also threw in a hand-cranked microfilm reader and Mutoscope along with a wind-powered generator to power the movie projector.[66]

Pessimism about civilization manifested itself above all in the heightened attention to language. These depositors prepared for the likelihood not only that their recipients might speak a language other than English but also that their most learned archaeologists might have lost, through historical rupture, even the ability to read that ancient language. One solution to the problem of undecipherable words was to use Isotype (International System of Typographic Picture Education), a language that Viennese émigrés were then promoting in part to facilitate socialism and internationalism.[67] Thomas Kimmwood Peters, a filmmaker Jacobs appointed as both professor of visual education and archivist of the crypt, commissioned three professional artists to pictographically narrate changing technologies of illumination, communication, and power generation from prehistoric to modern times (**fig. 6.11**).[68] Similarly, the partial sphere adorning its door (**fig. 6.8**), evoking the universally recognizable shape of a full sphere, communicated visually to future archaeologists that there was something hidden within. Another solution was to teach those archaeologists English through some kind of modernized Rosetta stone. Like the old professor in Jack London's *Scarlet Plague*, Harvey promised to place in the center of the Pyramid's main vault, inside a glass container, "a key book to the English language."[69] Jacobs and Peters went further and developed what they called a Language Integrator (**fig. 6.12**; also in foreground of **fig. 6.7**). An old Mutoscope machine connected to a record player, it would show a man holding up an object and pronouncing its name while the word appeared beneath. This speaking dictionary, placed near the entrance, was programmed to deliver three thousand commonly used English words. Peters derived that list from Basic (British American Scientific International Commercial) English, a simplified language introduced in the late 1920s by British linguist Charles Kay Ogden, who, like advocates of other lingua francas such as Esperanto and indeed Isotype, hoped it would bring about world peace. Such hopes were tied to apocalyptic fears; in H. G. Wells's 1933 novel

FIG. 6.11 George L. Carlson, *untitled* [1938?]. This popular illustrator, already well known (especially in Atlanta) for his dust jacket for *Gone with the Wind* (1936), was commissioned to produce one of the pictographs for the interior of the Oglethorpe Crypt. In a series of eight images that were "fused" into its enamel walls, he used the new and supposedly universal language of Isotype (International System of Typographic Picture Education) to convey to distant viewers the history of human communication, from cave paintings, gongs, drums, and smoke signals to semaphore, telegraphy, telephony, and radio. An additional sixteen pictographs by two other artists conveyed further aspects of "the story of man's rise from primitive pre-historic times to the complexity of modern civilization." Ironically, the artists had to revert to what was essentially cave painting to make this point. George Carlson collection of George Hagenauer.

FIG. 6.12 *Six Thousand Years From Now Some Puzzled Invader of What Is Now the State of Georgia May Break Into the Vault With Its Machine to Save the English Language.* Illustration from undated newspaper clipping, Crypt of Civilization filing cabinet, folder 23, Philip Weltner Library Archives, Oglethorpe University Special Collections.

The Shape of Things to Come, Basic English helped to unify the world (and sharpen mental acuity) after its devastation by war, economic collapse, and plague.[70] It would also presumably reinforce the hegemony of Anglo-Saxon culture.

Toward the Scientific Millennium

For all their pessimism about civilization, Ward, Harvey, and Jacobs did share Westinghouse executives' fundamental faith in science and technology. All three believed that science was the one thing still, in

Jacobs's word, "progressing."[71] "In the immediate past twenty years," the latter's archivist declared, "more important discoveries have been made than in all the six thousand years that have gone before."[72] Peters's sentiments were echoed at the dedication of the crypt's steel door in 1938 by the president of the Radio Corporation of America, David Sarnoff, who contrasted the piecemeal, "empirical methods" of the nineteenth century with the theoretical leaps of the twentieth, which will make "man the master of infinite new sources of power."[73] Jacobs himself envisioned the people of 8113 benefiting from "the infinite treasuries of the atom and producing new species of living things at will by means of X-ray and radio-activity."[74] Thus, while the "immediate" outlook was "dark," the "long future" looked bright, as science would prevail. In what might be called a helical philosophy of history, he saw the human race advancing in spirals, "retreating often but eventually arriving at a higher point of achievement." The spiral may now be circling back but soon will "point forward."[75]

Indeed, Jacobs envisaged his archaeological offerings to the future not just as historical sources but also as practical tools that, like Basic English in Wells's novel (or the encyclopedias in England's "The Last New Yorkers"), would assist those who might be struggling after the collapse of civilization to reestablish science and industry. Anticipating the survivalist handbooks and doomsday prepper kits of our day, the crypt included exact scale models and blueprints of various machines of transportation, communication, and production so that (in Peters's words) they could "be reconstructed in case civilization has gone back to primitive times."[76] Notably absent, however, were technologies of destruction—omitted, perhaps, to avoid another downturn of the spiral. Concerns that such gifts might prove fatal to future generations—dramatized most famously in Walter Miller's 1959 nuclear fable *A Canticle for Leibowitz*—were already evident in Arthur Stangland's "The 35th Millennium," published in Gernsback's *Wonder Stories* in 1931. Among the "marvelous inventions" that the people of AD 5000 sealed for "posterity" in a large, subterranean concrete chamber in the Ozarks—and that the hero discovers thirty millennia later—are a set of photon ray guns and microfilmed instructions for manufacturing further ray guns and long-range biological weapons. These prompt the leaders of that declining remnant of ancient Americans to undertake a war of aggression, while other "infernal war machines" fall into the hands of their enemies, prompting the hero to wish "he had never found the . . . chamber."[77] Apparently oblivious to such dangers, Westinghouse, a defense contractor since 1915, did deposit newsreels of battleships, tanks, and other "war machines" to demonstrate American "military prowess."[78]

As for Ward, he ultimately resisted Spengler's and others' conclusion that the collapse of Western civilization was inevitable. He saw his life's work as the quest to determine what might *save* civilization from decline and fall, a quest initiated at an early age by reading Edward Gibbon, which instilled in him an intense despair at all those "centuries . . . lost from human progress."[79] Unlike ancient civilizations, he reassured Professor Shambaugh, we have a trump card, namely, "the Scientific Method."[80] He argued in his writings that since its birth around 1600, this empirical procedure has struggled against religious superstition and is only beginning to be applied to social and political issues. But "if we get out of this quagmire, if some movement of better and more scientific minds can begin to control human activity, an age of marvelous progress and splendor is not far ahead."[81] By the time his first letter to the future is opened in 2131, he hoped, an "Age of Science" will have dawned. Ward thus viewed the present troubles merely as symptoms of what might be an "Epoch" of "Transition" between "Tradition" and "Science"—a period of both "Confusion and Clearness . . . of Inventive Progress and Moral Decline."[82]

Ward's vessel was to ensure or even accelerate that transition by facilitating a scientific approach to history itself. No longer can civilization be taken "for granted," he wrote to Shambaugh. It must be "purposely developed" by engaging, through his foundation, in the "fundamental study" of the preconditions for its progress—an appeal that echoed Harvey's own claim that civilization depended on the "study of civilization as a science" and that earlier civilizations fell because of ignorance of the factors causing their decline.[83] Historiography has failed so far to guide civilization, Ward argued, because it has depended on incomplete and unintended sources and is thus "very inaccurate and uncertain."[84] Nor will cornerstone deposits (and, he might have added, some earlier vessels) be much use, as they transmit "mere local curiosity and gossip across perhaps a hundred years." His vessels, by contrast, will preserve "serious, well-chosen and classified data," enabling future historians to objectively measure the progress (or regress) of civilization across a long time span and to identify "relative stages."[85] (And if civilization did collapse, that traumatic event could at least be redeemed through a kind of postmortem performed by future historians with the aid of the time vessels' contents.) These measurements would, furthermore, "rene[w]" posterity's sense of purpose.[86] Alluding to a recent electrical innovation, Ward described his foundation as a

"Historico-Social Thermostat" designed to prevent the uncomfortable highs and lows of civilization, those "zeniths of progress and nadirs of degeneracy," by rekindling "Science, . . . the best warming material ever known, . . . at right intervals." With this automatic regulation installed, "progress would go on without limits," century after century.[87] This notion of history as a science that could save *all* science went beyond even Konta's paeans to objective history. Ward's emphasis on positivist "data" rather than subjective messages represented a further drift away from the kind of affective engagements with the amateur and local historians of the future found in earlier vessels.

Buoyed by their faith in science (historical science included), time-vessel architects wagered their entire projects on the continuance of the capitalist status quo. Through the magic of compound interest, Ward explained, a donation of just ten thousand dollars, invested at a 4 percent rate in his Long-Range Endowment—a scheme he modeled on Ben Franklin's 1790 trust for Philadelphia's and Boston's artisans of 1990—would yield ten billion dollars by 2351.[88] Even the antiusury militant Harvey embraced capitalist interest in declaring that the proceeds from the sale of his books would be deposited in a trust fund to pay for groundskeeping and caretaking of the Pyramid.[89] A similar failure to imagine an open, postcapitalist future is evident with time vessels of the period that incorporated monetary gifts to the future inhabitants of their town in the form of a fund invested (with interest) in local banks. Bloomingdale's 1930 vessel contained bankbooks registering a seventy-five-dollar deposit that would purportedly approach two million dollars by its target date of 2130.[90] Such investments not only foreclosed the future but also substituted a merely financial transaction—one involving minimal financial sacrifice—for deeper ethical duties to posterity.

Belief in scientific progress also pushed Ward and Jacobs to the limits of their religion. As a student at Harvard, Ward had embraced the "rational faith" of Unitarianism as a refuge from the revivalist "threats and promises" of his childhood. But his loyalty to the radical Unitarian Francis Ellingwood Abbot, who had been banished in 1867 for his rejection of all scriptural dogma and embrace of evolutionary science, put him at odds with his congregations and hindered his career as a minister.[91] He continued thereafter to criticize Christianity for clinging to anthropomorphic visions of God and outdated ethical codes.[92] Jacobs's repudiation of superstition and his vision of a religion based on the scientific laws of evolution had provoked similar tensions with his fellow Presbyterians. His refusal to ban an evolutionist textbook from campus led some to withdraw their bequests and blacklist his university as "a center of heresy."[93] Undeterred, he developed a course called Cosmic History, which applied new discoveries in archaeology, anthropology, biology, and other sciences to spiritual questions about the origin of life and the fate of the earth.[94] We might even view their vessels' deep time spans as further repudiation of religious beliefs in the end time. At the same time, both men drew on theological rhetoric to describe the coming Age of Science. For Ward, it would be the "salvation" of humanity; for Jacobs, a millennial paradise, with disease, murder, and war eradicated and humanity transfigured into the "divine."[95]

In Praise of Acetate

The interwar vessels—especially Oglethorpe's and Westinghouse's—were to testify to the progress toward that scientific millennium not simply through their content but also through the media and preservation technologies they employed to transmit it. While their predecessors had drawn on ancient as well as modern media, Jacobs and Pendray aligned themselves almost wholly with the latter.[96] Chief among these was the synthetic chemical compound cellulose acetate. By 1925, after a long quest for a cheaper, less flammable substitute for cellulose nitrate, manufacturers began introducing acetate products, ultimately including gramophone records and film. Although acetate was not adopted in movie theaters for another two decades, as it could melt under their high intensity arc lights, it was immediately used for the smaller format promotional films screened in schools and other public settings, and it was particularly suitable for a cold, dark time vessel.[97] As many as 250 motion pictures—from fiction and newsreel to industrial and governmental films—ended up in the crypt, along with more than fifty gramophone records (many of them acetate). Peters's prior career as cinematographer for Pathé, Kinetoscope, Hale's Tours, and several Hollywood studios, proved crucial, as did his prior experience as a film collector.[98]

The crypt and capsule also deployed acetate as a material base for photographically preserving books and documents, thus fulfilling what Konta had dreamed of doing with cumbersome glass plates. Peters was again instrumental, having earlier invented the first 35-millimeter microfilm camera.[99] Loading an improved model with a "special long life" acetate film, which he fancifully claimed would last "at least 6000 years" thanks to a "vaporate" process that gave it a protective coating, he now used it to undertake the long, painstaking process of miniaturizing more than 800 books, totaling 640,000 pages at 360 pages per hour. (Even at that rate, Peters still had female students "running three eight-hour shifts" to complete the task.)[100] As if in competition, Westinghouse amassed a "Micro-File" of ten million words, which would "take an ordinary person more than a year to read," presuming they could endure the eyestrain of reading them through the enclosed microscope.[101] Peters and Pendray thus aligned themselves with the growing movement among librarians and archivists in the 1930s to microfilm their collections and sometimes even destroy the originals. Even critics such as Lewis Mumford endorsed this agenda. In *Technics and Civilization* (1934), he celebrated these "new forms of permanent record" (photography, film, and their offshoots) for their potential to reduce the "bulky" old forms (buildings, monuments, texts, etc.) to more compact, economical formats that would not impede "the free development of a different life in the same place. . . . It is no longer necessary to keep vast middens of material in order to have contact, in the mind, with the forms and expressions of the past."[102] Thus, whereas the MHRA had advocated microphotography as part of a broader program of archival preservation and devised time vessels to publicize that program, its successors deployed microfilm and vessels as technologies for rendering printed matter disposable.

The Crypt of Civilization served as a laboratory for preservation techniques involving new metal alloys as well as plastics. In case the acetate films and microfilms failed to survive five millennia, Peters drew on another of his inventions, metal film. Using a process he kept secret, he printed backups of some sound films, photographs, and texts onto thin, nickel strips. Unlike traditional photoengraving, which imprints images onto coated metal surfaces and thus remains subject to decay, he etched them directly into the metal, allegedly rendering them permanent. As opaque images, they required a specially devised projector that used mirrors to display their reflection (**fig. 6.13**).[103]

The crypt also pioneered new techniques of long-term media storage. In addition to a conventional copper strongbox for printed materials and glass bell jars for the figurines, stainless steel cylinders (double lined with asbestos and glass) were to hold the microfilm, with nickel foil between the acetate rolls (**fig. 6.14**). Stainless steel shelves were installed to accommodate these vessels-within-a-vessel. Peters, whose patented inventions included vacuum and rare gas devices, found a way to replace the air in those cylinders and bell jars—and in the crypt itself—with the inert gas helium (another idea copied by Westinghouse, albeit with nitrogen).[104] In fact, that idea had already occurred to Konta and indeed Harvey, whose admiration for modern technology—in particular, reinforced concrete and automobile highways—belies the stereotype of populists as luddites. The latter's plan was to enlist a glass factory to hermetically vacuum seal his books in heavy glass containers and to remove air also from the Pyramid's vaults.[105] These technologi-

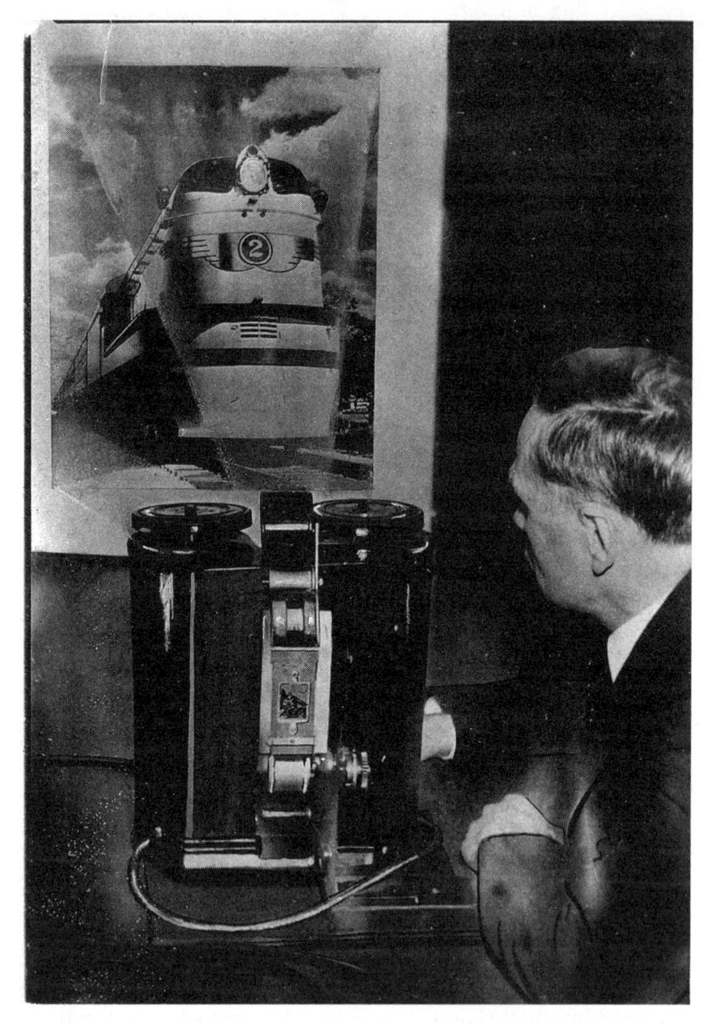

FIG. 6.13 "The Method by Which the Metal Pictures Are Projected." Peters demonstrating the "opaque type projector" he designed to screen his metal film. Illustration from T. K. Peters, "The Story of the Crypt of Civilization," *Bulletin of Oglethorpe University* 25, no. 1 (1940).

FIG. 6.14 "Glass, Metal, and Asbestos Receptacles." Illustration from T. K. Peters, "The Story of the Crypt of Civilization," *Bulletin of Oglethorpe University* 25, no. 1 (1940).

cal innovations were not merely means of conveyance. The crypt's ultimate meaning, Peters suggested, resided less in its contents than in its demonstration of breakthroughs in recording and preserving, which now allowed the present to be fully captured for posterity, thus proving modern civilization's superiority. As one set of pictographic images on the wall suggested, the crypt represented the culmination of an evolutionary trend (one endorsed, with some reservations, by Mumford) from the constraints of "primitive" communication to the long-range media of "modern civilization" (**fig. 6.11**). The medium, to borrow a phrase, *was* the message—a point Peters reinforced by depositing a film he had made about the making of the crypt.[106]

The Cult of the Expert: Technocracy and the Postdemocratic Future

Besides showcasing advances in science, such materials and procedures more specifically upheld the authority of the scientific "expert." Each of these time-vessel creators emphasized how specialists were supervising aspects of their project. For technical issues, Pendray turned to engineers, geologists, and chemists, including Kodak's research director, and for suggestions on content, he relied on "archaeologists, historians, and authorities in virtually every field of science, medicine and the arts."[107] Jacobs closely collaborated with the owner and editors of *Scientific American*, while Harvey claimed he would enlist a "paper expert" from New York to endorse his methods.[108] Both the Westinghouse and Oglethorpe vessels, moreover, boasted a stamp of approval from C. G. Weber, a researcher at the US Bureau of Standards who specialized in the long-term stability of motion picture film.[109] Ward, meanwhile, anticipated the endorsement of future scientific experts, in particular those representing the American Association for the Advancement of Science (AAAS). A proud fellow of that society, Ward charged it with overseeing the vaults' opening every two hundred years, reprinting the records, adding others, and, at the thousand-year point, "reconsider[ing]" the project's future. He empowered the AAAS also to grant researchers access to the vaults at other times.[110] Like Pendray, who invited the AAAS president, Albert Blakeslee, to speak at the capsule sealing, Ward clearly realized the prestige such an association could confer in the present. He introduced his idea at the AAAS annual meeting in 1932, securing a unanimous vote of approval from its council, and in 1937 he gave its officers a guided tour of the mausoleums.[111] Notably, while women had played crucial roles with the earlier time vessels, every one of these experts was male. Conscription of scientific experts served not just to legitimize but also to defeminize the time vessel.

Far from limiting scientific expertise to technical issues, however, these time-vessel architects believed that expertise should be brought to bear on the social and governmental realm. The heralding of scientists as society's saviors had reached its zenith in the early 1930s with Technocracy Inc., founded by a former engineer, Howard Scott. Yet although that movement had split and largely faded by 1936, these vessels suggest that the dream lived on.[112] Jacobs pictured the people of 8113 living under a government "administered by experts" as "medicine and dentistry and law" are already—and, in case they were not, he intended to deposit *ABC of Technocracy* for their perusal.[113] In his address at the Westinghouse sealing, Blakeslee similarly predicted that "governments will depend increasingly on expert knowledge—that we shall seek information before legislation rather than the reverse."[114] Indeed, the New York World's Fair was—on a larger scale—an exposition of what could be achieved through technocratic collaboration between science, engineering, corporations, and government.[115] The most stridently technocratic of time-vessel architects was Ward. Since the 1920s, he had published warnings that civilization could be saved only by scientists organized into "directing committees."[116] In the program for a symposium he organized in 1934 to promote his scheme—with Edward A. Ross, the famous sociologist and early advocate of technocracy, as keynote speaker—he again declared that politicians "solve no problems," that "institutions . . . are not keeping pace with the findings of social investigators," and that scientists alone can halt the "slump."[117] Several other prominent technocrats also endorsed his scheme, including sociologist William G. Ogburn and the president of the Carnegie Institution for Science, John C. Merriam.[118]

The politics of this diverse movement were complex. Technocracy did not exclude liberal viewpoints. It had emerged in part out of the writings of leading progressive reformers and intellectuals such as Thorstein Veblen and appealed to socialists such as H. G. Wells and Albert Einstein.[119] (The latter submitted a message for the Westinghouse capsule that appeared to call for technocratic planning of the "production and distribution of commodities.")[120] Ward himself espoused a variety of progressive causes in his writings, from women's rights, the protection of labor, and the municipalization of the economy to the elimination of business interests from politics and the introduction of direct democracy.[121] This was reflected in his proviso that the vessel was not for "the special benefit of any [future] individual or class" and was not to be "exploited or copyrighted"; it was for the people.[122] He also hoped, like Ehrich, that his scheme would instill among his own contemporaries a duty to posterity and in particular an ability to "tilt the mirror of [our] imagination" and "conceive our problems as Future Ages would look at them"—a method he termed *futuring*.[123]

Nevertheless, during the 1930s many reactionaries embraced technocracy as an alternative to democracy. Technocracy Inc.'s refusal to participate in the political process—along with such paramilitary accessories as the gray uniform, ancient symbols, and public saluting of their leader—smacked of fascism.[124] Ward appears to have been drawn to this aspect too. He denounced radicals, warned that the proletariat (which remains suspicious of science) will "wreak ruin," and declared that democracy had failed.[125] Jacobs strayed even farther from the political mainstream. In a 1936 address, he lamented the republic's degeneration into a democracy and urged the professionalization of politics and the disenfranchisement of all recipients of federal or state assistance. The lecture was widely condemned in the press, with the *Miami Herald* dismissing it as "a strange brand of Americanism, subversive as Fascism and Communism." Indeed, Jacobs publicly endorsed the fascist ideas of the French biologist Alexis Carrel, including the notion that "25% of our population simply cannot think" and therefore should not be allowed to vote.[126] Jacobs even treasured a newspaper clipping about his crypt that had been cut from the Nazi mouthpiece, *Göttinger Tageblatt*.

These antidemocratic tendencies were reflected in the administering of the vessels. Unlike their antecedents, they were not envisaged as open collaborations with the public. Contributions to Ward's vaults were by invitation only. He sent his prospectus to those (like Shambaugh) he considered "best informed regarding facts and literature within their . . . fields," requesting copies of their books, along with a brief biography and portrait.[127] And although public input was solicited for the Oglethorpe crypt and Westinghouse capsule, it was a pretense to gain exposure. *Scientific American* allowed Jacobs to ask its readers for suggestions but only in a "cautiously worded manner because we are bound to receive large numbers of crank letters which we do not wish to handle."[128] Perhaps for the same reason, Westinghouse opted for the less popular magazine, *Science* (published by the AAAS), placing a call there for the "thoughts" of "scientific men."[129] Even when inundated with letters, both Jacobs and Pendray reserved the right to choose what to deposit.[130] The associations that Harvey and Ward incorporated to raise funds for their vessels provided a similarly false impression of equality. The Pyramid Association was evidently a single-handed operation, while the locals who served on the Far-Reaching Foundation's "voluntary committee" did so under Ward's presidency. "I have always expected to work alone," the latter confessed, "and do my utmost to carry out these 'outstanding ideas,' because of the difficulty in getting them understood."[131]

The fanciful titles of these associations, moreover, were arguably a facade concealing their creators' ultimate goals of promoting and memorializing themselves. While self-interest also motivated earlier vessels, it remained in tension with civic rhetoric and popular participation. That tension had largely evaporated by the 1930s. A prime function of Ward's, Harvey's, and Jacobs's projects was to disseminate their own writings. All three had authored numerous books and pamphlets (with Ward's numbering at least seventy-one) and had established their own presses to publish them.[132] While these publications had limited circulation, their authors apparently hoped that, by depositing them in their vessels (and requesting they be reprinted), they would win posthumous acclaim. The assertion that a given book was to be sealed for posterity was also a useful sales tactic in the present, one that Harvey employed on his frontispieces.[133] Self-aggrandizement was also evident in Ward's decision to entomb himself, his wife, and close collaborators alongside his records—something even Harvey denied any intention of doing. Whereas the "attention" given a professor entombed "in an Iowa City cemetery" would gradually fade, he told Shambaugh, the interest given someone who founded such a scheme would "increase . . . with every decade and every century."[134]

Even at the events for these vessels, popular involvement was limited. The consecration of Ward's vaults, lowering of the Westinghouse capsule, dedication of the crypt door, and final sealing of the capsule and crypt (**figs. 6.8, 6.15**) were not local, participatory events but rather spectacles transmitted by mass media to national audiences. Paramount Studios filmed all three projects for newsreels that were

FIG. 6.15 Anthropologist Clark Wissler (*left*) and Westinghouse vice president David S. Youngholm pouring Electropast (a petroleum-based sealant concocted by Westinghouse) into the "Immortal Well," thus sealing the time capsule. Photograph, September 23, 1940. New York World's Fair 1939–1940 records, MssCol 2233, Manuscripts and Archives Division, New York Public Library.

dispatched across the country and abroad (to fifty million people, in Ward's estimate), while NBC broadcast the crypt dedication and the Westinghouse sealing live on the first nationwide radio network, in the process promoting the latter's line of transistor radios.[135] Even at a world's fair noted for interactive exhibits, the capsule engaged its audiences not as active collaborators but as passive spectators, peering through the mediating lens of a periscope. Indeed, public relations officers viewed these fairgoers as an aggregate audience analogous to those of mass media, and they defined success in terms of attendance figures.[136] The few present at the sealing of the hole, moreover, were denied the earlier concession of adding last-minute offerings because the capsule was already closed. Even more undemocratically, Ward appears to have sealed his vaults in 1936 without any ceremony.[137]

Bolstering the Aryan Race

These men held out hope for one particular field of expertise, namely, eugenics. Their predecessors, especially Mosher and Ehrich, had invoked this pseudoscience in vaguely expressed hopes that the "healthy" could be encouraged to reproduce (positive eugenics) and the "unfit" discouraged (negative eugenics). The 1930s time vessels, however, followed a trend toward the involuntary sterilization of the latter. "In our day we are doing much more to help the fit to be fitter," Ward wrote in one of the pamphlets promoting his vessels, "but nothing yet to hinder the unfit from swamping improvement." Writing to the future offered him a means to indulge his fantasy: "Within two generations society will not allow the lascivious, the passionate, the uncontrolled to reproduce," thus saving itself from "absolute destruction."[138] In Harvey's sketch of a utopia "Twenty Years Hence," he imagined the advances toward "a better race" made by this movement, now universally accepted and enshrined in laws (his utopia also featured a National Institution of Science that would find cures for diseases by performing "dangerous experiments" on prisoners).[139] Even Westinghouse's chairman, Robertson, exploited the capsule ceremony to promote eugenics by imagining it being opened by the "race of supermen and superwomen" that "the principle of breeding" will have achieved by 6939; in that world, he added, "the abnormal will have no place."[140] Such beliefs may even have inspired the capsule's contoured shape; streamlining was the architectural and design style that 1930s eugenicists employed to convey visions of a purified future.[141]

The organizers of the transmillennial vessels touted eugenics as a cure-all for a range of social problems. Ward and Harvey suggested it would reduce crime (the former advocating the sterilization of all "hereditary and incorrigible individuals *before* they have committed crime"), while Jacobs offered it as a solution to poverty. The latter predicted that there will have been "many New Deals" by 8113, but only "a world-wide adoption of the principles of eugenics" will "eliminat[e]" the "ill-clad, ill-housed and ill-fed."[142] Jacobs's and Ward's advocacy of eugenics contradicts historians who assume it found little support among ministers.[143]

Ward, Jacobs, and Harvey also appear to have been among those who embraced eugenics above all to bolster Anglo-Saxon supremacy against ethnic others. Even after the enactment of national quota laws in 1924, demanded for so long by the eugenicist-influenced Immigration Restriction League, Ward continued to warn of threats to his fellow "Nordics." In 1932 he was writing a book titled *Making America Thorobred*, which would show how the nation's "racial stock," already contaminated by slave importation, had been further tainted by immigration from eastern Europe and Asia, requiring "the application of Eugenical measures, including widespread sterilization"—a similar argument to that of Ross's *The Old World in the New* (1914).[144] Jacobs borrowed the title of another racial eugenicist book—Madison Grant's *The Passing of the Great Race* (1916)—to describe how those immigrants were overthrowing the "White, Protestant, Nordic-Aryan republic," spreading their "political sores, infections, and diseases" and taking over Northern cities.[145] Harvey appears to have shared these views, judging by his repeated citations of William Swinton's *Outlines of the World's History*, whose guiding assumption was that Caucasians are "the only truly *historical* race."[146] Xenophobic views even pervaded Pendray's science fiction. His "Yellow Peril" novel *Earth-Tube* (1929) narrated an Asian invasion of the Americas via a tunnel of indestructible metal they had dug through the earth.[147]

All three men reserved special opprobrium for Jewish immigrants. Harvey tapped into the populist variant of anti-Semitism, particularly in his 1894 novel *A Tale of Two Nations*, in which the financier Baron Rothe (a caricature of Rothschild) conspires from London to control the US Congress. Although the historian Richard Hofstadter considered Harvey's Judeophobia secondary to his Anglophobia, it became more virulent by the 1920s, especially in his writings on usury.[148] Ward's anti-Semitism, by contrast, derived from his conviction that Judaic ethics were impeding social solutions such as eugenics. In one letter to the future, he expressed his hope that "modern Nordics" would break free of that "ancient Semitic view."[149] Jacobs railed even more aggressively about Jewish "propaganda" and influence. In a 1938 editorial in *The Georgian,* he attributed their persecution in Germany to (among other things) their "ability to accumulate power and wealth." This resulted in the termination of his column, which prompted him to issue further denunciations of Jewish advertisers' power to silence their critics—although he continued to accept donations from financier Bernard Baruch and publicity from RCA president David Sarnoff.[150]

For Jacobs, WASP dominance in the South was imperiled not only by Jewish (and Catholic) immigrants but also by blacks. The "worst blow" in the nation's history, he wrote in his diary, came from "the pen of its worst presidential blunderer" in the form of the Emancipation Proclamation.[151] He expressed his nostalgia for the plantation era in his 1940 novel *Red Lanterns on St. Michaels,* a tale of Southern romance and heroism set against the backdrop of General Sherman's attack on Charleston, which he deposited in the crypt alongside a movie script (donated by David O. Selznick) for *Gone with the Wind.* His vision of the South's future was one from which blacks had been banished. He believed in encouraging the Great Migration because it was ridding the South of a "burden" on "our white civilization," and he urged black leaders to advocate repatriation to Africa as the only "solution" to the "race problem."[152] But with the outbreak of war, he now feared a further expansion of federal government that would undermine state's rights and impose racial equality on the South—"a denial of racial differences as absurd as a denial of differences among horses or dogs or anthropoid apes."[153] These beliefs were reinforced by Peters's designation of Confederate landmarks—the carvings on Stone Mountain (a model of which was in the crypt) and the Civil War battlefield of Kennesaw Mountain—as triangulation points allowing future explorers to locate the crypt.[154]

Pendray's views on race can be gleaned from "A Rescue from Jupiter," serialized in *Science Wonder Stories* in 1930. The nationalization of industry by socialist governments in the twenty-second century, while dissolving class distinctions and ushering in a "golden age" of "peace and plenty" for all, had exhausted all fossil fuels by the seventy-sixth century. The subsequent discovery of free atomic power by splitting water—and the weaponization of that technology—resulted in the earth's dehydration. Without clouds to protect them from the merciless sun, humans all became dark skinned within "four or five generations," which in turn prompted a valorization of blackness, with people resorting to "artificial stimulation and cosmetics" to darken their skin. Although Pendray appears to be questioning essentialist notions of races as fixed and distinct, he was in fact reviving an older geographical theory of environmental determinism, which adduced climate and topography to shore up racial categories and hierarchies. Such hierarchies are evident throughout the solar system. Whereas Jupiter's greater gravity and heat have bred a "brutal and selfish" race, those who inhabit its lighter moons (though descended from the same stock) are white, "slender and refined"—differences reflected in their respective music, art, and architecture. The light-skinned Jupiterians ultimately rescue the last two inhabitants left on earth thanks to their ability to translate the message the latter deposited in a metal-alloy cylinder placed in the mechanical hand of a black figure erected as a "last monument" to humanity (**fig. 6.16**). As they depart the dead earth, the Jupiterians reassure the earthlings that their moon will soon "make you white as we are."[155]

Time vessels themselves supposedly proved white progress and superiority. This logic, implicit in Konta's scheme, was now rendered explicit. According to Ward, concern for the future could only have developed at a higher stage of mental and social evolution. "Space is primary to human experience. It is an early faculty," he wrote, whereas temporal consciousness comes later.[156] Applying this evolutionary framework, he noted the "limited chronologies" of ancient civilizations, the gradual emergence of an interest in the past, and finally a growing consideration for the future, culminating in his idea of preserving data for future ages. He thus discounted the Pharaohs' tomb building as motivated not by "the good of the Future, but [by] royal honor and certain salvation in some hazy after-death existence." Furthermore, Egyptians were "racially and geographically outside the stream of the conquering Aryan career, so that their strange experience and example have been solemn curiosities, of small value in solving human life problems."[157] Boasts about the invention of the time capsule thus instantiated claims of white supremacy.

Lipstick and Underwear: Encapsulating "American Civilization"

These vessels' reactionary tendencies—their privileging of scientific and eugenic expertise, their curtailment of popular participation, their disparagement of racial and ethnic minorities—appear at odds with

FIG. 6.16 Frank R. Paul, illustration from Gawain Edwards [Edward Pendray], "A Rescue from Jupiter," *Science Wonder Stories* 1, no. 9 (1930): 774. The light-skinned hero from one of Jupiter's moons retrieves the metal-alloy cylinder from earth's last monument, thereby triggering an "automatic mechanism" causing the black sculptural figure to disappear back into the structure. James L. "Rusty" Hevelin Collection, Special Collections, University of Iowa Libraries.

their inclusive, democratic, even populist rhetoric about encompassing American civilization in the aggregate. Earlier vessels, while often claiming to be microcosmic, had prioritized the names and faces of politicians and professionals, or had focused on the narrow social community of a college or spa town or of a city's middle class. Even Konta, who embraced the New History, produced little more than autographed and cinematographed testimony of leading figures. The 1930s vessels, however, subordinated celebrity testimony and middle-class experience to the task of encapsulating American society as a whole.[158]

Ward, Jacobs, and Peters claimed that this striving for a complete "cross section" of the present—a term all three used—was inspired by their sense of the partial, biased nature of the archaeological record.[159] While researching ancient civilizations, they had each been struck by the dearth of sources about the lives of their common folk, which prompted concerns about a potential erasure of ordinary experience of the 1930s.[160] The need for a fuller account of "our age" drove Ward to document not just scientific inquiries but also social conditions, yielding a total of 2,500 books supplemented with pictures, charts, and tables—an expansion of the cornerstone tradition by a factor of a thousand, as he put it.[161] Through these materials, he wrote, "we here lay open to you the whole life of our day. . . . We render confession of our whole Civilization. . . . The whole collection will form, in compact and classified space, a synopsis of the life of our time."[162] Jacobs similarly described the crypt as an "index to our civilization," documenting our "life, manners, and customs," while Pendray vowed to embrace the full spectrum of "our thought, activity, and accomplishment"; indeed, in yet another plagiarism, the latter initially named the capsule "A Cross-Section of Our Time."[163] Whereas the centurial vessels were modest in their microcosmic tendencies, often acknowledging the inherent impossibility of their task or the inevitable need for selectivity, the 1930s vessels aspired to a more totalizing, all-encompassing microcosm, giving rise to a sheer exhaustive accumulation of books and pages.

This epistemological hubris also engendered lists reminiscent of Jorge Luis Borges's absurd Chinese encyclopedia. Pendray reduced contemporary civilization to a mere fifteen categories, ranging from "Where We Live and Work" to "Our Religions and Philosophies," while Ward needed thirty-six so as to include "Our Crimes" and "Our Domestic Animals."[164] To cover such a dizzying range of topics, they resorted to "general information" publications that might be considered microcosmic in their own right: the *World Almanac for 1938*, synoptic histories of the United States and the world, Sears Roebuck catalogs, and patent office inventories. Foremost among these was the *Encyclopaedia Britannica* (14th edition), which historian James Harvey Robinson described, in his entry on civilization, as a complete survey—a capsule, as it were—in "some thirty-five million words" of "what mankind has hitherto done and said." In a masterful shortcut, Pendray took care of sixty-three of his subtopics by simply microfilming the relevant essays from the *Britannica*.[165] Ward and Harvey resisted such shortcuts, dedicating extensive time and money to the Sisyphean task.

Since their inception in eighteenth-century France, encyclopedias have been inspired by civilization's perceived fragility. Denis Diderot viewed his as a hedge against "some catastrophe so great as to suspend the progress of science and interrupt the labors of craftsmen, and plunge a portion of our hemisphere in darkness once again."[166] Yet the intensified encyclopedic impulse of the interwar years was a symptom of, or even counterdiscourse to, the specter of total war, which engendered fears of wholesale loss of knowledge.[167] In *The Shape of Things to Come*, Wells envisioned a future "Encyclopaedic organization," based in Barcelona and staffed with seventeen million workers, that "accumulates, sorts, keeps in order and renders available everything that is known," thus creating a "collective Brain" or "Memory of Mankind."[168] On the eve of Pearl Harbor, Isaac Asimov expanded the idea into an *Encyclopedia Galactica*, a "universal compendium of knowledge" compiled by thousands of scientists in a remote corner of the galaxy, safe from "the turmoil of the Empire."[169] The shadow of future aerial warfare also stimulated visions of universal libraries (another Enlightenment ideal rendered absurd by Borges), now preserved through some kind of civilizational ark. The rocket that evacuates 103 people from earth before its destruction in Philip Wylie and Edwin Balmer's *When Worlds Collide* (1933) contains a "complete library" of books that usefully double up as "insulating material."[170]

This frenzied spirit of completism prompted the interwar time vessels to engage, even more fully than their predecessors, with material culture. The size of the vaults permitted larger objects and a larger number of them, with Harvey envisaging a collection of contemporaneous domestic and industrial items ranging "from the size of a needle and safety-pin up to a Victrola."[171] Yet there was also a qualitative shift away from genteel, bourgeois material culture (lace collars, silver spoons, silk fans, etc.) and an expansion of the commitment (initiated by earlier vessels) to the generic and mass produced. Westinghouse included about thirty-five "articles of common use," grouped under the categories of "convenience, comfort, health, safety" (e.g., eyeglasses);

children's toys; games; and fashion or vanity items (**fig. 6.17**).[172] The crypt offered an even broader range: domestic and business appliances (e.g., automatic toaster, Royal typewriter), toys (e.g., Lincoln logs), popular food and drink (e.g., chewing gum, beer), a ladies' purse filled with typical contents (e.g., lipstick, keys, hairpins), and a complete set of men's and women's clothing, including underwear.[173] To illustrate how, and by whom, the clothes were worn, Peters supplied miniature dolls representing people from "every walk of life and in various trades and occupations" dressed in the "appropriate costumes"—an idea Frederic Harrison had earlier suggested (**fig. 6.7**). These were inspired by the ushabti, figurines of domestic slaves that leading Egyptians included in their tombs in the belief that they would come alive in the afterlife, inadvertently providing rare testimony about the lower classes.[174]

The embrace of the commonplace was also evident in the vessels' catholic approach to arts and amusements. Their predecessors were, in varying degrees, culturally elitist; Detroit's mayor had commissioned essays on art, theater, classical music, and opera but none on the city's prominent ragtime and vaudeville scenes. The crypt, in contrast, gave

FIG. 6.17 "Items to go into time capsule." New York World's Fair 1939–1940 records, MssCol 2233, Manuscripts and Archives Division, New York Public Library.

equal (if not greater) emphasis to the "popular arts"—which had recently gained legitimacy as objects of intellectual scrutiny—as to the "high" arts ("poetical masterpieces," reproductions of "famous" paintings and sculptures, and "great" orchestral scores). Despite Jacobs's complaints that the mass media had "poisoned" the nation's "life blood" and spread Jewish propaganda (or perhaps to illustrate the point), his crypt preserved recordings of the Marx Brothers, Popeye (Fleischer Studios), and jazz musician Artie Shaw (born Arshawsky) along with Bing Crosby and entire radio programs such as NBC's *Lucky Strike Hour*, thus bottling samples of a notoriously ephemeral medium.[175] Westinghouse threw in further pop culture highlights, such as Slim Gaillard's hit song "Flat-Foot Floogee," Mickey Mouse cartoons, and an issue of *True Confessions*.[176] Following classic studies such as Johan Huizinga's *Homo Ludens: A Study of the Play Element in Culture* (1938) and Constance Rourke's *American Humor* (1931), both vessels emphasized popular leisure activities too. Instead of the genteel tea parties, charity balls, and needlework documented in Colorado's century chest, they preserved newsreels of football and baseball games, mass-produced board games, joke books, and other commercialized amusements.[177] Furthermore, they conceived of newsreel footage not simply as records of events, as Konta envisaged, but as documents of American "customs" and of "How we Appear, Talk and Act."[178]

Behind this celebration of the popular and the ordinary, however, lurked hidden agendas. One was a cultural nationalism that supplanted the localism of the centurial vessels and the internationalism of Konta's. "We are preserving records of American civilization only," Jacobs declared. "We can speak with authority about that; we cannot presume to estimate the culture of other nations. They will have to provide their own crypts—and perhaps, after this, they will."[179] American life was presented, moreover, as a complex yet distinctive and homogeneous whole—or, to use a concept then gaining favor among anthropologists, as a *culture*. Although E. B. Tylor had proposed that usage of the word in 1871, it was not fully established in American anthropology until the interwar years in the work of Franz Boas's students (such as Edward Sapir and Ruth Benedict) and others such as Yale professor and American Museum of Natural History curator Clark Wissler.[180] In fact, Wissler served as Pendray's chief advisor for the capsule and appears to have shaped it in multiple ways. He brought in four other ethnologists and anthropologists as consultants, which helped to overcome other experts' insistence that it should focus only on scientific knowledge and should disregard popular culture because no one in 6939 "will care about Howard Hughes or college football or fashion shows or Donald Duck or women's hats."[181] In letters to Pendray, Wissler also urged greater attention to the cultural realms of school and home. Yet he influenced the capsule, above all, through his conception of cultures as discrete and holistic. In *Man and Culture* (1923), he had introduced the idea of the "culture area" circumscribed by certain geographical barriers. While all humans, with their innate "biological equipment," share a "universal" "culture pattern" consisting of nine general categories (speech, material traits, art, knowledge, religion, family and social systems, property, government, and war)—a list that overlaps with Pendray's—each culture possesses its own distinctive pattern (or configuration) of "trait-complexes" shaped by the factors of environment and race and by processes of diffusion. Also a leading eugenicist, Wissler used the culture concept to shore up racial distinctions.[182]

In turn, Pendray's vessel (like Jacobs's) contributed to the diffusion of the culture concept beyond anthropology and even academia. Already, sociologists Robert and Helen Lynd had employed it in their best-selling *Middletown: A Study in American Culture* (1929), to which Wissler had written the foreword, and which was deposited in the crypt.[183] While the Lynds retained a critical perspective, other Americans—including some of the founders of the discipline of American Studies—now embraced the concept as a means to distinguish their "way of life" from that of totalitarian countries and to assert its resilience in the face of economic depression. As the historian Warren Susman showed, the popularization (or "acculturation") of the idea of culture in the 1930s gave rise to a "vast literature . . . [and] vast body of films, recordings, and paintings . . . that described and defined every aspect of American life." The transmillennial vessels thus paralleled broader efforts—from surveys such as *Recent Social Trends in the United States* and George Gallup's samplings of public opinion to various competitions, including one at the 1939 fair, to identify the "most typical" American family—to determine the national average.[184]

For all the appearance of statistical impartiality, social discriminations were built into these summations of American life. The competition at the 1939 fair failed to shortlist a single black family for its prize, while the Lynds excluded blacks from their study.[185] So, too, were the vessels racially exclusive. The only apparent acknowledgment of nonwhites was a "Negro doll" in the crypt and the Gaillard record in the capsule. Jacobs's choice of a white musician (Artie Shaw) to represent jazz, moreover, was symptomatic of how the growing acceptance of that genre was predicated on what Gunther Schuller called its "sanitization."[186] They also denied the gender, ethnic, and class conflicts within the nation in accordance with the notion of culture as holistic and consensual. Unlike earlier vessels, they neglected to document labor unions (then approaching their apogee) or women's struggles. Surely, "the work being done by women to advance the position of women," the former chairwoman of the National Woman's Party Jane Norman Smith wrote to Jacobs, "would be as interesting 6000 years hence as chewing gum and canned peaches!"[187] Rose Arnold Powell, a tireless campaigner for the commensurate memorialization of women's rights leaders, was similarly incensed that Pendray had not asked a single "woman . . . to make [a] statement on [the] civilization of our day" and urged him (in vain) to contact octogenarian Carrie Chapman Catt.[188]

These 1930s vessels can be further differentiated from more critical studies of American culture (such as the Lynds's) by their preference for the term *civilization* rather than *culture*. Although sometimes used interchangeably, the two had distinct, even opposite, connotations. In the first volume of his best-selling series, *The Story of Civilization* (1935), historian Will Durant, who endorsed Ward's scheme, was typical in defining civilization against culture. Whereas culture existed even in Paleolithic times, civilization emerged only in certain regions and was a product of certain economic and social conditions, in particular urbanization, and it retained its etymological associations with the political status of *civitas* (belonging to the city) and the moral status of *civilité* (or refinement). It also implied a linear narrative of progress unfolding from East to West—the kind of evolutionary *telos* from which Boas hoped the culture concept would liberate anthropologists.[189] To be sure, as Robinson emphasized in his *Encyclopaedia Britannica* essay on the subject, a less normative and more pluralistic definition of *civilization*—as that which is not biologically inherited—had emerged, rendering it applicable to nonurbanized peoples; indeed, the MHRA had applied the term to American Indians. But he, too, ultimately reinforced its dominant associations with material, scientific, and technological progress advanced by an intellectual elite and encapsulated by the encyclopedia itself.[190] Both Robinson and Durant viewed civilization as a positive force, whereas those alarmed by the spread of "artificial," "industrial," or "machine" civilization looked to "genuine," "organic," and "spiritual" culture generated by artists or by the people.[191]

The vessels' bias toward civilization rather than culture was evident not only in their founders' repeated usage of that term but also in their contents. Eschewing the handcrafted artifacts that emerging folk-culture scholars (many of them female) were analyzing, they selected only the latest commercial products. Those products were not preused but shipped straight from the company—with Woolworth supplying the crypt with a "complete set of samples" of its wares, along with their prices.[192] No attempts were made to document the actual uses people made of them or the subjective meanings they projected onto them. Instead, they were offered as evidence of the positive influence of manufacturing advances on national life. Westinghouse, for instance, deposited "specimens of modern cosmetics" to show how American women became "the most beautiful, most intelligent, and best groomed of all the ages."[193] The progress of civilization was also signified through the sheer quantity of stuff. At the capsule sealing, Wissler referred to the objects as "baggage" and claimed that human progress can be viewed in terms of a measurable increase in it. In a subsequent article in *Natural History*, he asserted a hierarchy of peoples based on their "average baggage load," with Australian Aborigines at the bottom, Plains Indians (with their domesticated "draught animals") at "intermediate levels," and modern civilization at the top. Indeed, our increased load was in turn stimulating even greater demand for "mechanical devices to carry and house" it.[194]

Burnt Film and Torn Stockings: Countervisions of the Capsule

The technocratic and corporate appropriation of the time vessel, however, did not go unchallenged. Some criticisms came from celebrities invited to speak or contribute. At the Westinghouse capsule's sealing-in, Gene Tunney, the retired heavyweight boxing champion and—perhaps unbeknown to Pendray—admirer of the socialist George Bernard Shaw, spoke out against the American Dream of a "white-collar job" and predicted "greater respect and appreciation 5,000 years hence for the mechanic."[195] The vessels' ideological message was also contradicted by some of their contents. Two of the nine paintings chosen by Westinghouse to epitomize the arts were by artists with socialist affiliations. Pablo Picasso was represented by his recent *Guernica* and Otto Dix by his expressionist portrait *Dr. Mayer-Hermann* (1926), which transforms electrical lights and metal instruments of the kind produced by Westinghouse into the stuff of nightmares (**fig. 6.18**). Another choice, Charles Sheeler's iconic precisionist view of Ford's River Rouge plant, *American Landscape* (1930), while often assumed to be sympathetic to industrial modernity, has also been viewed as a critique of a system

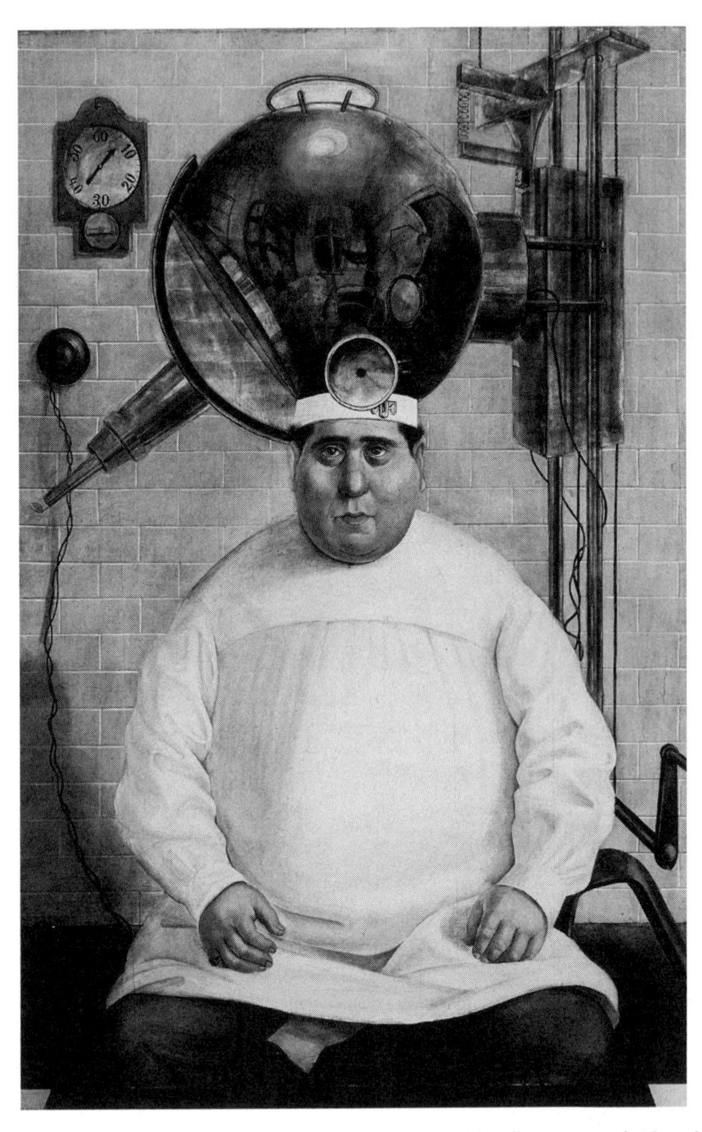

FIG. 6.18 Otto Dix, *Dr. Mayer-Hermann*, oil and tempera with collage on wood, 58″ × 39″, 1926. A reproduction of this portrait exemplifying the harsh realism of the *Neue Sachlichkeit* (New Objectivity) movement was deposited in the Westinghouse Time Capsule. Museum of Modern Art, New York, Gift of Philip Johnson.

that was devaluing and displacing humanity.[196] The most unambiguous countervoice in the capsule, though, was Thomas Mann's. Having emigrated to the United States on the eve of the fair, the German novelist, socialist, and outspoken antifascist was selected, along with the physicists Albert Einstein and Robert Millikan, to submit a brief message to the future. Mann began by denouncing "the idea of the future as a 'better world'"—the premise of the entire fair—as "a fallacy of the doctrine of progress." He also contradicted Westinghouse's technocratic creed by issuing a defense of cultural humanism and saluting those "brothers of the future [who are] united with us . . . in this endeavor."[197]

There were also criticisms of the vessels from without. Some were minor, questioning the location or precautions, or taking issue with certain selections; one letter in the *New York Times* complained that the choice of Picasso, among others, would give a "monstrously distorted idea . . . of our day!"[198] Others were more sustained and far-reaching, most notably "The Time Capsule," published in *Astonishing Stories* in 1941 (**fig. 6.19**). Written pseudonymously by Roger Hoar, a Massachusetts state senator and prominent campaigner for unemployment benefits and women's suffrage, this short story challenged the techno-utopian optimism of both Westinghouse and Gernsback's other authors by narrating a dismal future of relentless barbarism and warfare. Charting the unanticipated twists of that open future, it also suggested multiple ways in which the Westinghouse deposit might *fail* to benefit posterity. Its site became not a beacon of scientific reason but a "place of worship" for a primitive tribe of "beast men" attracted by the mysterious monument installed there after the fair (a granite marker was actually installed later that year). A repressive oligarchy that conquered New York circa 2900 commandeered the adjoining "Time Museum," which permanently housed the replica of the capsule and its contents, and thereby restored industry, along with "rigidly defined" "social and economic distinctions." That regime was supplanted by a new civilization that enslaved the beast-men and then by another thousand years of barbarism. Finally, a "pitiful little remnant" of whites from the west coast arrives in Flushing on the appointed date—their "handmade leather garments" and "crude tools" puncturing the Westinghouse chairman's eugenicist dream of a "better race." Lacking the "scientific curiosity" of "archaeologists," they believe the capsule is a magical force, a "chrysalis" that would appear by itself and release the "Spirit of Civilization." When it fails to do so, they start digging. After "many days" of relentless labor, they retrieve and eventually manage to open

FIG. 6.19 The reception, in AD 6938, of the Westinghouse Time Capsule by a "group of white savages" armed with "crude stone axes." Illustration from Ralph Farley [Roger Sherman Hoar], "The Time Capsule," *Astonishing Stories* 2, no. 4 (1941): 26. Courtesy: Special Collections Research Center, Syracuse University Libraries.

the capsule yet are unable to "recognize" a "single object." Particularly mysterious are the "long strips of narrow silvery tape," one of which catches fire. Believing the smoke represents the long-awaited "Spirit of Civilization," they burn the remainder on their campfire "to the accompaniment of appropriate incantations by their priests." Civilization, the story bleakly concludes, "had returned for a brief glimpse of the world of man, before deserting it forever."[199]

Corporate co-optation of time vessels, moreover, did not foreclose their reclamation for more public-minded ends. In 1938, the New York Public Library issued a call for anonymous New Yorkers of all classes and occupations, including "subway motormen [and] housewives," to submit their diaries to their manuscript division. "Thousands of New Yorkers," the director wrote, "have the diary-keeping habit. Yet they modestly leave the final deposition of their records of experiences to chance." He offered them the option of having their memoirs sealed for a specified number of years, thus replicating the time capsule in all but name. Publicizing the initiative, the pulp magazine *Argosy* declared it "adequate compensation" for "those millions of humble citizens who didn't get in to the Time Capsule at the World's Fair (unless by way of the New York telephone book)," thus foregrounding the library's implicit critique of how Westinghouse privileged the scientific expert and literary celebrity.[200]

Westinghouse's scheme was also subject to communist appropriation. While denouncing the Time Capsule for publicity mongering and for including "only 'bourgeois' concepts of what would be valuable and interesting 5,000 years hence," the official mouthpiece of the Soviet government, *Izvestia*, grasped its underlying political potential. It announced that a "bigger and better" one, recording "the development of Socialism and Communism," would be placed in the fountain of the Palace of the Soviets, the colossal structure designed—but never completed—in the center of Moscow. There was also a plan to duplicate the capsule in other parts of Russia.[201]

The corporate embrace of the time vessel, furthermore, did not preclude the dispatching of transtemporal messages by *individuals*. On the contrary, E. L. Doctorow's semiautobiographical novel *World's Fair* (1985) shows how the Westinghouse's capsule might have *stimulated* such personal efforts. His alter ego, Bronx schoolboy Edgar Altschuler, had been captivated by the gleaming cylinder since seeing its burial in a Movietone newsreel. When he sees the actual vessel shortly before the fair closed, as the runner-up of a competition to find the Typical American Boy, it loses some of its gloss as his father exposes its ideological biases, raising the question of why there was "nothing in the capsule about the great immigrations that had brought Jewish and Italian and Irish people to America or nothing to represent the point of view of the workingman. 'There is no hint from the stuff they included that America has a serious intellectual life, or Indians on reservations or Negroes who suffer from race prejudice. Why is that?'" Yet, despite—or, indeed, because of—these troubling absences, Edgar resolves to make his own version. In the novel's final scene, he fabricates a capsule not from new high-tech alloys but from recycled scraps: a cardboard mailing tube lined with tinfoil "methodically collected from the insides of cigarette packs and gum wrappers." Rather than filling it with impersonal, brand-new products, he offers up some of his own well-used and highly prized possessions: his Tom Mix decoder badge, a harmonica handed down from his older brother, and a rocket ship whose paint had worn off along with a handwritten essay about Franklin Roosevelt he had written for school. Whereas Westinghouse emphasized the durability of American textiles, Edgar submits one of his mother's torn silk stockings, rescued from the trash. Dispensing with the fanfare of gongs and speeches or fancy stone markers, he and his friend simply bury the tube in Claremont Park and then stomp on the ground to seal it up, scattering "some leaves and crumbs of dirt . . . for camouflage."[202]

SEVEN

Breaking the Seal: The Vicissitudes of Transtemporal Communication

In the hopes and visions of their depositors, time vessels would awaken to dramatic scenes of jubilation and gratitude. Anna Deihm imagined the Century Safe's opening as the climax of the bicentennial celebrations; Kansas Citians pictured the "expressions of wonder" on historians' faces as "they rummage" through their box; and Thornwell Jacobs described "prominent men from all over the world" convening in the eighty-second century to "witness the breaking of the [crypt's] seal" while "radio-newspaper headlines" and "television sight-and-sound receivers" trumpet this "story of international importance and significance" to all humankind.[1]

Utopian and science fiction may have fueled such fantasies of reception. In Alvarado Fuller's 1890 time-travel novel, *A.D. 2000*, the US president summons his entire cabinet for the opening of the late nineteenth-century time vessel, and afterward they adjourn to the War Department to "carefully read" its messages.[2] Even though only two people are present to open the vessel in George Allan England's "The Last New Yorkers," they make up for it with expressions of quasi-religious enthusiasm. "An air of mystery, of long expectancy seemed brooding everywhere; it seemed almost as though the spirit of the past were waiting to receive them . . . 'Waiting!' breathed [Allan] . . . , 'all these long centuries—for us! For you, Beatrice, for me! And we are here, at last!'"[3] Some characters find the anticipation psychologically unendurable. In a *Munsey's* story from 1894, the narrator—an aging American professor—has come into the possession of a bronze box sealed in 1389 by a forebear in England and handed down from father to first-born son. As the quincentennial target date approaches, he experiences insomnia, rapid aging, an inability to lecture, hallucinations, and ultimately insanity culminating in his consignment to an asylum.[4]

Needless to say, the fate of the vessels in this book—and of time capsules in general—has hardly comported with such fantasies. Far from being treasured through the generations, their temporal passage has often been marked by neglect and indifference. As we saw, Henry Cogswell's gift to San Francisco encountered verbal and physical abuse until 1904, when it was judged an obstruction and removed to Washington Square Park. Deihm's Century Safe also suffered the ignominy of removal. Its sealing in Statuary Hall in 1879 provoked a senator to immediately introduce a legislative provision prohibiting the exhibiting of any "work of art or manufacture other than the property of the United States" in the Capitol's prominent spaces. As Congress had not formally accepted Deihm's offering, the safe was banished to the "inferior location" of a hallway, while President Garfield's promise to restore it to Statuary Hall was nullified by an assassin's bullet.[5] Subsequently relegated to an even more remote spot beneath the steps of the east entrance, it silently witnessed the inaugurations of several presidents, each failing to add their signature to its album. In that spot, it was exposed to the elements, causing the ironwork to grow rusty and pockmarked, the paintings, inscriptions, and ornaments to erode, and the handle and dial mechanism to break—in contrast to Fuller's fictional safe, which remained "in as good condition" as when it was sealed.[6]

Such neglect and indifference can precipitate a worse problem, namely, the mislaying and forgetting of time vessels altogether. Deihm's was rediscovered in 1962 and relocated to a dusty storage closet to await the bicentennial "amid ladders and cartons of light bulbs," but others have been less fortunate.[7] The sixty-year bronze chest sealed by the Lower Wall Street Business Men's Association in 1914 and entrusted to the New-York Historical Society fell into oblivion as a result of being uncataloged and later consigned to offsite art storage in Chelsea, and thus it missed its date with destiny. While that vessel was eventually rediscovered in the late 1990s and belatedly opened (by myself and others) in 2014, others remain missing.[8] My efforts to track down the twin hundred-year vessels that the Modern Historic Records Association sealed in 1913 have proved fruitless. As for Konta's prototype vessel, the New York Public Library confirmed receipt in its 1913 bulletin but is now unable to locate it (so, too, the collection of diaries it solicited in 1938).[9] A similar fate befell the iron box dispatched on the centennial of Lyndon, Vermont, in 1891; its bicentennial recipients searched in vain through the town vault, the bank, and the public library and began to wonder whether it had ever been deposited.[10] In addition, there are unknown unknowns: those numberless time capsules whose very existence (let alone location) has been forgotten. Of the tens of thousands of capsules buried across the United States, the vast majority—according to the International Time Capsule Society, which offers to keep a registry of new deposits—are either lost or forgotten.[11]

Time vessels are prone to undershoot as well as overshoot their target date. In embedding his Memorial Safe in the lobby of Chicago's city hall, Charles Mosher failed to account not only for mayoral forgetfulness (there was no ceremony to mark the quarter century in 1901) but also for accelerated architectural obsolescence. When the building was demolished in 1908, the safe was opened and its contents removed and photographed for the newspapers. (Kansas City's century box was similarly endangered when Convention Hall was demolished in 1936, only to be safely installed in the new auditorium later that year, behind a metal plate).[12] Perhaps realizing their mistake, Chicago's city officials resealed Mosher's albums in a large trunk and delivered them to the new city hall three years later.[13] But in 1915, a "curious clerk" succumbed to the temptation to reopen it, at which point it was decided that the photographs were "of no use or value while in the vault" and were handed over to the Chicago Historical Society.[14] The Memorial Safe's reception was thus premature and unauthorized—far from the grand, ceremonial opening Mosher envisaged. Other vessels have fallen prey to theft. During the installation of air conditioning in the Paris Opera House in 1989, it was discovered that the basement vault—not due to be opened for several more years—had been breached and the gramophone and several of the records stolen.[15]

Despite these hazards, most of the vessels described in this book did fulfill their mission and reached their target date intact. The half-century box in which the Ancient and Honorable Artillery Company had embedded their one-hundred-year vessel was opened at Faneuil Hall on the specified day in 1930—possibly the earliest on-time reception of a time capsule.[16] Deihm's Century Safe was reinstalled in Statuary Hall for President Gerald Ford to open in the presence of members of Congress and guests on July 1, 1976 (**fig. 7.1**). Three years later, San Francisco officials invited mayor Dianne Feinstein to officiate the opening of Cogswell's P.O. Box. In the meantime, they called in a crane to lift Ben Franklin's "battered and bullet holed" pewter body—his collarbone "cracked" by "numerous attempts to lasso him from his perch"—to reveal the hundred-pound lead casket, thus dispelling rumors that thieves had stolen it from the pedestal during the 1904 relocation.[17] Other successful vessels of the period include the bottle sealed by Civil War veterans in Chicago's Palmer House Hotel in 1879 (opened 1979), the century box in Faneuil Hall (1881–1980), and the Cleveland centennial box (1896–1996). Each of the centurial vessels was delivered on time, too, as was the Oklahoma City century chest (1913–2013). And in 2010, the first volume of Mark Twain's autobiographical manuscript was published exactly a hundred years after his death, having been sequestered in the Bancroft Library as he had instructed.[18] These successes might be attributed to their consignment to enduring institutions, calling into question our later tendency to bury them in the ground. But luck was also involved. The Mount Holyoke box had been

FIG. 7.1 William Fitz-Patrick, untitled photograph (print from Tri-X panchromatic negative, July 1, 1976). President Gerald R. Ford removes the album, "Photographs of Our Great American People," from Deihm's Century Safe after its relocation to the US Capitol's Statuary Hall, Washington, DC. Architect of the Capitol George S. White stands to his side, while congressmen, staffers, and guests look on from the wings. Courtesy Gerald R. Ford Library.

lying forgotten in the college archives, its inscription shrouded in dust, until a librarian happened on it in 1988.[19] Duren Ward's mausoleum vaults in Denver were also apparently forgotten until my own discovery of them. Now, for better or worse, they remain on course—with the Oglethorpe and Westinghouse projects—for more remote destinations.

Although these openings were well attended and widely reported, certain unexpected complications did arise. Vessels that had been soldered rather than screwed shut were difficult to crack. Mount Holyoke's groundskeepers had to come to the rescue with their gardening tools, while several men spent hours working on the Faneuil Hall box with "hammers, chisels, and a drill."[20] Those with locks presented their own problems. Chicago's mayors had evidently neglected to pass on the combination code to Mosher's vault, necessitating the summoning of a safebreaker in 1908, while the Smithsonian had no record of the key Deihm apparently deposited there.[21] Nor could any of her direct heirs be located, all her children having predeceased her.[22] After a lengthy search, the Capitol's archivists found the spare key in the custody of a Florida retiree. This precipitated a further difficulty vessels can encounter: legal dispute over their ownership. The Floridian claimed custody by virtue of his possession of the key only to be trumped by a congressional resolution that belatedly accepted Deihm's gift and thereby rendered it US property, allowing it to be moved back to Statuary Hall.[23] A similar dispute ensued between the California Historical Society and the San Francisco Public Library over the rights to Cogswell's box; it was eventually resolved by the city's attorney in favor of the former.[24] The timing of the openings was another issue. Repudiating the calendrical convention for turns of the century, the millennium was reckoned to arrive in 2000 not 2001, so the Kansas City and Detroit vessels were a year late for the celebrations. Recipients also brought the date forward to suit their schedules. Detroit's box was opened a day early to avoid a conflict with New Year's Day football (its depositors, an official declared, failed to anticipate how "football has come into our lives"). Deihm's was opened three days early, and Cogswell's two months early.[25] Confirming the latter's curse on anyone who prematurely unsealed it, the heavens opened just as the ceremony began, causing one umbrellaless participant, a descendent of the dentist, to curse his forbear for "wreak[ing] his vengeance on us."[26]

When the vessels were finally breached, audiences often expressed disappointment. Unrealistic expectations had accumulated in the long countdown to the ceremonies. Some even speculated that Deihm's safe would reveal a million dollars in gold, a human skeleton, or a long-lost original version of the Liberty Bell.[27] A Capitol official even suggested finding "a discreet and dependable man" to inspect the safe ahead of time to prevent embarrassment.[28] Before turning the key, President Ford tried to defuse these expectations by reminding spectators that "no safe is big enough to contain the hopes, the dreams, the energies of our people."[29] But as he retrieved various albums, the sense of anticlimax was palpable. One staffer complained that "Mrs. Deihm did not seem to have a very good concept of what would be important and interesting a hundred years later."[30] Ford himself did not take the contents (not to mention, his female opponents) seriously; he held up a framed photograph of President Rutherford Hayes's wife and called out to the Democratic representative and women's rights advocate Lindy Boggs, "Lindy! I have a picture of a chairperson! I don't have any indication of her name but she looks mighty pretty" (**fig. 7.2**).[31] Nor did he carry out the solemn, ceremonial signing of autographs that Deihm had requested. Mosher's photographs, exhibited that same year at the Chicago Historical Society, elicited similar apathy. They do not "seem especially 'artful,'" observed one reviewer. "The project is awash in a sea of stolid Midwestern faces . . . a fairly grim and unattractive lot." Evidently, it was "a business venture."[32]

FIG. 7.2 William Fitz-Patrick, untitled (print from Tri-X panchromatic negative, July 1, 1976). President Ford holds up a silver-framed photograph of Mrs. Rutherford Hayes, which he has just removed from the Century Safe. Courtesy Gerald R. Ford Presidential Library.

The recipients' indifference toward Deihm's and Mosher's offerings is surprising given the upsurge during the mid-1970s of nostalgia not just for all things 1776 but also for the lost world of 1876. Gore Vidal's novel *1876* was a best seller that year, Victorian furnishings were in vogue, and the Smithsonian's blockbuster exhibition of more than thirty thousand artifacts and machines from the Philadelphia exposition, restored to "like-new appearance," offered visitors a chance to "return to the Centennial."[33] This nostalgia did not extend to the Century Safe for a number of reasons. Despite the superficial similarities between the centennial and bicentennial—both celebrations were promoted as an antidote to chronic recession, postwar fatigue, and political scandal—they played out quite differently. Whereas the former centered on a single city and event (the Philadelphia exposition), the latter was dispersed across the nation. The decision to decentralize stemmed from a lack of funding (Congress was unwilling to grant the necessary appropriations, and cities were hamstrung by their own fiscal crises), a dwindling interest in staging world's fairs, a growing preference for more locally oriented and ethnically inflected events, and officials' concerns that oppositional and minority groups, already mobilizing for protests and boycotts through the People's Bicentennial Commission, would hijack any large-scale celebration. Thus, the backdrop of a spectacular bicentennial exposition that Deihm and Mosher envisioned never transpired. The official events that did take place were designed to be more inclusive and vernacular and to eschew any "high moral tone" or "formal rituals" such as dignitaries signing autograph albums.[34] But above all, the Century Safe's lukewarm reception—along with the decision not to top it up for the tricentennial—reveals a further erosion of faith in progress.[35] Victorian Americans had resolved their anxieties about the imminent future by investing hopes in the distant future, but the experience of Vietnam, Watergate, the oil crisis, and stagflation now appeared to obstruct such utopian vistas.

The great hopes of past depositors, moreover, appeared to place a burden on the present, making time-vessel openings an occasion for soul searching rather than celebration. Particularly stark was the gulf between predictions of the limitless economic and demographic growth of cities and the realities of urban decline. By the time it opened its century box, Kansas City had been decimated by deindustrialization and "white flight," its population down to 450,000 from a peak of over half a million in 1970. In the same period, Detroit had plummeted from over 1.5 million to 951,270—well short of the four million confidently predicted by one contributor to its vessel—and had its credit rating reduced to junk status. (Of the centurial vessel cities, only Colorado Springs has continued to grow at a steady rate and looks set to overtake Kansas City and Detroit by the next census.)[36] Late-Victorian dreams of unending urban and technological progress were also belied by the four-day power outage that had paralyzed Detroit six months earlier.[37]

To be sure, recipients noted evidence of *social* progress since the vessels were sealed. That San Francisco's and Kansas City's were opened by their first female mayors was celebrated as a repudiation of white male control of the political establishment. The audience "erupted" in "laughter" when Mayor Kay Barnes of Kansas City read the presumptuous salutation of her predecessor's letter, "Dear Sir."[38] Those jeers, though, masked the fact that the suffragette contributors' hopes had hardly been borne out, with women remaining a minority in Congress: 3.7 percent in 1979, 13.6 percent in 2001, and still only 19.8 percent in 2018.[39] Williamina Fleming's letter in the Harvard chest, with its laments about gender inequalities in academia and the workplace, served as a reminder of further unfulfilled promises. Similarly, while the civil rights activist David Augustus Straker would presumably have been gratified to know that a black mayor (Dennis Archer) would unveil his message in Detroit in 2000, racial conditions in that city fell far short of his vision of a postracial paradise. In the very month of the opening, an investigation was initiated into alleged civil rights violations by the Detroit Police Department.[40] Depositors who had envisaged progress in the implementation of temperance and eugenics would also have been disappointed. After the San Francisco ceremony (which failed to include a reading of the message denouncing "liquor interests"), participants "retired to nearby taverns to hoist a toddy or two in memory of all the things Dr. Cogswell didn't stand for."[41]

Furthermore, although the reading out of letters and holding up of photographs appeared to grant the posthumous recognition their depositors had sought, in reality the fame conferred by a time capsule is decidedly short lived.[42] After the ceremony, those documents are typically consigned to archival oblivion. Occasionally, the receptacle and a selection of its contents are put on display, yet hardly in the prominent fashion that Mosher imagined for his photographs in that vast Memorial building on Chicago's lakefront.[43] Ignoring citizens' requests for Deihm's gift to be displayed in perpetuity, the Architect of the Capitol subsequently dispatched the safe to a storage warehouse and its contents to the basement archives.[44] Although an archive might appear a natural home for retired vessels, it potentially undermines their identity and coherence. Despite the long-established principle of "archival integrity," archivists have sometimes dispersed time-capsule collections, as if to punish their heterodox approach to historical documents. Mosher's portraits were not restored to the safe (as he stipulated) but merged into the Chicago History Museum's larger collection of cabinet cards, while Cogswell's photographs, artifacts, and manuscripts (after remaining uncataloged for more than two decades) were transferred to their "proper" divisions, with some items now missing.[45] Reordering documents, often alphabetically, and inserting them into sterile manila folders and archival boxes has further stripped them of their origin and meaning as transtemporal missives. Similarly, the recent digitization of the Colorado and Detroit documents, while useful for remote historical research, cannot reproduce the affective responses that physical traces of the past can spark in their presence.

Even rendered as archival data, the contents of time capsules have failed to interest historians. Deihm's and Cogswell's materials, among others, remained unconsulted until my visit in 2010.[46] If historians were to consult time-vessel collections, they would be unlikely to find "smoking guns," that is, primary documents that directly overturn the historical record. As for vessels' published documents (books, commercial photographs, newspapers, etc.), most have remained extant in libraries around the country, as that critic of the Harvard chest had predicted. Even the Brady photographs in the Century Safe are redundant; thanks to the Library of Congress's 1954 purchase of ten thousand negatives from his nephew's estate, now known as the Brady-Handy Collection, his portraits of Washington politicians (and other notables) have largely survived. Wariness of prepackaged archives, especially those ostensibly assembled to memorialize the elite, exonerate capitalists, or trumpet a city's commercial prospects, may also have contributed to this historiographical neglect of time vessels.

Despite this litany of neglect, however, time vessels constitute a crucial source for historians albeit not along the lines for which they were intended. They bring to light crucial debates from the Gilded Age and Progressive Era, debates over how one should, and to what extent one can, historicize the social and political conflicts of one's own time; over the politics of memorialization (who is entitled to be memorialized and who or what institutions should control the process of memorialization); over the implications of the expansion of print culture, the introduction of new media technologies, and the acceleration of architectural obsolescence for the longevity of cultural memory; and over the possibilities and shortcomings of new approaches to history. The vessels, through their messages and their specifications, allow us to trace mounting anxieties about the future of the state and of "modern civilization" itself along with the changing cause of those anxieties (from the conflict between capital and labor to the specter of total war). They also unveil the hopes that past individuals harbored for various futures, some of which presumed a radical deviation from the assumed direction of history. And above all we see how utopian wishes—for the triumph of Christian Science, temperance, or eugenics, or conversely the eradication of class distinctions, gender inequalities, racial discrimination, or capitalism itself—were rendered concrete through the ritual depositing of material artifacts, the targeting of a specific date in the future, and the addressing of actual, embodied individuals. We thus glimpse efforts to forge a physical and affective bond with future members of their movement or social group. And yet, as many of those time-vessel messages were addressed more generally to "future historians," we are *all* interpellated by such transtemporal expressions of hope and are thus conscripted into a historical imperative to redeem—or in some cases thwart—those utopias. William James's assertion that time capsules leave little for the future historian to do could hardly be further from the truth.

Epilogue: The Time Capsule's Futures

Whatever the prospects for individual time capsules, the future of the practice itself now appears in doubt. At a British Academy forum in 2007 on the topic of posterity, one discussant noted that time capsules have gone "out of fashion," a verdict echoed by another recent scholar.[1] Certainly there have been signs of decline since the 1970s. Given the decentralization of the US bicentennial celebrations noted above, there was little interest in assembling a new capsule representing the nation as a whole—besides a metal container deposited in the National Archives, a smaller collection of financial souvenirs cobbled together by the Treasury Department, and boxes of signatures of "rededication" gathered around the country by the Bicentennial Wagon Train (inexplicably, the latter were stolen en route to Valley Forge, where President Gerald Ford was to have buried them, and they remain missing to this day).[2] This pattern persisted with the abandonment of plans for a five-hundred-year time capsule to mark the Columbus quincentenary in 1992.[3] And when the proposal for a "large-scale" capsule for the millennium failed to attract corporate sponsors, one of its advocates quipped, "Are time capsules becoming a thing of the past?"[4]

If interest in time capsules is indeed diminishing, it might be a reaction to the corporate variant introduced by Westinghouse in 1939. As early as 1953, the University of Iowa's head librarian was privately questioning why "people who are otherwise relatively sane in their thoughts and actions spend time and money" on projects like Westinghouse's when libraries "have all the information about our time that the future will want." He concluded that time capsules were "a publicity stunt and not much else."[5] In the postwar years, corporations consolidated their co-optation of the time capsule. Westinghouse continued to lead the way with Time Capsule II, buried at the 1964–1965 New York World's Fair ten feet north of its predecessor, also for opening in 6939. This second installment followed the same corporate-industrial logic. It consisted of a missile-shaped container machined out of a new Westinghouse alloy (Kromarc) and stocked with a sample of scientific wonders embodying "the progress of those remarkable years" since the 1939 fair: an electric toothbrush, a credit card, a bikini, and, perhaps to take the edge off the presumed stupefaction of its recipients, tranquilizer pills. Decisions on the contents were again made by a committee of "experts" rather than by ordinary people. This corporate practice for promoting American products was itself exported, arguably reaching its zenith with the nested capsules organized by Panasonic at the 1970 World Expo in Osaka.[6]

Close associations with the military-industrial complex may also have contributed to a backlash against time capsules. Reflecting the national defense backgrounds of several of its committee members (most notably Vannevar Bush), Westinghouse's second capsule, like the fair as a whole, gave particular prominence to two major developments of the Cold War. The first was atomic energy, represented by deposited items such as a relic from Enrico Fermi's original nuclear reactor. Instead of testifying to the horrors of Hiroshima, Time Capsule II emphasized the benefits of the atom, for example through a film of the opening of the world's first nuclear power station. Even if the capsule resembled a nuclear missile about to be loaded into its silo, its emphasis on technologies of food preservation such as freeze-dried food and plastic wrap and survival tools such as a pocket radiation monitor evoked that embodiment of atomic optimism, the fallout shelter.[7] The specter of the mushroom cloud thus did not fundamentally transform time capsules; their deployment to project visions of the ultimate triumph of technology and technocracy, already initiated by Westinghouse and Oglethorpe in the 1930s, was merely expanded. That techno-futurism was finally burst in the 1970s by a growing antinuclear movement, a reactor meltdown at Three Mile Island, and an outpouring of science fiction confronting the implications of nuclear apocalypse. An earlier classic, Walter L. Miller's *A Canticle for Leibowitz* (1961), had already questioned the wisdom of preserving knowledge of nuclear power for future civilizations, allowing them to repeat the cycle of devastation and regression.[8]

The other product of the military-industrial complex, rocket science, was represented in Westinghouse's second capsule through such deposits as a fragment from America's first manned spacecraft, Aurora 7.[9] Five years later, this affinity between time capsules and the Space Race spawned a further offshoot: the space-time capsule. In inscribing the names of members of Congress and NASA administrators along with "goodwill messages" from four American presidents (and a further sixty-nine world leaders) on a coin-size silicon disc sealed in an aluminum case and affixed to the section of Apollo 11's lunar module that remains on the moon, NASA effectively appropriated the time capsule as a monument to an American triumph over the Soviets.[10] While that capsule rests in the Sea of Tranquility, others are hurtling through the galaxy either in the form of radio signals—such as astronomer Frank Drake's Arecibo transmission of 1974, a binary-encoded dispatch to a specific cluster in the constellation of Hercules—or as attachments to space probes. Pioneer 10 (1972) and Pioneer 11 (1973), the first artificial objects to attain the required velocity to escape the Solar System, each bear a plaque engraved with pictorial messages for potential extraterrestrial recipients. For Voyager 1 and 2 (1977), now entering interstellar space, the idea was expanded into a larger collection of 116 photographs and diagrams as well as audio recordings of musical selections, spoken greetings, and terrestrial sounds—from a horse and cart to a fighter jet, and from a human kiss to a whale song—all encoded in analog form on a gold-plated record. The organizer of all four projects, scientist Carl Sagan, acknowledged the decisive inspiration of being taken as a four-year-old to see the original Westinghouse capsule.[11]

We may also be witnessing a reaction to the temporal hubris of such projects. What has been called the Golden Age of the time capsule—the half century that followed the conception of the Oglethorpe crypt—was characterized by ever more ambitious time spans.[12] Space-time capsules, in particular, presented the possibility of exceeding even the five or six millennia envisaged by Pendray and Jacobs. The Arecibo radio message will take twenty-five millennia to reach its target—and a further twenty-five to get any confirmation of receipt. And although the Pioneer and Voyager messages lacked a target date—even after the probes' batteries expire and their instruments shut down circa 2050, they will sail on indefinitely—the 1976 launch of the LAGEOS-1 satellite for geodynamic measurements of the earth allowed Sagan to design a timed message, as its intermediate circular orbit will cause it to reenter earth's atmosphere in 8.4 million years.[13] Seeking to outdo that, a payload of images was designed for NASA's Cassini mission to Saturn to be left on its frozen moon Titan sealed inside a durable material (an "artificial fossil") to await the sun's red giant phase, which would generate the thawing that might allow life and ultimately an intelligent species to evolve there. It was thus effectively a time capsule with a target date of "5 or 6 billion years hence."[14] Transcending calendrical scales, such deep-time messages are often accompanied by elaborate dating devices. Both the Apollo and LAGEOS-1 plaques contain a depiction of the earth's continents to give future recipients a point of comparison from which to estimate the lapse of time since they were launched.[15]

By the 1970s, faith in the possibility of communicating across vast chasms of time (and space) was displaying signs of increasing strain. Naive assumptions about the future intelligibility of contemporary artifacts and messages—satirized in earlier decades by Edgar Allan Poe, John Ames Mitchell, and Mark Twain, among others—came under renewed attack from science fiction writers. Arthur C. Clarke's "History Lesson" (1949) presented the absurdity of aliens from Venus trying to reconstruct the long-extinct human race from the film reel of a Mickey Mouse cartoon that a last remnant had deposited in a mountain, while Robert Nathan's *The Weans* (1960) and David Macaulay's *Motel of the Mysteries* (1979) playfully exposed the unreliability of modern artifacts as archaeological evidence.[16] Yet perhaps the most trenchant challenge came from the semioticians and archaeologists whom the US Department of Energy (DOE) enlisted in 1983 to devise markers that could deter, for at least ten thousand years, future civilizations from disturbing underground repositories of radioactive waste. Rather than endorsing the markers, these scholars questioned the reliance on signs—verbal or visual—as they can be removed, become incomprehensible, or even attract the curious. Instead, they suggested perpetuating knowledge of the site through existing institutions such as libraries and archives, through invented rituals and legends—lived memories, fostered by an "atomic priesthood"—or conversely, by abstaining from any kind of transtemporal communication, concealing the waste beneath a silent landscape.[17]

The time capsule was vulnerable not only to semiotic critique but also to a new skepticism, expressed by philosophers such as Michel Foucault, toward the totalizing claims of modern encyclopedic forms in general.[18] During their golden age, time capsules' orchestrators made grandiose boasts about their synoptic encompassing of the present through ever-larger collections of data—whether measured in words, pages, or linear feet of microfilm. There was also an expansion of their focus from "American civilization" to "Western" or "modern civilization," and ultimately to the entire species. In aspiring to represent humanity—through samples of diverse musical cultures, languages, and photographs of buildings and faces from around the world—the Voyager Golden Record succumbed to the kind of universalist essentialism and "sentimental humanism" that critics like Susan Sontag deplored in photographer Edward Steichen's earlier "Family of Man" exhibit.[19] Its depiction of a "typical" woman as smaller than a man attracted feminist criticism, while its photographic captions revealed a Eurocentric bias in naming Westerners such as Jane Goodall but not, for instance, "Dancer from Bali" or "Man from Guatemala." Increasing wariness of speaking for the other, along with internal rivalries, prompted NASA to abandon its "Portrait of Humanity," a single, carefully posed photograph of fourteen individuals of various races and ages that had been planned for the 1997 Cassini mission.[20]

The post-1970s decline of the far-reaching, all-encompassing time capsule, however, has not entailed a decline of time capsules in general. That decade in fact witnessed a proliferation and vernacularization of the practice as the bicentennial prompted smaller groups (schools, local communities, etc.) and individuals to make their own ad hoc deposits, often in their backyard.[21] This new populist embrace of the tradition has spawned a cottage industry of companies—Time Capsules, Inc., Heritage Time Capsules, Future Packaging and Preservation, to name a few—offering a range of durable receptacles and even custom engraving. It has also spurred a new pedagogy of the time capsule. The Smithsonian, the Library of Congress, and ultimately an International Time Capsule Society (founded 1990) all offer amateurs advice on how to improvise a homemade container, what to include in it, and where to deposit it.

The time capsule's resurgence has gained further impetus from the internet revolution. Various websites have made it even easier to upload, encrypt, and "seal" messages and audiovisual fragments for the future. Life's Time Capsule (lifestimecapsule.com) and myLegacy.org, among others, allow one to designate multiple recipients, select any target date (up to infinity), and upload larger amounts of data—if one opts for the more expensive pricing plans, which range up to $9.95 per month or a onetime fee of $499.95.[22] For those preferring not to pay, there are time-delayed email forwarding services such as Future Post-box (futurepostbox.com) or FutureMe (futureme.org), the latter now boasting more than seven million letters to posterity. And with Facebook's "On This Day" feature (introduced 2015), users can receive a kind of "time capsule" daily and automatically without the need for any prior act of transmission.

However, the internet's limitations have arguably done more to reinvigorate the time capsule. Greater awareness of the ephemerality of digital memories—the fact that they can be easily overwritten (intentionally or accidentally), hacked, or rendered inaccessible by the rapid obsolescence of formats and storage devices—may have generated the growing interest in older, more stable mnemonic devices such as the time capsule. There are also concerns about the privacy, ownership, and transferability of memories that we entrust to social media corporations. Meanwhile, the so-called digitization of everything, which has inundated our hard drives and personal clouds with a torrent of images and songs, has generated nostalgia for the analog, the tactile, and the manageable.[23] One online instructor who has illuminated the art of the digital time capsule now offers the course "Creating Tangible Time Capsules," which involves depositing a few printed photographs with "letters, drawings and all sorts of stuff" in boxes.[24] The practice may even appeal to the growing "slow movement" as a negation of digital communication's instantaneity and its tendency to accelerate daily life.[25]

This time-capsule revival is not without its problematic aspects. One concern is its appropriation as a vehicle for self-commemoration. In its period of emergence, the time vessel retained a balance between its (often self-interested) architect and the larger community he or she was obliged to enlist as collaborators. Now, containers are deposited in backyards, attics, and wall spaces as monuments to their individual creators, sometimes even targeted to their own offspring or their older selves. Digital capsules, too, betray a certain solipsism, evident in the names of the websites (mytimecapsule.net, hearfromme.com, etc.), which appears related to the larger preoccupation (fostered by social media and wearable devices) with archiving oneself, curating one's preferences, and "life-logging." This privatization of the future can again be traced back to the 1970s. The bicentennial "craze for [time capsule] burials," complained the geographer David Lowenthal a year later, "appeared to be animated less by a desire to show future generations present-day artifacts than to achieve immortality for personal effects, if not for one's person. Like the ancient Pharaohs, some Americans seemed determined to take it all with them."[26] During that same decade, Andy Warhol appropriated the term *time capsule* to describe the cardboard boxes, eventually numbering 610, in which he jettisoned whatever private ephemera were cluttering his studio: fan letters, photographs, fingernail clippings, acne medication, as well as art. The diverse materials—recently unpacked and meticulously cataloged—share the common denominator of (and derive their value from) Warhol's own persona.[27]

In more recent years, corporations have capitalized on this self-commemorative impulse. While earlier vessels may have appealed to ordinary citizens' desire to immortalize themselves, offering them an opportunity to address their progeny, they typically subordinated those personal messages to the collective documentations of their social world. By contrast, the capsules launched by recent corporations tend to lack any larger purpose beyond the sheer accumulation of personal messages—a kind of outsourcing of the labor of transtemporal communication. For its fifteenth anniversary in 2006, Yahoo! invited web visitors to contribute to a "digital anthropology" archive for the year 2020 by uploading intimate snapshots, video or audio clips, or words in response to questions such as "What do you love?" or "What makes you sad?" The 170,857 autoethnographic submissions—which, in social media fashion, could be viewed and commented on by anyone during the open period—were simultaneously sealed online, buried as hard copy at Yahoo! headquarters, deposited in the Smithsonian's Center for Folklife and Cultural Heritage, and laser projected through space-time from an ancient pueblo in New Mexico.[28] Google and Paramount devised their own crowdsourced capsule as a marketing scheme for the movie *Interstellar* (directed by Christopher Nolan, 2014), soliciting amateur videos from across the globe that convey our life on earth—"our hobbies, routines, families, achievements"—to "future generations."[29] The ploy has also been used to enlist public interest in the now privatized business of space exploration. Several upcoming missions to explore robotically the surface (or subsurface) of the Moon or Mars will be "kickstarted" by individuals who purchase access to "memory boxes" that the robots will leave there.[30] One veteran of the Voyager project has denounced a similar venture (the collecting of signatures for the Cassini spacecraft) as a "high-tech tombstone in space," and as such, an abnegation of the duty to convey a serious message.[31]

Recent collections of personal ephemera—whether sealed online, in cardboard boxes, in backyards, or beyond the earth—also tend to eschew other commitments made by the earlier vessels, such as that of filtering and ordering. The committee of citizens, carefully selecting and cataloging items, has been supplanted by an indiscriminate, catchall approach, or "data dump"—a reversion, in effect, to the cornerstone tradition. If there is any selectivity, it tends to be based on popular vote—as in the Earth Tapestry project, which is polling people around the world on which natural or man-made landmarks are to be archived on the moon by a robotic spacecraft—or else by computer algorithm—as in Jonathan Harris's 10 × 10 project, which "encapsulate[s]" moments in time by autonomously displaying a collage of the one hundred photographs recurring most frequently on international news sites.[32] This emphasis on contemporaneous display hints at another lapsed commitment, that of confidentiality. Following Oglethorpe's and Westinghouse's example, recent capsules often divulge their contents—with the one sealed in 1994 in the federal building of Juneau, Alaska, remaining fully visible through a plate-glass observation window.[33] The sustained attention to the question of modern media's durability, exemplified by earlier figures such as Konta, also appears to have diminished. The depositing of discs and hard drives in capsules or on celestial bodies begs the question of readability. Digital time capsules, too, betray a naivete about the life span of websites, servers, cloud services, and email addresses—let alone the retrievability of passwords. (There are, of course, exceptions, such as the Memory of Mankind, an ongoing project to imprint scientific, political, and cultural documents and personal messages onto ceramic tiles and deposit them in an old salt mine in the Austrian Alps—a fulfillment of one of the MHRA's dreams).[34]

More significantly, recent time capsules—especially the vernacular or digital variety—lack the temporal depth and openness of their predecessors. If the standard hundred-year time span stemmed from the assumption that vessels were for the unborn, the new interest in preserving time capsules for one's later self has generated shorter intervals. One capsule manufacturer revealed that "many of his customers were planning to open their capsules in a mere 10 to 25 years."[35] Digital capsules are typically even shorter, in accordance with the temporal phenomenon of "internet time." Students and faculty at MIT sealed one in 1999 for five years (there is no record that anyone remembered to open it), while FutureMe has announced that the most popular time span, chosen by 44 percent of its visitors, was just one year.[36]

These reduced time spans reflect and reaffirm the general foreshortening or "forfeit" of the future in the postindustrial, neoliberal era: the renunciation of utopian thinking and grand planning, especially after 1989; the temporal myopia or "short-termism" of many politicians; the recycling of past futures in science fiction; the adoption of lightweight, provisional materials in architecture; or the rejection of the Romantic idolization of posthumous reception by postmodern artists who strive instead for a "Warholian contemporary notoriety."[37] This diminished regard for posterity was again first widely noted in the 1970s in critiques by prominent liberal intellectuals. The bicentennial provided historian Henry Steele Commager with an occasion to decry how far Americans had fallen from the founders' "commitment to posterity" judging by the ongoing destruction of the environment, the stockpiling of nuclear weapons, the deepening of the national debt, and the disinvestment in education.[38] This growing "inability to identify with posterity," argued another historian Christopher Lasch that same year, was a symptom of a larger "culture of narcissism." In the retreat from political turmoil, the "prevailing passion" is to "live for the moment . . . [and] for yourself."[39]

It would be hasty, however, to write off the time capsule altogether as a debased tradition. Vernacular variants continue to represent a challenge to professional archival practices, in particular in their embrace of an ever broader swath of material culture. Contributors' inclusion of personal artifacts also indicates an expanded awareness of the historicity of their everyday lives, personal memories, and aesthetic preferences. If, as Carl Becker famously noted, "Everyman [is] his own historian," now everyone is potentially their own archivist.[40] Such commemorations of the local and the ordinary, moreover, suggest an ongoing disengagement from the national memory that official institutions and ceremonies seek to construct. In some cases, these implicit challenges become overt expressions of opposition. For the new capsule deposited at the foot of Cogswell's Franklin statue in 1979, locals contributed such countercultural items as a pair of Levi's, Armistead Maupin's *Tales of the City*, a Gay Freedom Day Parade poster, a poem by Lawrence Ferlinghetti, and a marijuana joint.[41] More recently, the practice was appropriated by those alarmed by Donald Trump's election. To convert fear into hope, the women's rights group MADRE launched a one-year time capsule and invited supporters to reaffirm—visually or verbally—their political principles, thus rehistoricizing a moment of crisis as a moment of "rebirth of our movements for peace, rights and justice."[42]

As the product of an international organization, the MADRE project further points to time capsules' expanded potential in the digital age to nurture an emerging sense of global citizenship. Although crowdsourcing offers corporations a cheap and quick, ostensibly democratic and culturally tactful method for devising time capsules, it can also foster a genuinely transnational perspective on our shared futures. To mark the week in 2011 that the global population (as estimated by the United Nations) reached the ominous milestone of seven billion, the *New York Times* invited readers across the world to submit a photograph that would represent "what the world looks like today" and then presented a keepsake box containing one hundred of them to twenty babies in a maternity ward in India, the presumed country of the seven billionth person, for them to open in twenty years.[43] The two-week window for entries evoked a related phenomenon of recent years: the depiction of a single day, or even minute, in the life of the world through a global crowdsourcing of amateur and professional photographs and videos. These collective self-portraits are typically exhibited online as interactive mosaics, such as the *Times'* own "Moment in Time" of 2010, or edited into feature-length documentaries, such as *Life in a Day* (directed by Ridley Scott and Kevin Macdonald, 2010) or the more critical *One Day on Earth* (directed by Kyle Ruddick, 2012), and are described, only in the loose sense, as time capsules (the raw footage from every single country in the world, Ruddick explains, has been "archived . . . as a time capsule" of that day "for years to come").[44] Yet one such archive—A Day in the World (2012), organized by the foundation Expressions of Humankind—has been deposited in a stainless steel time capsule in a disused mine in Sweden, while another—The Earth Pyramid—is to be deposited for a thousand years in a pyramid constructed out of materials from around the world, with a capstone cast from bullet shells from multiple warzones. More than just vehicles for armchair travel and "Family of Man" fantasies, these latter projects were conceived by nonprofits (often in conjunction with NGOs and UN agencies) as a means to promote mutual understanding and deep reflection on global problems.[45] Time capsules have thus preserved an affiliation with utopian internationalism that dates back to Konta's.

If these capsules remain somewhat anthropocentric, others have reached beyond the human, thus harking back to the depositing of seeds and acorns in earlier vessels such as Cogswell's. Various biological banks, or "biobanks," have been established to systematically preserve specimens (or DNA samples of) plants and animals that are endangered by climate change, pollution, biotech monoculture, and other human activities. One of them, the ongoing Time Capsule Program of Japan's National Institute for Environmental Studies, began with a plan to implant containers of specimens and atmospheric samples sixty-five feet into the Antarctic ice to help scientists in the year 3000 research climate change or resuscitate species. It has since constructed its own cryogenic facility, the Environmental Time Capsule building, for medium-term deposits.[46]

These and other efforts or proposals to suspend threatened species in liquid nitrogen, however, have drawbacks. By holding out hopes of species resurrection (or "de-extinction"), biocapsules—like the nuclear waste markers—might encourage public complacency or a sense of having done enough for posterity and thus thwart efforts to tackle climate change itself. Cryopreservation is "far from the best way to save ecosystems and could easily fail," naturalist Edward O. Wilson has written. "It is at best a last-ditch operation."[47] The inclusion of human DNA in some of these biocapsules—and indeed on interplanetary capsules such as "the Immortality Drive," a microchip containing DNA sequences of forty humans, delivered to the International Space Station in 2008, in case there is a global catastrophe—raise further questions. What lengths should we go to perpetuate our own species? What are the ethical implications of projecting a human into a posthuman world? And whose DNA would we choose for the purpose? (The choice of mainly white businessmen and celebrities for the Immortality Drive reinscribes the long-standing connection between time capsules and eugenics.) Perhaps better to construct, as one architectural student has proposed, a cenotaph to the human race, which would at least force us to confront our species' self-destructive tendencies (although that, too, could promote resignation).[48]

Besides offering species backups, time capsules may generate deeper responses to climate change. A series of capsules, buried in botanical gardens on four continents on World Environment Day in 1994, consist of pledges to the people of 2044 to actively address the problem.[49] Even more ambitious is the Clock of the Long Now. Conceived by the computer scientist and inventor Danny Hillis, named by the musician Brian Eno, and elaborated by the environmentalist Stewart Brand, this monumental, thermal-powered clock (now under construction) will be installed inside a remote mountain in west Texas and will mark the passing of each year for ten millennia. As an embodiment of deep time, this transmillennial device is to inspire long-term thinking and responsibility, above all in the generation that designs and constructs it. The clock will be supplemented with a library containing private messages to the future, a registry of other time capsules, and a record of where political leaders have stood on issues such as climate change.

Brand, the creator of *The Whole Earth Catalog*, believes it will "do for thinking about time what the photographs of Earth from space have done for thinking about the environment."[50]

Taken together, such projects may indicate an inchoate sense of what time-capsule architect Louis Ehrich called *posteritism*. Indeed, the Long Now Foundation, established to oversee the clock and other projects, is one of a number of new institutions that are challenging short-termism; others include the World Future Council, the Institute for the Future, and the Intergenerational Foundation. It is also related to philosophers' and legal theorists' efforts, prompted by nuclear and environmental threats, to establish a basis for the notion that future generations have rights in the present—a notion that cannot rely on traditional ideas of community, reciprocity, or rationality, let alone instinct.[51] With the emergence of our technological capacity to destroy the earth, wrote Hans Jonas, a pioneer of this field, "our duty to posterity assumes entirely new dimensions, embraces entirely new objects, and now even includes the responsibility for there going to *be* . . . an indefinite posterity for man on earth."[52] In this context, the launching of a time capsule for distant human recipients may constitute the ultimate expression of existential hope: that we may foster the various conditions (biological, cultural, political, etc.) that would enable it to reach its target.

For time capsules to serve these lofty goals, they need to be perpetually reinvented. A negative effect of their formalization in the 1930s and their proliferation since the 1970s has been a dilution of their capacity to stir the temporal imagination as they devolved into empty rituals, repeated formulaically on standard occasions. One source for new variations has been the art world. Buried containers have long fascinated conceptual artists, a notable example being Sol LeWitt's *Buried Cube Containing an Object of Importance but Little Value* (1968), a deadpan documentation of his act of burying a stainless steel cubic box (which he called a time capsule) containing an unidentified artwork of his, thereby precluding its commodification by the art market.[53] There have also been efforts to create future-oriented art. Robert Smithson considered long-term processes of entropy when creating earthworks such as *Partially Buried Woodshed* and *Spiral Jetty* (both 1970). Other works have engaged with the "slow time" of tree growth, such as *7000 Oaks* (1982), Joseph Beuys's eco-minded planting of trees around Kassel, Germany, as sculptural "monuments" for three hundred years hence; or *Ash Dome* (1977), a ring of ash trees that David Nash, in defiance of warnings of imminent nuclear war and economic collapse, planted and sculpted in a secret location in Wales for the year 2000.[54]

More recently, Trevor Paglen curated *The Last Pictures* (2012), a collection of one hundred photographs imprinted on a small silicon disc and launched on a telecommunications satellite into geosynchronous orbit, so that (unlike LAGEOS-1) it will circle the earth "potentially for more than a million years and perhaps even indefinitely." Repudiating Westinghouse-style ambitions to offer "a grand representation of humanity" but without succumbing to postmodern cynicism about long-term communication, he selected photographs to convey a specific theme: contemporary civilization's suicidal tendencies. Whatever the likelihood of their being recovered from the ring of "dead spacecraft ruins" girdling the earth (and Paglen was more realistic than others), the photographs—issued also in book format—would inspire contemporary audiences to confront ecological collapse head-on.[55]

Finally, these archival, conceptual, and land art tendencies come together in the Future Library (2014–), by artist Katie Paterson, which consists of unread book manuscripts submitted by one author each year, starting with Margaret Atwood. Installed in wooden boxes in a special room in the Oslo Public Library, they will finally be unveiled and published in 2114. Echoing Beuys's *7000 Oaks*, this "100-year artwork" involved the planting of a thousand trees in a nearby forest (on inalienable public land) to supply the paper for these publications. Lured initially by the mystery of an ark of secret books, we gradually become aware of the deeper lessons: the need to care for that forest and to recognize the mutual imbrication of nature and culture. In the words of the second author, David Mitchell, it represents "a vote of confidence in the future. Its fruition is predicated upon the ongoing existence of Northern Europe, of libraries, of Norwegian spruces, of books and of readers."[56]

Acknowledgments

For assistance with research, I would like to thank staff at the following archives: Ancient and Honorable Artillery Company of Massachusetts, Architect of the Capitol Curatorial Division, California Historical Society, Chicago History Museum, Colorado College Special Collections, Detroit Historical Society, Harvard University Archives, Iowa State Historical Society, Kansas City Museum, Mount Holyoke College Archives and Special Collections, New-York Historical Society (where I spent a memorable year as an NEH fellow; thanks to Valerie Paley, Margi Hofer, Ted O'Reilly, Nina Nazionale, Michael Ryan, Marilyn Kushner, and others), New York Public Library, New York University Archives, Oglethorpe University Philip Weltner Library Archives, San Francisco Public Library, Senator John Heinz History Center, and University of Iowa Special Collections. I am also grateful to those who scanned and sent me materials from private collections, especially George Hagenauer.

This book began in 2008 as a lecture presented at a University of Chicago symposium marking the retirement of my former advisor Neil Harris, whose work continues to inspire me. Questions from audience members there helped me hone my argument. So, too, at Columbia University's Temple Hoyne Buell Center for the Study of American Architecture, the New-York Historical Society, York University's School of the Arts, Media, Performance and Design, and, closer to home, in the history and American studies departments at the University of Iowa.

This book profited greatly from the insightful feedback

of scholars who read individual chapters or conference papers from it: Steven Hoelscher, Aleksandra Kaminska, Kim Marra, Alexis McCrossen, Glenn Penny, John Durham Peters, Lauren Rabinovitz, John Raeburn, and Steve Warren. I am particularly grateful to those who read the entire manuscript: Paula Amad, Travis Vogan, and the anonymous readers for the University of Chicago Press.

Early portions of the manuscript appeared as "Encapsulating the Present: Material Decay: Labor Unrest, and the Prehistory of the Time Capsule, 1876–1914," *Winterthur Portfolio* 45, no. 1 (2011): 1–28, and "Posing for Posterity: Photographic Portraiture and the Invention of the Time Capsule, 1876–89," *History of Photography* 38, no. 4 (2014): 331–55. Thank you to Amy Earls at *Winterthur* and Luke Gartlan at *History of Photography* and to their anonymous readers.

Numerous others contributed indirectly to the book through exchanges over coffee at Prairie Lights or Cortado, at conferences and archives, or in emails. I would like to thank Thomas Allen, Gyorgy Baics, Ross Barrett, Steve Choe, Susan Crane, Brian Durrans, Craig Eley, Naomi Greyser, Joseph Heathcott, Tom Mix Hill, Shelley Hornstein, Linda Kerber, Joni Kinsey, Kathy Lavezzo, Steve Moga, Brian Murphy, Tom Oates, Michele Pierson, Roland Racevskis, Matt Smerdon, Carl Smith, Harry Stecopoulos, Jim Throgmorton, and Dave Wittenberg. Daniel Yezbick provided crucial information relating to George Carlson's artwork for the Oglethorpe Crypt of Civilization, and Trudy Peterson answered my queries regarding the history of archives, as did Harriet E. Smith regarding Mark Twain's autobiography.

Iowa City has been an ideal place to produce this book. Three graduate students here—Paul Bartels, Erica Stein, and Steven Williams—worked as research assistants in the early stages. Other students—graduate and undergraduate—sparked ideas in seminars through their responses to materials relating to temporality and memory. A special thank you to Craig Eley, Allison Wanger, and Larissa Wehrnyak; and to Eric Johnson, who, as an Arkansas native, first alerted me to Coin Harvey's Pyramid at Monte Ne. Thanks also to all my colleagues in the departments of American studies and history at the University of Iowa. I am truly fortunate to have had such supportive chairs as Susan Birrell, Lisa Heineman, Kim Marra, Horace Porter, Lauren Rabinovitz, and Landon Storrs and equally supportive associate deans in Raul Curto and Joe Kearney. The University of Iowa also provided a subvention to help fund the reproduction of images.

At Chicago, I owe a particular debt of gratitude to my editor, Tim Mennel, for his expert advice at every stage and for his astute close reading of the manuscript toward the end. His assistant Rachel Kelly was patient and helpful, especially with all the images. And Steve LaRue's copyediting was deft and meticulous.

Above all, I must thank my family in London—my parents Rosemary and Tony Yablon and my big sister Emma Parlons, brother-in-law Jeremy, niece Sophie, and nephew Will—for supporting and inspiring me in so many ways. Their visits and my returns always helped to restore flagging spirits. So, too, did visits and calls to my parents-in-law Peter and Marie and all the other Amads of Melbourne, Australia—especially during Ashes victories. But it is to Paula Amad that I owe the largest debt. She has been my smartest reader, my wisest counselor, my guiding light, and the love of my life. And finally, Max—who was born around the same time as this project, ensured I took breaks from it, and gleefully clicked the send button to put an end to it—this book is dedicated to you. I always imagined it as a kind of time capsule for you to read one day. As for your own homemade coffee-tin time capsule, I hereby pledge never to peek inside.

Abbreviations

ARCHIVES AND SPECIAL COLLECTIONS

AOC	Office of the Curator, Architect of the Capitol, Washington, DC
BL	Bancroft Library, University of California, Berkeley
CHM	Chicago History Museum, Chicago, Illinois
CHS	California Historical Society, San Francisco
CSC	Centennial Safe Collection, Office of the Curator, Architect of the Capitol, Washington, DC
CSCCC	Colorado Springs "Century Chest" Collection, 1901, Ms 0349, Colorado College Tutt Library, Special Collections (all dated late July/early August 1901)
DCBC	Detroit "Century Box" Collection, 2001.061, Detroit Historical Society (all dated December 30, 31, 1900, or January 1, 1901)
HDCTCC	Henry D. Cogswell Time Capsule Collection, 1847–1879, California Historical Society, San Francisco
HUC	Harvard University "Chest of 1900," 1899–1900, HUA 900.xx, Harvard University Archives (all dated March 1901)
KCCBC	Kansas City "Century Box" Collection, Kansas City Museum (all dated January 1, 1901, unless otherwise specified)
LC	Library of Congress, Washington, DC
MHC	Box sealed by Mount Holyoke Class of 1900, Mount Holyoke College, Archives and Special Collections
MHRA	Modern Historic Records Association
NYPL	New York Public Library, New York City
NYHS	New-York Historical Society, New York City
OCCCC	Century Chest Collection, First Lutheran Church of Oklahoma City

OU	Philip Weltner Library Archives, Oglethorpe University, Atlanta, Georgia
RCB	First Congregational Church of Rockland's "Century Box," in the possession of Reverend Alan Copithorne, Rockland, Maine (all dated December 29–31, 1901)
SFPL	San Francisco Public Library, San Francisco, California
SHSI	State Historical Society of Iowa City
TJC	Thornwell Jacobs Collection, Oglethorpe University, Atlanta, Georgia
UISC	University of Iowa Special Collections, Iowa City
WC	Westinghouse Collection, Senator John Heinz History Center, Pittsburgh

OTHER

NYT	New York Times
GPO	Government Printing Office

Notes

INTRODUCTION

1. See G. E. Pendray, "Time Capsule Ceremony" (typewritten schedule), box 5, Clark Wissler Collection, Ball State University Special Collections. On the granite monument, see "Westinghouse Time Capsule," *Science* 94, no. 2437 (1941): 251. Reference to gong in [Pendray], *Story of the Westinghouse Time Capsule* (East Pittsburgh, PA: Westinghouse Electric and Manufacturing Company, [1939]).
2. Thornwell Jacobs, "Today—Tomorrow," *Scientific American* 155, no. 5 (1936): 260. Knute Berger describes the Westinghouse Time Capsule as "the world's very first time capsule" in "New York's Sacred Meadow: The Vital Legacy of the Westinghouse Time Capsules," http://www.nywf64.com/weshou14.html.
3. William E. Jarvis, *Time Capsules: A Cultural History* (Jefferson, NC: McFarland, 2003), 155. Robert Ascher describes Westinghouse's as the "first really elaborate time capsule," in "How to Build a Time Capsule," *Journal of Popular Culture* 8, no. 2 (1974): 242. On the enduring reputation of Westinghouse's Time Capsule, see Lester A. Reingold, "Capsule History," *American Heritage* 50, no. 7 (1999): 92.
4. Ittai Weinryb, ed., *Ex Voto: Votive Giving across Cultures* (New York: Bard Graduate Center, 2016), 1. Jarvis traces the time capsules' antecedents backs to antiquity in *Time Capsules*, 82–96. On cornerstone deposits in the United States, see Neil Harris, *Building Lives: Constructing Rites and Passages* (New Haven, CT: Yale University Press, 1999), 12–31. On deposit rituals in non-Western cultures, see, for example, Douglas R. Mitchell and Judy L. Brunson-Hadley, eds., *Ancient Burial Practices in the American Southwest: Archaeol-*

ogy, Physical Anthropology, and Native American Perspectives (Albuquerque: University of New Mexico Press, 2004), esp. 129–31.

5. For an early instance of an American inventor using such an envelope, see "Transactions of the American Institute," in *Documents of the Assembly of the State of New York* (Albany: Benthuysen, 1861), 664.
6. Although journalists often apply the term *time capsule* to cornerstone deposits, this book retains its specificity (i.e., the target date). There are of course precursors, such as manuscripts sealed in libraries for a fixed number of years; see below, 100.
7. Alexis de Tocqueville, *Democracy in America*, ed. J. P. Mayer (1835–40; New York: Harper Perennial, 1988), 507, 631, 207–8.
8. See Barbara Buckner Higginbotham, *Our Past Preserved: A History of American Library Preservation, 1876–1910* (Boston: G. K. Hall, 1990).
9. See Aleida Assmann, *Cultural Memory and Western Civilization: Functions, Media, Archives* (New York: Cambridge University Press, 2011), 154–68.
10. See, for example, John Bodnar, *Remaking America: Public Memory, Commemoration, and Patriotism in the Twentieth Century* (Princeton, NJ: Princeton University Press, 1993); Len Travers, *Celebrating the Fourth: Independence Day and the Rites of Nationalism in the Early Republic* (Amherst: University of Massachusetts Press, 1997); and Lynn Spillman, *Nation and Commemoration: Creating National Identities in the United States and Australia* (New York: Cambridge University Press, 1997).
11. Benedict Anderson, *Imagined Communities: Reflections on the Origin and Spread of Nationalism* (London: Verso, 1983), 22–36; Alexis McCrossen, *Marking Modern Times: A History of Clocks, Watches, and Other Timekeepers in American Life* (Chicago: University of Chicago Press, 2013), 89–142; Michael O'Malley, *Keeping Watch: A History of American Time* (New York: Viking, 1990), 55–144.
12. Thomas M. Allen has led the way with *A Republic in Time: Temporality and Social Imagination in Nineteenth-Century America* (Chapel Hill: University of North Carolina Press, 2008), esp. 17–58.
13. Allen, 6–11, summarizes (and critiques) that thesis.
14. The few studies of the time capsule are from other disciplines and, while invaluable, do not focus on its pre-1938 development. Jarvis (a library scientist) only briefly mentions Mosher's and Deihm's 1876 time capsules (*Time Capsules*, 45–46, 114–16), while anthropologist Brian Durrans illuminates recent deposits in "Time Capsules as Extreme Collecting," in *Extreme Collecting: Challenging Practices for 21st Century Museums*, ed. Graeme Were and J. C. H. King (New York: Berghahn Books, 2012), 181–202. Ascher discusses time capsules from the perspective of archaeology in "How to Build a Time Capsule."
15. See especially Johann Gustav Droysen, *Outline of the Principles of History* (1858; Boston: Ginn, 1893), 18–21, 51; Marc Bloch, *The Historian's Craft*

(1949; New York: Knopf, 1954), 60–64; and R. G. Collingwood, *The Idea of History* (Oxford: Oxford University Press, 1946), 138, 257.

16. Ascher, "How to Build a Time Capsule," 245, 249–50.
17. Such distinctions appear in Bodnar, *Remaking America*, 13–20, and in Aleida Assmann, "Four Formats of Memory: From Individual to Collective Constructions of the Past," in *Cultural Memory and Historical Consciousness in the German-Speaking World Since 1500*, ed. David Midgley and Christian J. Emden (Bern: Peter Lang, 2004), 25. See, also, Maurice Halbwachs, who, despite viewing individual memories as deeply social, distinguished between those of individuals and families and those of larger social groups, in *On Collective Memory*, trans. Lewis A. Coser (first published 1925 by F. Alcan [Paris]; Chicago: University of Chicago Press, 1992), 52–166.
18. Leo Braudy, *The Frenzy of Renown: Fame and Its History* (New York: Vintage Books, 1997), and Loren Glass, *Authors Inc.: Literary Celebrity in the Modern United States, 1880–1980* (New York: New York University Press, 2004), focus on how the *famous* viewed the subject of fame.
19. Although "prospective memory" is an emerging field in psychology, there has been little effort to apply it to cultural memory, though see Jan Assmann's *Cultural Memory and Early Civilization: Writing, Remembrance, and Political Imagination* (1992; New York: Cambridge University Press, 2011), 45–47, 53–54, 149–50.
20. Raymond H. Geselbracht, "The Origins of Restrictions on Access to Personal Papers at the Library of Congress and the National Archives," *American Archivist* 49, no. 2 (1986): 144. I am using *archives* loosely here to refer to both manuscript collections and the public archives that emerged after 1900.
21. See, for example, *Archives, Documentation, and Institutions of Social Memory*, ed. Francis X. Blouin Jr. and William G. Rosenberg (Ann Arbor: University of Michigan Press, 2006).
22. On nineteenth-century historical societies' and museums' accumulation of artifacts with historical associations, see Teresa Barnett, *Sacred Relics: Pieces of the Past in Nineteenth-Century America* (Chicago: University of Chicago Press, 2013), esp. 29–49.
23. On the challenge posed by purpose-built archives of modern media, see Paula Amad, *Counter-Archive: Film, the Everyday, and Albert Kahn's Archives de la Planète* (New York: Columbia University Press, 2010).
24. Despite Oliver Wendell Holmes's 1859 dream of photographic archives (see below, 46), it was probably not until 1889 that a library, Denver Public Library, launched one; *Encyclopedia of Library History*, ed. Wayne A. Wiegand and Donald G. Davis Jr. (New York: Garland, 1994), 48. The Library of Congress did not establish a phonograph record collection until 1903 and did not fully embrace film until 1939 (it accepted paper prints of films from 1893 onward only for purposes of copyright); see Anthony

Slide, *Nitrate Won't Wait: A History of Film Preservation in the United States* (Jefferson, NC: McFarland, 1992).

25. For a reminder that "data is not knowledge, and data storage is not memory," see Abby Smith Rumsey, *When We Are No More: How Digital Memory Is Shaping Our Future* (New York: Bloomsbury, 2016), 12.
26. Bernard Stiegler, "Memory," in *Critical Terms for Media Studies*, ed. W. J. T. Mitchell and Mark B. N. Hansen (Chicago: University of Chicago Press, 2010), 66–68.
27. John Higham, *History: Professional Scholarship in America* (1965; Baltimore: Johns Hopkins University Press, 1989), 148, 151; Bonnie G. Smith, *The Gender of History: Men, Women, and Historical Practice* (Cambridge, MA: Harvard University Press, 2000), 130–31, 148–49, 206.
28. Higham, *History*, 158–59; Smith, *Gender of History*, 9, 122, 133, 137–38, 150. There were, to be sure, calls for a "new history" at the turn of the century (see below, 203), but few heeded them; Peter Burke, "The New History: Its Past and Future," in *New Perspectives on Historical Writing*, ed. Peter Burke (University Park: Pennsylvania State University Press, 2001), 7. Those who did continued to neglect social history, women's history, and vernacular sources.
29. For the work of a pioneering historian of the everyday, see Lucy Maynard Salmon, *History and the Texture of Modern Life: Selected Essays*, ed. Nicholas Adams and Bonnie G. Smith (Philadelphia: University of Pennsylvania Press, 2001). On the "boom" in local history societies and projects in the late nineteenth century, see Carol Kammen, *On Doing Local History* (Walnut Creek, CA: AltaMira, 2003), 15, 17. On efforts to collect material artifacts in this period, see Thomas J. Schlereth, "Material Culture Studies in America, 1876–1976," in *Material Culture Studies in America: An Anthology*, ed. Thomas J. Schlereth (Oxford: AltaMira, 1999), 9–15.
30. Smith, *Gender of History*, 119, 135–36, 146.
31. William James, letter to William Coolidge Lane, March 21, 1908, in "Information about the box" folder, box 62, HUC.
32. That distinction was famously codified by Hilary Jenkinson in *A Manual of Archive Administration* (Oxford: Clarendon, 1922), 106 ("the Archivist is not, and ought not to be, an Historian"), but has a longer history; see Margaret Procter, "Consolidation and Separation: British Archives and American Historians at the Turn of the Twentieth Century," *Archival Science* 6, no. 3/4 (2006): 361–79.
33. Susan Crane offers a rare critique of this tendency in "Writing the Individual Back into Collective Memory," *American Historical Review* 102, no. 5 (1997): 1372–85.
34. See Assmann, *Cultural Memory*, 196.
35. See, for example, Michael Kammen, *Mystic Chords of Memory: The Transformation of Tradition in American Culture* (New York: Vintage Books, 1991).

36. Eugène Minkowski, *Lived Time: Phenomenological and Psychopathological Studies*, trans. Nancy Metzel (1933; Evanston, IL: Northwestern University Press, 1970), 80; Martin Heidegger, *Being and Time*, trans. Joan Stambaugh (1927; Albany, NY: SUNY Press, 2010), esp. 292–97.
37. See Fredric Jameson, "Progress versus Utopia; or, Can We Imagine the Future?," *Science Fiction Studies* 9, no. 2 (1982): 147–58. Elizabeth Grosz defines an "open future" (and thus implies its opposite) in *Time Travels: Feminism, Nature, Power* (Durham, NC: Duke University Press, 2005), esp. 110.
38. See Elizabeth Grosz, "Thinking the New: Of Futures Yet Unthought," in *Becomings: Explorations in Time, Memory, and Futures*, ed. Elizabeth Grosz (Ithaca, NY: Cornell University Press, 1999), 15–16, and Grosz, *The Nick of Time: Politics, Evolution, and the Untimely* (Durham, NC: Duke University Press, 2004), 260.
39. See Barbara Adam and Chris Groves, *Future Matters: Action, Knowledge, Ethics* (Leiden: Brill, 2007), quotation on 11; and Elena Esposito, *The Future of Futures: The Time of Money in Financing and Society* (Cheltenham, UK: Edward Elgar, 2011).
40. Minkowski, *Lived Time*, 100.
41. See François Weil, *Family Trees: A History of Genealogy in America* (Cambridge, MA: Harvard University Press, 2013), 143–79.
42. John Durham Peters, *Speaking into the Air: A History of the Idea of Communication* (Chicago: University of Chicago Press, 1999).
43. Gillian Beer, "Imagining Posterity, Then and Now," *International Literary Quarterly* no. 5 (November 2008), http://www.interlitq.org/issue5/gillian_beer/job.php#.
44. See Michael Warner, "Irving's Posterity," *ELH* 67 (2000): 773–99.
45. Ernst Bloch, *The Principle of Hope*, trans. Neville Plaice, Stephen Plaice, and Paul Knight, 3 vols. (1959; Cambridge, MA: MIT Press, 1995), 2:623, 1:145–46.
46. See, for example, Avner De-Shalit, *Why Posterity Matters: Environmental Policies and Future Generations* (London: Routledge, 1995); Wilfred Beckerman and Joanna Pasek, *Justice, Posterity, and the Environment* (New York: Oxford University Press, 2001); Christopher Groves, *Care, Uncertainty and Intergenerational Ethics* (New York: Palgrave Macmillan, 2014); and, on posterity's economic rights, David Willetts, *Pinch: How the Baby Boomers Took Their Children's Future—And Why They Should Give It Back* (London: Atlantic Books, 2011).
47. Carl Becker, *The Heavenly City of the Eighteenth-Century Philosophers* (New Haven, CT: Yale University Press, 1932), 130–37.
48. Becker, 137–58. On the Romantics' "cult of posterity," see Andrew Bennett, *Romantic Poets and the Culture of Posterity* (New York: Cambridge University Press, 1999).
49. David Lowenthal, "Stewarding the Future," *Norwegian Journal of Geography* 60, no. 1 (2006): 17.

50. Hans Jonas, "Ontological Grounding of a Political Ethics: On the Metaphysics of Commitment to the Future of Man," in *The Public Realm: Essays on Discursive Types in Political Philosophy*, ed. Reiner Schurmann (Buffalo, NY: SUNY Press, 1989), quotation on 154.
51. David Lowenthal, *The Past Is a Foreign Country* (New York: Cambridge University Press, 1985), xvi–xvii.
52. Harold A. Innis, *The Bias of Communication* (1951; Toronto: University of Toronto Press, 2008), esp. 33–60.
53. Henry Steele Commager, "Commitment to Posterity: Where Did It Go?," *American Heritage* 27, no. 5 (1976): 5–6; Christopher Lasch, "The Narcissist Society," *New York Review of Books*, September 30, 1976, 5–12, and Lasch, *The Culture of Narcissism: American Life in an Age of Diminishing Expectations* (1978; New York: W. W. Norton 1991), 51.
54. Innis, *Bias of Communication*, 82–83.
55. John B. Thompson, *The Media and Modernity: A Social Theory of the Media* (Stanford, CA: Stanford University Press, 1995), 219–25.
56. Lowenthal, "Stewarding the Future," 17–18.

CHAPTER ONE

1. Details regarding Deihm's Century Safe (aka Centennial Safe) from J. S. Ingram, *The Centennial Exposition, Described and Illustrated* (Philadelphia: Hubbard Bros., 1876), 620. On Mosher's exhibit, see "Photography in the Great Exhibition," *Philadelphia Photographer* 13, no. 150 (1876): 184. The only secondary source specifically on Mosher is Larry A. Viskochil, "Chicago's Bicentennial Photographer: Charles D. Mosher," *Chicago History* 5 (Summer 1976): 95–104. Visitor numbers from Thomas J. Schlereth, "The 1876 Centennial: A Model for Comparative American Studies," in *Artifacts and the American Past*, ed. Thomas J. Schlereth (Nashville, TN: American Association for State and Local History, 1980), 133.
2. Deihm, standardized printed letter, December 15, 1876, CSC.
3. Estimates of the number of Mosher photographs vary; this figure from "To Be Opened A.D. 1976," *Chicago Tribune*, May 19, 1889, 11.
4. Allan Sekula, "The Traffic in Photographs," *Art Journal* 41 (1981): 18–19. Even the archiving of photographs—for policing, medical, colonial, or industrial purposes—is a kind of circulation, albeit one restricted to "experts."
5. Review of *The Pencil of Nature*, by Henry Fox Talbot, *Athenaeum* 904 (1845): 202.
6. "1876 Century Safe Lost in Oblivion," *Washington Post and Times Herald*, September 12, 1954, Centennial Safe—Correspondence folder, Art and Reference Files (hereafter A&RF), AOC.
7. "Charles D. Mosher: Veteran Photographer Succumbs to Attack of Apoplexy," undated newspaper clipping, Charles D. Mosher Papers, 1876–1915, CHM (hereafter Mosher Papers).

8. Requests for copies in Mosher Papers; prices in Charles D. Mosher, *Catalogue of Memorial Photographs of Prominent Persons Whose Likenesses Will Appear in Memorial Halls at the Second Centennial, 1976* (Chicago: Mosher, 1887), 7, Mosher Papers; he claimed to provide the memorial portraits "without charges" (*Catalogue of Memorial Photographs*, 2).
9. *National Republican* [Washington] (1879), quoted in C. F. Deihm, ed., *President James A. Garfield's Memorial Journal* (New York: C. F. Deihm, 1882), 198. Mosher referred to his award on the back of his photographs.
10. Charles D. Mosher, "Voices from the Craft," *Philadelphia Photographer* 13, no. 146 (1876): 45.
11. Archibald Wilberforce, ed., *The Capitals of the Globe* (New York: Peter Fenelon Collier, 1893), 510; George Brown Goode, *First Draft of a System of Classification for the World's Columbian Exposition* (Washington, DC: GPO, 1893), 654; both refer to the Columbian Exposition.
12. See Bruno Giberti, *Designing the Centennial: A History of the 1876 International Exhibition in Philadelphia* (Lexington: University Press of Kentucky, 2002).
13. Mosher, *Catalogue of Memorial Photographs* (1887).
14. The term is Tony Bennett's, from *The Birth of the Museum: History, Theory, Politics* (London: Routledge, 1995), 59–88. See also Neil Harris, *Cultural Excursions: Marketing Appetites and Cultural Tastes in Modern America* (Chicago: University of Chicago Press, 1990), 56–81.
15. Deihm, *Garfield's Memorial Journal*, 196. On vitrines, see Giberti, *Designing the Centennial*, 118–39.
16. Joan Brown, "Mrs. Deihm's Centennial Safe: History Lives in a Tiffany Inkwell," *Washington Star*, June 19, 1977; "1876 Century Safe Lost in Oblivion," *Washington Post and Times Herald*, September 12, 1954, 5, Centennial Safe—Correspondence folder, A&RF, AOC.
17. Deihm, *Garfield's Memorial Journal*, 196, 199.
18. Giberti, *Designing the Centennial*, 24 (emphasis in original). See also Michael Kammen, *Mystic Chords of Memory: The Transformation of Tradition in American Culture* (New York: Vintage Books, 1991), 135–36; and Schlereth, "1876 Centennial," 139–40.
19. Kammen makes that inference in *Mystic Chords*, 138, as does Lynn Spillman in *Nation and Commemoration: Creating National Identities in the United States and Australia* (New York: Cambridge University Press, 1997), 70, and Michael Bellesiles in *1877: America's Year of Living Violently* (New York: New Press, 2010), 18–19.
20. Editorial, *United States Centennial Welcome* 1, no. 1 (1876): 2, in box 3 (large) of miscellaneous materials, CSC.
21. "Bicentennial Opens Up New Interest in Time Capsules," *NYT*, June 21, 1976, 62; "Nineteen Hundred and Seventy-Six," *Independent House* [Woodbridge, NJ], June 15, 1876, 4; *State of Ohio: General and Local Laws*, vol. 74 (Columbus: Nevins and Myers, 1877), 535.
22. "Address of Leroy B. Gaston," *Old Folks' Record* 1 (1875): 478.

23. "Joint Resolution on the Celebration of the Centennial in the Several Counties or Towns," *Statutes of the United States of America Passed at the First Session of the Forty-fourth Congress, 1875–76* (Washington: GPO, 1876), 211. Quotation from *Red Cloud Chief* [Red Cloud, Nebraska], May 4, 1876, 2.
24. John W. Forney, quoted in Spillman, *Nation and Commemoration*, 79–80.
25. "Washington in 1976," *Washington Star-News*, May 12, 1876, 1.
26. Walt Whitman, *Democratic Vistas: The Original Edition in Facsimile*, ed. Ed Folsom (Iowa City: University of Iowa Press, 2010), lix, lxi, 3, 60–61, 135.
27. Kammen, *Mystic Chords*, 136. See also Giberti, *Designing the Centennial*, 24, and Spillman, *Nation and Commemoration*, 79.
28. "The Old Year," *NYT*, December 31, 1876, 6.
29. Dee Brown, *The Year of the Century: 1876* (New York: Scribner, 1966), 346–47.
30. Whitman, *Democratic Vistas*, 4, 70, 71, 73.
31. Whitman, 3, 33, 36–37, 40–41.
32. "The Old Year," 6.
33. The Remington No. 1 went into production in 1874 but was fully launched at the exposition; see "A New Writing Machine," *National Stenographer* 3, no. 4 (1892): 187–89. Brian Dolan, *Inventing Entertainment: The Player Piano and the Origins of an American Musical Industry* (Lanham: Rowman and Littlefield, 2009), 42. Ernest Freeberg, *Age of Edison: Electric Light and the Invention of Modern America* (Harmondsworth: Penguin, 2013), 18. Brown, *Year of the Century*, 133.
34. See James Carey, *Communication as Culture: Essays on Media and Society* (New York: Routledge, 1989), 201–30.
35. John Durham Peters, *Speaking into the Air: A History of the Idea of Communication* (Chicago: University of Chicago Press, 1999), 94–101, 142–44; Jeffrey Sconce, *Haunted Media: Electronic Presence from Telegraphy to Television* (Durham, NC: Duke University Press, 2000), 21–58; Molly McGarry, *Ghosts of Futures Past: Spiritualism and the Cultural Politics of Nineteenth-Century America* (Berkeley: University of California Press, 2008), 14, 20.
36. George H. Whitman, in *Ceremony at the Sealing of the Century Box by the Ancient and Honorable Artillery Company in Faneuil Hall, Boston* [. . .] (Boston: Alfred Mudge, 1882), 36.
37. Carolyn Marvin, *When Old Technologies Were New: Thinking about Electric Communication in the Late Nineteenth Century* (New York: Oxford University Press, 1988), 87–97.
38. Charles Barnard, "Kate: An Electro-Mechanical Romance," and William John Johnston, "A Centennial-Telegraphic Romance," in *Lightning Flashes and Electric Dashes*, ed. W. J. Johnston (New York: Johnston, 1877), 53–62 and 100–111, respectively. Michel Chion elaborates on the "acousmatic" in *Sound: An Acoulogical Treatise*, trans. James A. Steintrager (1998; Durham, NC: Duke University Press, 2016), 131–49.

39. Marvin, *When Old Technologies Were New*, 89–91.
40. David M. Henkin, *The Postal Age: The Emergence of Modern Communications in Nineteenth-Century America* (Chicago: University of Chicago Press, 2006); Richard R. John, *Spreading the News: The American Postal System from Franklin to Morse* (Cambridge, MA: Harvard University Press, 1995).
41. "Nineteen Hundred and Seventy-Six," 4; Antiquarian Postoffice, discussed below (chap. 2).
42. Charles Sanders Peirce, *The Essential Peirce: Selected Philosophical Writings*, vol. 2, *1893–1913* (Bloomington: Indiana University Press, 1998), 7–8.
43. Charles D. Mosher, *Catalogue of Memorial Historical Photographs of Prominent Men and Women and Souvenirs for the Second Centennial, 1976, By C. D. Mosher, for his Memorial Offering to Chicago* (Chicago: Mosher, 1883), 23; Mosher, *Catalogue of Memorial Photographs* (1887), 6.
44. On the physionotrace, see Wendy Bellion, "Heads of State: Profiles and Politics in Jeffersonian America," in *New Media: 1740–1915*, ed. Lisa Gitelman and Geoffrey Pingree (Cambridge, MA: MIT Press, 2003), 31–60; on human taxidermy, see Edward L. Schwarzschild, "Death-Defying/Defining Spectacles: Charles Wilson Peale as Early American Freak Showman," in *Freakery: Cultural Spectacles of the Extraordinary Body*, ed. Rosemarie Garland Thomson (New York: New York University Press, 1996), 82–96, quotation on 87; on embalming, see Gary Laderman, *Rest in Peace: A Cultural History of Death and the Funeral Home in Twentieth-Century America* (New York: Oxford University Press, 2003), 6, 14–15.
45. John Moring, *Early American Naturalists: Exploring the American West, 1804—1900* (Lanham, MD: Taylor Trade, 2005), 171–86; Quentin R. Skrabec Jr., *H. J. Heinz: A Biography* (Jefferson, NC: McFarland, 2009), 71.
46. Teresa Barnett, *Sacred Relics: Pieces of the Past in Nineteenth-Century America* (Chicago: University of Chicago Press, 2013), 25–26, 39–40, 42.
47. Patricia West, *Domesticating History: The Political Origins of America's House Museums* (Washington, DC: Smithsonian, 1999), 1–37.
48. Scroll (and protographs) in CSC; Deihm deposited further autographs in the form of an album titled "Citizens' Autographs: United States 1876." Mosher's photographs in cabinet card photographs collection, CHM; see also the two ledger books—a photograph register and an autographs volume—in boxes 7 and 8, Mosher Papers. "Eminent men" quoted in Deihm, *Garfield's Memorial Journal*, 196.
49. Tamara Plakins Thornton, *Handwriting in America: A Cultural History* (New Haven, CT: Yale University Press, 1996), 87; Edgar Allan Poe, "A Chapter on Autography," *Graham's Magazine* 19 (November 1841): 224–34.
50. Barbara McCandless, "The Portrait Studio and the Celebrity," *Photography in Nineteenth-Century America*, ed. Martha Sandweiss (Fort Worth, TX: Amon Carter Museum, 1991), 55.
51. McCandless, 65. Such claims did not entail a literal realism; Mosher, as we will see, employed retouching. Yet such interventions were believed

to access a deeper truth; see Miles Orvell, *The Real Thing: Imitation and Authenticity in American Culture, 1880–1940* (Chapel Hill: University of North Carolina Press, 1989), 90, 95.

52. See, for example, W. H. Lipton, "Anatomy, Phrenology, and Physiognomy, and Their Relation to Photography, No. 1," *Philadelphia Photographer* 14, no. 158 (1877): 53–54.
53. Mosher, *Catalogue of Memorial Photographs* (1887), 2, back cover.
54. Mosher, 27; "A Gigantic Plan: The Coming Union Memorial Home," *Inter-Ocean*, n.d., scrapbook, 149, box 8, Mosher Papers.
55. Mosher, *Catalogue of Memorial Historical Photographs* (1883), 23.
56. Mosher, 24; Charles D. Mosher, *A Scrap-Book and Half-Hour Chit-Chats with the President and Lawmakers at Washington* (Chicago: Mosher, 1892), 11, Mosher Papers.
57. Susan A. Crane, "The Pictures in the Background: History, Memory, and Photography in the Museum," in *Memory and History: Understanding Memory as Source and Subject*, ed. Joan Tumblety (New York: Routledge, 2013), 123–28.
58. Alan Trachtenberg, *Reading American Photographs: Images as History, Mathew Brady to Walker Evans* (New York: Macmillan, 1990), 43.
59. On Brady's Washington studio and his photographing of politicians, see Mary Panzer, *Mathew Brady and the Image of History* (Washington, DC: Smithsonian, 2004), 93, 99. Edward Anthony and J. M. Edwards were apparently the first, in 1843, to photograph every member of Congress; McCandless, "Portrait Studio," 55–56.
60. Trachtenberg, *Reading American Photographs*, 32; see also 43, 48.
61. Trachtenberg, 38, 49.
62. McCandless, "Portrait Studio," 54, 55; Trachtenberg, *Reading American Photographs*, 37, 38.
63. Trachtenberg, *Reading American Photographs*, 41; Panzer, *Mathew Brady*, 43–45.
64. "Mosher's Reception," undated newspaper clipping and untitled article by "Junius," undated newspaper clipping, scrapbook, 146, box 8, Mosher Papers.
65. Charles Edwards Lester, preface to Mathew Brady, *Gallery of Illustrious Americans* (New York: Brady, D'Avignon, Lester, 1850).
66. Lester, preface; McCandless, "Portrait Studio," 57.
67. See p. 131, below.
68. "Brady's Gallery," *Harper's Weekly* 7, no. 359 (1863): 722.
69. "A Broadway Valhalla: Opening of Brady's New Gallery," *NYT*, October 6, 1860, 4.
70. Horace Traubel, *With Walt Whitman in Camden*, 3 vols. (New York: Mitchell Kennerley, 1906–14), 3:553.
71. David Brewster, *The Stereoscope: Its History, Theory, and Construction* (London: Murray, 1856), 181.

72. Panzer, *Mathew Brady*, 65, 20.
73. Jeana K. Foley, "Recollecting the Past: A Collection Chronicle of Mathew Brady's Photographs," in Panzer, *Mathew Brady*, esp. 190–91; McCandless, "Portrait Studio," 62; and Robert Wilson, *Mathew Brady: Portraits of a Nation* (New York: Bloomsbury, 2013), 210–11.
74. Elizabeth E. Siegel, "'Miss Domestic' and 'Miss Enterprise', or How to Keep a Photograph Album," and Sarah McNair Vosmeier, "Picturing Love and Friendship: Photograph Albums and Networks of Affection in the 1860s," in *The Scrapbook in American Life*, ed. Susan Tucker, Katherine Ott, and Patricia P. Buckler (Philadelphia: Temple University Press, 2006), 253–55 and 208, respectively.
75. Charles D. Mosher, *Half-an-Hour's Chit-Chat with My Friends, Photography the Subject* (Chicago: Mosher, 1873), 27–28, CHM. Mosher, *Catalogue of Memorial Historical Photographs* (1883), 25; Mosher, *Scrap-Book*, 46.
76. Mosher sold his celebrity photographs both directly and through dealers; Mosher, *Catalogue of Memorial Photographs* (1887), 7. On the mixing of family and celebrity photographs, see Siegel, "'Miss Domestic,'" 253.
77. Mosher, "Improvement in photograph-albums," U.S. Patent 169,186, filed October 26, 1875.
78. "The Photographic Society," *Illustrated London News*, January 22, 1859, 83.
79. James M. Reilly, *The Albumen and Salted Paper Book: The History and Practice of Photographic Printing, 1840–1895* (Rochester, NY: Light Impressions, 1980), 101–3.
80. M. Carey Lea, "An Examination into the Circumstances under Which Silver Is Found in the Whites of Albumen Prints," *Photographic News* 10, no. 415 (1866): 394; "The Care of Pictures—IV: Photographs," *Cassell's Household Guide* 1 (1877): 377.
81. Walter Benjamin, "Little History of Photography," in *Selected Writings*, vol. 2, *1931–1934*, ed. Michael W. Jennings et al. (Cambridge, MA: Belknap Press of Harvard University Press, 2005), 507, 514, 515, 517, 518, 519, 527.
82. Joe Nickell, *Real or Fake: Studies in Authentication* (Louisville: University Press of Kentucky, 2009), 34.
83. Benjamin, "Little History," 515, 517, 518.
84. Mosher, *Half-an-Hour's Chit-Chat*, 30–32.
85. Elizabeth Cady Stanton to Mosher, Mosher Exhibit Manuscripts, box 8, Mosher Papers.
86. Benjamin, "Little History," 519.
87. Mosher, *Catalogue of Memorial Photographs* (1887), 5; see also his cabinet cards' backmarks. Deihm, *Garfield's Memorial Journal*, 197.
88. See Neil Harris, "The Gilded Age Revisited: Boston and the Museum Movement," *American Quarterly* 14, no. 4 (1962): 545–66, esp. 548, 561–62. The Boston Public Library's founders spoke of the "need to meet the future no less than the present demands of the public" in *Second Annual Report of the Trustees of the Public Library* no. 74 (1854): 15. Mount Vernon

quotation from "Editor's Table," *Southern Literary Messenger* 24 (May 1857): 394; see also West, *Domesticating History*, 1–37. Frederick Law Olmsted, "The Value and Care of Parks," in *American Environmentalism: Readings in Conservation History*, ed. Roderick Nash (New York: McGraw-Hill, 1990), 51. Nevertheless, Yellowstone National Park was originally established in 1872 to prevent private acquisition of the land; only in the 1880s did antiutilitarian notions of preserving wilderness for posterity win out; see Roderick Nash, *Wilderness and the American Mind* (1965; New Haven, CT: Yale University Press, 2001), 101, 108–21.

89. George Perkins Marsh, *Man and Nature; or, Physical Geography as Modified by Human Action* (New York: Scribner, 1864), 327–28.
90. Marsh, 327; N. H. Egleston, ed., *Arbor Day: Its History and Observance* (Washington, DC: GPO, 1896); "The Sanitary Value of Trees," *Scientific American* 54, no. 24 (1886): 375; S. B. McCracken, ed., *Michigan and the Centennial* (Detroit: Detroit Free Press, 1876), 172–74.
91. Lewis Mumford, *The City in History: Its Origins, Its Transformations, and Its Prospects* (New York: Houghton Mifflin Harcourt, 1961), 420.
92. David Lowenthal, "Stewarding the Future," *Norwegian Journal of Geography* 60, no. 1 (2006): 17.
93. Hilary Ballon, *The Greatest Grid: The Master Plan of Manhattan, 1811–2011* (New York: Columbia University Press, 2012), 135.
94. Carl Becker, *The Heavenly City of the Eighteenth-Century Philosophers* (New Haven, CT: Yale University Press, 1932), 129, 131, 137, 140, 142, 149.
95. See Michael Lienesch, *New Order of the Ages: Time, the Constitution, and the Making of Modern American Political Thought* (Princeton, NJ: Princeton University Press, 1988), 161–64.
96. Douglass Adair, *Fame and the Founding Fathers: Essays by Douglass Adair*, ed. Trevor Colbourn (New York: W. W. Norton, 1974), 3–26.
97. Andrew Bennett, *Romantic Poets and the Culture of Posterity* (New York: Cambridge University Press, 1999).
98. See Stanley N. Katz, "Republicanism and the Law of Inheritance in the American Revolutionary Era," in *Property Rights in the Colonial Era and Early Republic*, ed. James W. Ely Jr. (New York: Garland, 1997), 157–86.
99. Alexis de Tocqueville, *Democracy in America*, ed. J. P. Mayer (1835–40; New York: Harper Perennial, 1988), 53.
100. Thomas Jefferson to James Madison, September 6, 1789, in *Jefferson: Political Writings*, ed. Joyce Appleby and Terence Ball (New York: Cambridge University Press, 1999), 596.
101. See Peter Fritzsche, *Stranded in the Present: Modern Time and the Melancholy of History* (Cambridge, MA: Harvard University Press, 2004); and see Nick Yablon, "Time and Space," in *A Cultural History of Memory in the Nineteenth Century*, ed. Susan A. Crane (London: Bloomsbury, forthcoming).
102. See Jefferson to George Whythe, January 16, 1796, Thomas Jefferson Papers, LC; and Nick Yablon, *Untimely Ruins: An Archaeology of American*

Urban Modernity, 1819–1919 (Chicago: University of Chicago Press, 2009), 179–80.

103. "Editorial Notes and Gleanings," *National Magazine* 12 (February 1858): 180.
104. Harold A. Innis, *The Bias of Communication* (1951; Toronto: University of Toronto Press, 2008), 172.
105. Innis, 4, 33–60, 64, 76.
106. Mosher, *Catalogue of Memorial Historical Photographs* (1883), 53; "Shades of the Departed," *Inter-Ocean*, n.d., Mosher scrapbook, 148, box 8, Mosher Papers.
107. Signed scroll in CSC.
108. *Libraries of the City of Chicago* (Chicago: Chicago Library Club, 1905), 27.
109. "Mosher's National Photographic Art Gallery," *Chicago Tribune*, March 6, 1881, 5; "Shades of the Departed"; and see Barbara Buckner Higginbotham, *Our Past Preserved: A History of American Library Preservation, 1876–1910* (Boston: G. K. Hall, 1990).
110. Oliver Wendell Holmes Sr., "The Stereoscope and the Stereograph," *Atlantic Monthly* 3, no. 20 (1859): 748.
111. Bonnie G. Smith, *The Gender of History: Men, Women, and Historical Practice* (Cambridge, MA: Harvard University Press, 2000), 105–16.
112. Craig Robertson, "Mechanisms of Exclusion: Historicizing the Archive and the Passport," in *Archive Stories: Facts, Fictions, and the Writing of History*, ed. Antoinette Burton (Durham, NC: Duke University Press, 2006), 73–75; Victor Gondos Jr., *J. Franklin Jameson and the Birth of the National Archives, 1906–1926* (Philadelphia: University of Pennsylvania Press, 1981), 3, 5, 7.
113. Gondos, *Franklin Jameson*, 3, 5; Donald R. McCoy, "The Struggle to Establish a National Archives in the United States," in *Guardian of Heritage: Essays on the History of the National Archives*, ed. Timothy Walch (Washington, DC: National Archives and Records Administration, 1985), 1, 2.
114. Albert Sidney Bolles, "Safes," in *Industrial History of the United States: From the Earliest Settlements to the Present Time* (Norwich, CT: Henry Bill, 1878), 280–85. These claims were obviously overstated, as the Chicago Fire revealed.
115. Mary Jane Moore, "U.S. Capitol's Queerest Safe Must Be Cracked" (typed manuscript, 1944), 2, in Centennial Safe—Correspondence folder, A&RF, AOC; "For Posterity: Ten Thousand Photographs of Prominent Chicagoans Being Prepared for Future Generations," *Chicago Tribune*, February 28, 1885, 7, in scrapbook, 142, box 8, Mosher Papers; "Chicago Has Relic Safe," *NYT*, July 14, 1907, 12.
116. "Notes and Queries," *Harper's Weekly* 1, no. 1 (1857): 11.
117. "After 200 Years, Sleuths Seek Capitol Cornerstone," *Chicago Tribune*, September 17, 1993, N6.
118. Backmark of Mosher cabinet card, CHM; see also below, p. 68.

119. "Centennial Safe," undated typed manuscript, in Centennial Safe—Correspondence folder, A&RF, AOC.
120. Deihm, *Garfield's Memorial Journal*, 197; "Shades of the Departed"; and Mosher, *Catalogue of Memorial Photographs* (1887), 7.
121. Deihm, *Garfield's Memorial Journal*, 196.
122. Deihm, letter "To the People of These United States," February 22, 1879, Centennial Safe folder, A&RF, AOC.
123. Bernard Stiegler, "Memory," in *Critical Terms for Media Studies*, ed. W. J. T. Mitchell and Mark B. N. Hansen (Chicago: University of Chicago Press, 2010), 66–68.
124. Innis, *Bias of Communication*, 34. Innis does suggest that paper subsequently stripped architecture of this role (24).
125. Mumford, *City in History*, 99.
126. Max Page, *The Creative Destruction of Manhattan, 1900–1940* (Chicago: University of Chicago Press, 2001), esp. 21–68, 111–44; Randall Mason, *The Once and Future New York: Historic Preservation and the Modern City* (Minneapolis: University of Minnesota Press, 2009).
127. William Dean Howells, *Their Wedding Journey* (Boston: Osgood, 1872), 37.
128. Tappan, Lyon, and Hone, cited in Yablon, *Untimely Ruins*, 107–46; Walter [*sic*] Whitman, "Tear Down and Build Over Again," *American Whig Review* 2, no. 5 (1845): 536.
129. See Yablon, *Untimely Ruins*, 135–40.
130. [James Fenimore Cooper], *Gleanings in Europe. Italy: By An American*, 2 vols. (Philadelphia: Carey, Lea, and Blanchard, 1838), 2:153, 155, 156; 1:220.
131. "Foreign Buildings at the Centennial," *American Architect and Building News*, March 25, 1876, 101.
132. "Vandals at the Fair," *Chicago Tribune*, January 8, 1894, 11; see also Ross Miller, *American Apocalypse: The Great Fire and the Myth of Chicago* (Chicago: University of Chicago Press, 1990), 244–50.
133. Giberti, *Designing the Centennial*, 73–74.
134. Tappan, quoted in Yablon, *Untimely Ruins*, 111.
135. Deihm, *Garfield's Memorial Journal*, 197. See Yablon, "'Land of Unfinished Monuments': The Ruins-in-Reverse of Nineteenth-Century America," *American Nineteenth Century History* 13, no. 2 (2012): 153–97.
136. Edgar Allan Poe, "Mellonta Tauta" (1850), in *Edgar Allan Poe: Poetry and Tales*, ed. Patrick F. Quinn (New York: Library of America, 1978), 882–84; on New York's aborted Washington Monument, see Yablon, "Land of Unfinished Monuments," esp. 154, 164–65.
137. Yablon, *Untimely Ruins*, 7, 41–42, 56–57.
138. Tocqueville, *Democracy in America*, 283, 469–70.
139. Montgomery Schuyler, "The Brooklyn Bridge as Monument" (1883), in *American Architecture: Studies by Montgomery Schuyler* (New York: Harper, 1892), 71–72, 76, 78–80. John Ames Mitchell, *The Last American: A Fragment from the Journal of Khan-li* (New York: Frederick A. Stokes and Brother, 1889), 36.

140. Whitman, "Crossing Brooklyn Ferry," in *Leaves of Grass* (Boston: Thayer and Eldridge, 1860), 380, 382, 383.
141. Whitman, 382. See Paul A. Orlov, "On Time and Form in Whitman's 'Crossing Brooklyn Ferry,'" *Walt Whitman Quarterly Review* 2 (Summer 1984): 12–21; Roger Gilbert, "From Anxiety to Power: Grammar and Crisis in 'Crossing Brooklyn Ferry,'" *Nineteenth-Century Literature* 42 (December 1987): 339–61; Folsom in Whitman, *Democratic Vistas*, xxxiv.
142. Whitman, "Crossing Brooklyn Ferry," 388.
143. Michael Lienesch, "The Constitutional Tradition: History, Political Action, and Progress in American Political Thought, 1787–1793," *Journal of Politics* 42, no. 1 (1980): 4–5, 17–18.
144. "Strike among the Iron Workers," *NYT*, June 3, 1876, 6; "The American Lazzaroni," *Chicago Tribune*, August 5, 1876, 4.
145. Henry George, "Presidential Speech at San Francisco, 1876," in *Henry George: Collected Journalistic Writings*, vol. 1, *The Early Years 1860–1879*, ed. Kenneth C. Wenzer (Armonk, NY: M. E. Sharpe, 2003), 92, 101.
146. M. L. Holbrook, "Topics of the Month," *Herald of Health* 61 (August 1876): 273.
147. Mosher, *Catalogue of Memorial Photographs* (1887), 6.
148. Deihm, *Garfield's Memorial Journal*, 196; "Shades of the Departed."
149. W. H. D., "How Many More?," *United States Centennial Welcome*, March 15, 1876, 7, box 3, CSC.
150. Deihm, *Garfield's Memorial Journal*, 197.
151. Walter Benjamin, "On the Concept of History," in *Selected Writings*, vol. 4, *1938–1940* (Cambridge, MA: Belknap Press of Harvard University Press, 1996), 395.
152. The century is not, of course, a universal unit of time; on the sixty-year cycle in Chinese culture; see Achim Mittag, "Time Concepts in China," in *Time and History: The Variety of Cultures*, ed. Jörn Rüsen (New York: Berghahn, 2007), 47.
153. Barbara Adam and Chris Groves, *Future Matters: Action, Knowledge, Ethics* (Boston: Brill, 2007), 11–15, 57–58.
154. "The Centennial Autograph Album," *United States Centennial Welcome*, March 15, 1876, 1, box 3, CSC.
155. [Deihm], "For United States Centennial Exposition" (circular, 1876), in Charles Henry Hardin Papers, 1875–1877, Office of Governor, Record Group 3.22, Missouri State Archives, Jefferson City.
156. Mosher, *Scrap-Book*, 30, 31. Reference to National Historical Photographer in Mosher, *Catalogue of Memorial Photographs* (1887), 5 (emphasis added).
157. "The Centennial Autograph Album," 1.
158. Mosher, *Catalogue of Memorial Historical Photographs* (1883), 5.
159. Rev. S. Domer, *The Self-Assertion of Our Citizenship: A Sermon Preached on the Last National Thanksgiving Day in the First Century of the American Republic, in Foundry Church, Washington, DC* (Washington: Bayne, 1875), 9, 11–12, 13, 15.

160. Mosher, *Scrap-Book*, 6–7, 8, 11, 51.
161. Mrs. S. H. Stevenson to Mosher, November 24, 1873, box 8, Mosher Papers. T. H. Huxley, "On the Geographical Distribution of the Chief Modifications of Mankind," *Journal of the Ethnological Society of London* 2, no. 1 (1870): 404–12. On the 1869 project, see Ann Maxwell, *Picture Imperfect: Photography and Eugenics, 1870–1940* (Brighton: Sussex Academic, 2010), 29–34.
162. Francis Galton, "Photographic Chronicles From Childhood to Age," *Fortnightly Review* 37, no. 181 (1882): 31. See also Maxwell, *Picture Imperfect*, 79–107.
163. Mosher, *Catalogue of Memorial Historical Photographs* (1883), AA. On eugenics' slow emergence in the United States, see Maxwell, *Picture Imperfect*, 7–10, 109, 111–46.
164. *United States Centennial Welcome* and *Our Second Century*, box 3, CSC; see endorsements in Deihm, *Garfield's Memorial Journal*, 200. As seen below, the bank president did not perform this task.
165. Mosher, *Catalogue of Memorial Historical Photographs* (1883), 33. See Marshall Field to Mosher (ca. 1881), Mosher Exhibit Manuscripts, box 8, Mosher Papers.
166. Mosher, *Catalogue of Memorial Historical Photographs* (1883), 48. On rising militancy in Chicago, see Richard Schneirov, *Labor and Urban Politics: Class Conflict and the Origins of Modern Liberalism in Chicago, 1864–97* (Urbana: University of Illinois Press, 1998); on strikes at the McCormick plant in 1873, see 45n59; on the failed strike at Armour's in 1879, see 107–8.
167. Mosher, *Catalogue of Memorial Photographs* (1887), 25.
168. Mosher's prices appear on the backmark of some of his cabinet cards, CHM. On the bourgeois ritual of the visit to the portrait photographer, see John Tagg, *The Burden of Representation: Essays on Photographies and Histories* (Amherst: University of Massachusetts Press, 1988), 37, 43; on the cabinet card's prestige, see 50.
169. Mosher, "Specialties," in *Philadelphia Photographer* 14, no. 168 (1877), unpaginated advertising supplement; Mosher, *Half-an-Hour's Chit-Chat*, 26.
170. See Oscar Lewis, *Mug Books: A Dissertation Concerning the Origins of a Certain Familiar Diversion of Americana* (Indianapolis, IN: Grabhorn, 1934).
171. Mosher's photographs of the convention, SSF—Political Conventions—1880—Republican, Prints and Photographs Division, LC, and political endorsement in scrapbook, box 8, Mosher Papers.
172. "The Centennial Safe," *United States Centennial Welcome*, June 15, 1876, in box 3 (large) of miscellaneous materials, CSC. Gano Dunn, *Peter Cooper (1791–1883): "A Mechanic of New York"* (New York: Newcomen, 1949), 19.
173. W. H. D., "How Many More?," *United States Centennial Welcome*, March 15, 1876, 7, box 3, CSC.

174. William A. Richardson, *Public Debt and National Banking Laws of the United States* (Washington, DC: Morrison, 1873), CSC.
175. Moore, "U.S. Capitol's Queerest Safe," 3.
176. Moore, 3, 4; "1876 Century Safe Lost in Oblivion," *Washington Post and Times Herald*, September 12, 1954, 2, Centennial Safe—Correspondence folder, A&RF, AOC.
177. "For Posterity: Ten Thousand Photographs of Prominent Chicagoans Being Prepared for Future Generations," *Chicago Tribune*, February 28, 1885, 7, in scrapbook, 142, box 8, Mosher Papers.
178. Gretchen Ritter, *Goldbugs and Greenbacks: The Antimonopoly Tradition and the Politics of Finance in America, 1865–1896* (New York: Cambridge University Press, 1999), 52.
179. Stephen J. Rosenstone, Roy L. Behr, and Edward H. Lazarus, *Third Parties in America* (Princeton, NJ: Princeton University Press, 1984), 65.
180. Deihm, *Garfield's Memorial Journal*, 197. Elizabeth Thompson's materials (not mentioned in Deihm's official inventory) in box 2, CSC.
181. Elizabeth Rowell Thompson, *The Figures of Hell; or, The Temple of Bacchus* (privately printed, 1878), 40, 49, CSC. On her support of suffrage, see *Selected Papers of Elizabeth Cady Stanton and Susan B. Anthony*, ed. Ann D. Gordon, 6 vols. (New Brunswick, NJ: Rutgers University Press, 1997–2012), 4:11, 12, 23. For her biography, see Richard C. S. Trahair, *Utopias and Utopians: An Historical Dictionary* (Westport, CT: Greenwood, 1999), 399, and *Notable American Women, 1607–1950*, ed. Edward T. James, Janet Wilson James, and Paul S. Boyer, 3 vols. (Cambridge, MA: Harvard University Press, 1971), 3:453–54.
182. James, James, and Boyer, *Notable American Women*, 3:453; Trahair, *Utopias and Utopians*, 18.
183. Elizabeth Thompson and William Saunders, *Prospectus of the American Worker's Alliance for the Advancement of Educational, Industrial, Cooperative, and Social Reform* (Washington, DC: The Alliance, 1879), 5 box 2, CSC.
184. Dr. Marsh [Arthur Merton], *The Crowned Republic; or, The New Demands of Scientific Knowledge, Association, and Industry* (Boston: The Workers, 1879), 1, 2, 5, 6, 7, 8, 9, 11.
185. Mosher, *Catalogue of Memorial Photographs* (1887), 23; Mosher, *Catalogue of Memorial Historical Photographs* (1883), 10, 20, 22, 30. Mosher, "A Gigantic Plan: The Coming Union Memorial Home," *Inter-Ocean*, n.d., news clippings scrapbook, 149, box 8, Mosher Papers.
186. Mosher, *Catalogue of Memorial Historical Photographs* (1883), 15, 21.
187. Mosher, *Scrap-Book*, 15–18, 25, 28, 34.
188. McCandless, "Portrait Studio," 58, 62–63; Panzer, *Mathew Brady*, 18.
189. Trachtenberg, *Reading American Photographs*, 30–33.
190. Citizen's Autograph Album, CSC.
191. Philip Quilibet [pseud.], "Safe Celebrity," *Galaxy* 22, no. 1 (1876): 125.
192. Mosher, *Scrap-Book*, 22–24, 28, 49.

193. Jennifer E. Manning, *African American Members of the United States Congress: 1870–2011* (Congressional Research Service Report, 2011), 2.
194. On that exclusion, see Kathleen Ann Clark, *Defining Moments: African American Commemoration and Political Culture in the South, 1863–1913* (Chapel Hill: University of North Carolina Press, 2006), 119–26.
195. See William D. Andrews, "Women and the Fairs of 1876 and 1893," *Hayes Historical Journal* 1 (1977): esp. 174.
196. *Statue of Miss Frances E. Willard Erected in Statuary Hall of the Capitol Building at Washington* (Washington: GPO, 1905), AOC.
197. William B. Tyler, "Profiles of the Founders," Thomas Thompson Trust, http://thomasthompsontrust.org/id1.html; Gordon, *Selected Papers of Elizabeth Cady Stanton*, 5:331.
198. Donald B. Marti, *Women of the Grange: Mutuality and Sisterhood in Rural America, 1866–1920* (Westport, CT: Greenwood, 1991); Thompson cited in Marti, 20. *The Crowned Republic*, the book she signed and apparently deposited, also declared that "all persons, of either sex, and of all races . . . have the same classes of rights" (Marsh, *Crowned Republic*, 8).
199. Letters from Stanton, Anthony, and Frances E. Willard, Mosher Exhibit Manuscripts, box 8, Mosher Papers.
200. "Centennial Safe Closed," *NYT*, February 23, 1879, 1; Mosher, *Catalogue of Memorial Historical Photographs* (1883), 35.
201. Mosher, *Scrap-Book*, 50, 24.
202. Moore, "U.S. Capitol's Queerest Safe," 5, 3, and "Washington's Birthday," *Washington Post*, February 24, 1879, 2, Centennial Safe folder, A&RF, AOC.
203. *New York Tribune*, February 25, 1879, quoted in Charles E. Fairman, *Art and Artists in the Capitol of the United States of America* (Washington, DC: GPO, 1927), Centennial Safe folder, A&RF, AOC.
204. "Observance of the Day: Arrangements Made for the Celebration of Feb. 22," *Chicago Tribune*, February 22, 1889, 8.
205. "To Be Opened A.D. 1976," *Chicago Tribune*, May 19, 1889, 11.
206. *National Republican*, quoted in Deihm, *Garfield's Memorial Journal*, 198.

CHAPTER TWO

1. "Formal Presentation of the Cogswell Fountain," *Daily Evening Bulletin*, June 16, 1879, col. A; "Cogswell Fountain: Dedicated Yesterday in the Presence of a Great Crowd," *Daily Alta California* (hereafter *Alta*), June 15, 1879, 1. Weight in "Ben Franklin's Downfall," "Time capsules" folder, San Francisco Ephemera Collection SF SUB COLL, SFPL.
2. Cogswell, letter "to the Public . . . ," carton 1, Cogswell Papers, 1846–1960, Mss 84/61c, BL (hereafter Cogswell Papers).
3. See Thomas J. Schlereth, "Material Culture Studies in America, 1876–1976," in *Material Culture Studies in America*, ed. Thomas J. Schlereth

(Oxford: AltaMira, 1999), 14. Universities did not emphasize material culture in US history courses until 1932; Michael Kammen, *Mystic Chords of Memory: The Transformation of Tradition in American Culture* (New York: Vintage Books, 1991), 343.

4. Cogswell, diary entry, June 1, 1879, box 2, Cogswell Papers.
5. Cogswell, letter to "the Historical Society of 1979" (June 1, 1879), 1, 6, 7, box 2, folder 24, HDCTCC. Poorhouse and size of fortune mentioned in "Henry D. Cogswell Biographical Research Notes," [1959?], 1, 2, MS 690, CHS. Golden molar mentioned in Frederick C. Moffatt, "The Intemperate Patronage of Henry D. Cogswell," *Winterthur Portfolio* 27, no. 2/3 (1992): 123. Robert D. Lundy, "The Polk Street Background of *McTeague*," in *Frank Norris: McTeague, A Story of San Francisco: Authoritative Text, Contexts, Criticism*, ed. Donald Pizer (New York: W. W. Norton, 1997), 267. On Franklin's epitaph, see Leo Braudy, *The Frenzy of Renown: Fame and Its History* (New York: Vintage Books, 1997), 369.
6. See Cogswell, diary entry, June 6, 1879, box 2, Cogswell Papers. On the fountains' European-inspired designs, see Moffatt, "Intemperate Patronage," 126, 133.
7. Box 1, folder 11; carton 1, folder 87; GS Box 057; box 1, folder 17; and box 2, folder 28, HDCTCC.
8. Box 1, folders 16, 17, HDCTCC.
9. Annette B. Weiner, *Inalienable Possessions: The Paradox of Keeping-While-Giving* (Berkeley: University of California Press, 1992).
10. Cogswell, "The Cogswell Fountains," *Daily Evening Bulletin*, June 22, 1886, 4.
11. Heinrich Schliemann, *Troy and Its Remains* (New York: Scribner, Welford, and Armstrong, 1875).
12. See, for example, *Elite Directory for San Francisco and Oakland* (San Francisco: Argonaut, 1879).
13. Phelps's prospectus is in carton 2, folder 135, HDCTCC. The book was eventually published as *Contemporary Biography of California's Representative Men*, 2 vols. (San Francisco: Bancroft, 1882).
14. On the law suits, see Barbara Allen Babcock, *Woman Lawyer: The Trials of Clara Foltz* (Stanford, CA: Stanford University Press, 2011), 81–82.
15. "Dr. Cogswell's Drinking Fountain," undated newspaper clipping, box 2, folder 28, HDCTCC.
16. Cogswell, letter "to the Public . . ."; diary entries for May 24, 30, and 31, box 2, Cogswell Papers.
17. Cogswell, letter "to the Public . . ."; "Dr. Cogswell's Drinking Fountain."
18. Cogswell, "[Initial] Communication to the Board of Supervisors [final draft]," n.d., box 1, folder 18, HDCTCC; "Cogswell Fountain: Dedicated Yesterday," 1.
19. Frank Soulé, John H. Gihon, and James Nisbet, *The Annals of San Francisco* (New York: Appleton, 1855), 244; *Compendium of the Tenth Census*

(Washington, DC: GPO, 1880), pt. 1, p. 452. Boston's population reached a quarter million by the 1880 census.

20. Cogswell, letter to "the Historical Society of 1979" (June 1, 1879), 1–2, box 2, folder 24, HDCTCC (emphasis added).
21. Cogswell, "[Initial] Communication to the Board of Supervisors [final draft]."
22. Fredric Jameson, *Postmodernism, or The Cultural Logic of Late Capitalism* (Durham, NC: Duke University Press, 1991), 284.
23. Peter Fritzsche, *Stranded in the Present: Modern Time and the Melancholy of History* (Cambridge, MA: Harvard University Press, 2004), esp. 16, 51.
24. Charles Baudelaire, *The Painter of Modern Life and Other Essays*, trans. Jonathan Mayne (London: Phaidon, 1964), 14; Mark Twain and Charles Dudley Warner, *The Gilded Age: A Tale of Today* (Hartford, CT: American, 1873).
25. Carol Kammen, *On Doing Local History* (Walnut Creek, CA: AltaMira, 2003), 12.
26. "Introductory," in *Papers of the California Historical Society* 1, pt. 1 (1887): xx; "California Historical Society 1852–1922," *California Historical Society Quarterly* 1, no. 1 (1922): 9–22; quotation on 13. *San Francisco Directory for the Year Commencing April, 1879*, ed. Henry G. Langley (San Francisco: Francis, Valentine, 1879), 1102.
27. See, for example, the two eighteenth-century colonial documents that CHS published in 1874. Henry L. Oak, "Catalogue of Books and Pamphlets in the Bancroft Library" (manuscript, 1879), carton 24, folder 17, MSS B-C 7, BL.
28. Martin Kellogg, "The Local Units of History," *Papers of the California Historical Society* 1, pt. 1 (1887): 1, 6, 12.
29. Fernand Braudel, *On History*, trans. Sarah Matthews (Chicago: University of Chicago Press, 1980), 3.
30. Book on mining in carton 1, folder 88; book on shipbuilding in carton 2, folder 112; railroad timetables in carton 1, folders 80–81; laundry price list in box 2, folder 26; water rates in carton 1, folder 90; quarantine laws in carton 2, folder 158; fire ordinances in box 2, folder 29; cable car booklet, carton 1, folder 98; menu in carton 2, folder 132; bathing house advertisement in carton 2, folder 129; hot springs prospectus in carton 2, folder 150; geysers guidebook in carton 1, folder 96; hymns, sermons, and tracts in carton 2, folder 159, carton 1, folder 91, carton 2, folder 154; homeopathy manual in box 2, folder 29; patent medicine advertisement in carton 2, folder 162; poetry anthology in carton 1, folder 83; libretto in carton 1, folder 63; Hutchinson program in carton 2, folder 113, HDCTCC.
31. S. N. D. North, *History and Present Condition of the Newspaper and Periodical Press of the United States* (Washington, DC: GPO, 1884), 77.
32. Philip J. Ethington, *The Public City: The Political Construction of Urban Life in San Francisco, 1850–1900* (Berkeley: University of California Press, 2001), 21, 24.

33. Josiah Royce, *California: A Study of American Character; From the Conquest in 1846 to the Second Vigilance Committee in San Francisco* (Boston: Houghton Mifflin, 1886).
34. James Ford Rhodes, "Newspapers as Historical Sources," in *Historical Essays* (New York: Macmillan, 1909), 86.
35. Kellogg, "Local Units of History," 7–8.
36. See Paula Amad, *Counter-Archive: Film, the Everyday, and Albert Kahn's Archives de la Planète* (New York: Columbia University Press, 2010), 178–79.
37. Lucy Maynard Salmon, *The Newspaper and the Historian* (New York: Oxford University Press, 1923), xli; on the continued resistance of historians, see xxxix, xl.
38. Fine arts collection, CHS; see also accession notes, 79-1-1 to 79-1-72. Deihm did include an inkwell and pens but merely for the signing of the autograph albums at the opening.
39. Teresa Barnett, *Sacred Relics: Pieces of the Past in Nineteenth-Century America* (Chicago: University of Chicago Press, 2013), 19–26. See also William L. Bird, Jr., *Souvenir Nation: Relics, Keepsakes, and Curios from the Smithsonian's National Museum of American History* (New York: Princeton Architectural Press, 2013), 7, 9, 33; Schlereth, "Material Culture Studies in America," 1–75; Kammen, *Mystic Chords*, 146–49.
40. Barnett, *Sacred Relics*, 18–25, 107.
41. Estimate from "The Cogswell Fountain," *Daily Evening Bulletin*, June 10, 1879, col. C.
42. See Henry George, *Progress and Poverty: An Inquiry into the Cause of Industrial Depressions and of Increase of Want with Increase of Wealth the Remedy* (1879; New York: D. Appleton, 1886), 248.
43. Henry Fountain, "Date Seed of Masada Is Oldest Ever to Sprout," *NYT*, June 17, 2008, F3.
44. As Robespierre's remains lay mixed with those of his supporters (relocated to the Catacombs), it seems unlikely to be his tooth.
45. Newspapers in carton 1, folders 30–61, HDCTCC.
46. F.L. Gordon, letter, oversize box 1, folder 177, HDCTCC.
47. Mary P. Ryan, *Civic Wars: Democracy and Public Life in the American City during the Nineteenth Century* (Berkeley: University of California Press, 1997), 225–58.
48. *Housekeeper's Encyclopedia* prospectus in carton 1, folder 86; *Cottage Hearth* in carton 1, folder 85, HDCTCC; needlework artifacts, lace collar, and pressed flower in the fine arts collection, CHS.
49. *Gem* in carton 1; Cooper, letter with poem in box 2, folder 26; Cooper, letter to Cogswell and wife in box 1, folder 15, HDCTCC.
50. Pamphlet in carton 1, folder 77, HDCTCC. On Gordon and Foltz's efforts, see Babcock, *Woman Lawyer*, 47–57.
51. Laura De Force Gordon, *Great Geysers of California, and How to Reach Them* (San Francisco: Bacon, 1877), frontispiece, carton 1, folder 96, HDCTCC.

52. Cogswell, note, box 1, folder 20, HDCTCC.
53. "Henry D. Cogswell Biographical Research Notes," [1959?], 4, MS 690, CHS.
54. Babcock, *Woman Lawyer*, 79–84.
55. *Souvenirs of San Francisco* (Philadelphia: Janentzky, [1879?]), carton 2, folder 155, HDCTCC.
56. Peter Bacon Hales, *Silver Cities: Photographing American Urbanization, 1839–1939* (1984; Albuquerque: University of New Mexico Press, 2006), 167; Rebecca Solnit, *River of Shadows: Eadweard Muybridge and the Technological Wild West* (New York: Viking, 2003), 161, 167.
57. Gunther Barth, *Instant Cities: The Rise of Modern City Culture in Nineteenth-Century America* (New York: Oxford University Press, 1980), 80.
58. Hales, Solnit, and others see Muybridge as catering to contemporaneous patrons and customers.
59. Hales, *Silver Cities*, 151–52, 167, 171.
60. Eadweard Muybridge, letter to William Kelby, May 13, 1897 (emphasis added), NYHS archives.
61. G. W. P., "Physiognomy of Cities," *American Whig Review* 6, no. 2 (1847): 234–35.
62. Herodotus, *The History of Herodotus*, trans. G. C. Macaulay, 2 vols. (London: Macmillan, 1890), 2:149.
63. "Cogswell Fountain," *Alta*, June 15, 1879, 1; blessing in box 2, folder 24, HDCTCC.
64. Cogswell, diary entry, June 14, 1879, box 2, Cogswell Papers.
65. Cogswell, letter to "the Historical Society of 1979" (June 1, 1879), 2, box 2, folder 24, HDCTCC. On the lack of public amenities in San Francisco before 1900, see Ryan, *Civic Wars*, 204.
66. "Formal Presentation of the Cogswell Fountain," *Daily Evening Bulletin*, June 16, 1879, col. A; Cogswell disavows temperance affiliation in "The Cogswell Fountains," *Daily Evening Bulletin*, June 22, 1886, 4. Cogswell, diary entry, June 14, 1879, box 2, Cogswell Papers; on Murphy, see Jack S. Blocker, David M. Fahey, and Ian R. Tyrrell, *Alcohol and Temperance in Modern History: An International Encyclopedia* (Santa Barbara: ABC-CLIO, 2003), 107–9, 426–27.
67. Moffatt, "Intemperate Patronage," 126, 133, 136.
68. Moffatt, 124–25; *Our Head Light* in carton 1, HDCTCC.
69. Based on Richard Mendelson's estimate of one hundred thousand saloons in America by 1872, in *From Demon to Darling: A Legal History of Wine in America* (Berkeley: University of California Press, 2009), 31.
70. Letters in box 1, folder 18, and box 2, folder 25, HDCTCC.
71. *Municipal Reports for the Fiscal Year 1880–81* (San Francisco: George Spaulding, 1881), 262.
72. Cogswell, letter to "the Historical Society of 1979," 2.
73. Moffatt, "Intemperate Patronage," 124.

74. See Nan Alamilla Boyd, *Wide-Open Town: A History of Queer San Francisco to 1965* (Berkeley: University of California Press, 1995), 45–46; and Gilman M. Ostrander, *The Prohibition Movement in California, 1848–1933* (Berkeley: University of California Press, 1957).
75. Cogswell, letter to "the Historical Society of 1979," 2 (emphasis in original).
76. John J. Rumbarger, *Profits, Power, and Prohibition: Alcohol Reform and the Industrializing of America, 1800–1930* (Albany, NY: SUNY Press, 1989), esp. 57–68.
77. Cogswell, letter to "the Historical Society of 1979," 1 (postscript).
78. Cogswell, 6.
79. Benjamin E. Lloyd, *Lights and Shades in San Francisco* (San Francisco: A. L. Bancroft, 1876), 451. On Cogswell's "remarkable frugality," see "More Gold to Enrich the Heirs," *San Francisco Call* [hereafter *Call*], February 21, 1902, 9.
80. See Peter R. Decker, *Fortunes and Failures: White-Collar Mobility in Nineteenth-Century San Francisco* (Cambridge, MA: Harvard University Press, 1978), 163–65.
81. Ryan, *Civic Wars*, 188; Decker, *Fortunes and Failures*, 167. Terrence J. McDonald cites rankings in *The Parameters of Urban Fiscal Policy: Socioeconomic Change and Political Culture in San Francisco, 1860–1906* (Berkeley: University of California Press, 1996), 66.
82. "Eloquent Statistics: American Industrial and Commercial Triumphs," undated newspaper clipping, box 2, folder 28, HDCTCC.
83. Decker, *Fortunes and Failures*, 239, 194.
84. Decker, 229–30.
85. Guillermo Prieto, *San Francisco in the Seventies: The City as Viewed by a Mexican Political Exile*, trans. and ed. Edwin S. Morby (1877; San Francisco: J. H. Nash, 1938), 8.
86. Walton Bean, *California: An Interpretive History* (New York: McGraw-Hill, 1968), 220; Decker, *Fortunes and Failures*, 160.
87. S. P. Dewey, *The Bonanza Mines of Nevada: Gross Frauds in the Management Exposed* [. . .] (San Francisco: privately printed, 1878), carton 1, folder 88, HDCTCC.
88. Antone Otto, message, box 2, folder 27, HDCTCC. On the drought, see Kevin Starr, *Americans and the California Dream, 1850–1915* (New York: Oxford University Press, 1973), 132.
89. Alexander Saxton, *The Indispensable Enemy: Labor and the Anti-Chinese Movement in California* (Berkeley: University of California Press, 1975), 106; William Issel and Robert W. Cherny, *San Francisco, 1865–1932: Politics, Power, and Urban Development* (Berkeley: University of California Press, 1987), 125.
90. Peter Tamony, "Hoodlums and Folk Etymology," *Western Folklore* 28, no. 1 (1969): 44–48.

91. Cogswell, letter to "the Historical Society of 1979," 3.
92. Saxton, *Indispensable Enemy*, 70, 108, 159; Ira Brown Cross, *A History of the Labor Movement in California* (Berkeley: University of California Press, 1935), 31, 132.
93. Henry George, "What the Railroad Will Bring Us," *Overland Monthly* 1, no. 4 (1868): 302.
94. George, *Progress and Poverty*, 9, 130–31, 246–47, 305.
95. Saxton, *Indispensable Enemy*, 16.
96. The *Alta* was the exception.
97. Jean Pfaelzer, *Driven Out: The Forgotten War against Chinese Americans* (Berkeley: University of California Press, 2008), 77; Michael Kazin, "The July Days in San Francisco, 1877: Prelude to Kearneyism," in *The Great Strikes of 1877*, ed. David O. Stowell (Urbana: University of Illinois Press, 2008), 149–51.
98. Ethington, *Public City*, 271–77; Bean, *California*, 285; Cross, *History of the Labor Movement in California*, 130.
99. Cross, *History of the Labor Movement in California*, 107.
100. Ethington, *Public City*, 244, 279; Issel and Cherny, *San Francisco*, 128.
101. A "minor riot" took place a month before the vessel ceremony; see "Yesterday's Election," *Alta*, May 8, 1879, 1, carton 1, HDCTCC.
102. "Macaulay's Prophecy," *Argonaut*, May 31, 1879, 7, carton 1, HDCTCC. On American appropriations of Macaulay's warnings, see Nick Yablon, *Untimely Ruins: An Archaeology of American Urban Modernity, 1819–1919* (Chicago: University of Chicago Press, 2009), 147–89.
103. "Pause! Read! Reflect! The New Constitution in Full," *Alta*, April 13, 1879, 1, carton 1, and "Is San Francisco in Earnest?," *Alta*, April 13, 1879, 3, carton 1, HDCTCC.
104. Henry George, "The Kearney Agitation in California," *Popular Science Monthly* 17 (August 1880): 446.
105. Cogswell, diary entry, May 7, 1879, box 2, Cogswell Papers; "Yesterday's Election," *Alta*, May 8, 1879, 1, carton 1, HDCTCC. On the women's rights clauses, see Babcock, *Woman Lawyer*, 53–55.
106. Robert Fogelson, *America's Armories: Architecture, Society, and Public Order* (Cambridge, MA: Harvard University Press, 1989); on the growth of the San Francisco police and the arrest of vagrant children, see Ryan, *Civic Wars*, 218.
107. "Various Methods of Combating Revolution," *Monitor*, June 5, 1879, carton 1, HDCTCC.
108. The dental college (which never materialized because of disagreements with the university) and the women's shelter are documented in box 1, folders 5–13, and box 2, folder 21, HDCTCC, respectively. Sarah B. Cooper, "The Kindergarten in Its Relation to Industrial Education," in *Eighth Annual Report of the Golden Gate Kindergarten Association* (San Francisco: Spaulding, 1887), 30–31.

109. Cogswell, letter to "the Historical Society of 1979," 3.
110. Cogswell, "[Initial] Communication to the Board of Supervisors [final draft]," n.d.; Cogswell, "[Second] Communication to the Board of Supervisors," n.d., 2, both box 1, folder 18, HDCTCC.
111. His women's home was on Broadway, between Kearny and Montgomery Streets; Cooper's first kindergarten was at 116 Jackson Street.
112. Lloyd, *Lights and Shades*, 78.
113. Lloyd, 298.
114. Saxton, *Indispensable Enemy*, 115; Kazin, "July Days in San Francisco," 138, 142, 154.
115. Kazin, "July Days in San Francisco," 151, 154, 156.
116. Ethington, *Public City*, 265, 312; George, "Kearney Agitation," 440, 442.
117. "Communism and Its Remedy," *Berkeley Advocate*, September 5, 1878; "Preparing to Deliver the City to Kearney" and "Why the New Constitution was Adopted Here," *San Francisco Real Estate Circular* 14, no. 7 (May 1879), all carton 1, HDCTCC.
118. C. T. Hopkins, *Common Sense Applied to the Immigrant Question* (San Francisco: Turnbull and Smith, 1869), 3, 20–22, carton 2, folder 112, HDCTCC.
119. Hopkins, 22, 26, 46, 48 (emphasis in original).
120. Hopkins, 60, and app.
121. Cogswell, diary entry, July 3, 1879, box 2, Cogswell Papers.
122. Cogswell, (draft) letter to the board of supervisors, n.d., carton 1, Cogswell Papers.
123. Jeff Jarvis, "The 100-Year War," *Examiner*, March 2, 1979, 5, box 1, HDCTCC.
124. "A Rat's Tale," *Los Angeles Herald*, March 18, 1884, 5; "Dr. Cogswell's Fountains," *Call*, July 21, 1898, 6; Moffatt, "Intemperate Patronage," 140.
125. "His Soul Was Sad," *Call*, March 2, 1893, 8.
126. "Dr. Cogswell's Fountains," 6; "Image Breakers," *Call*, January 3, 1894, 8.
127. Moffatt, "Intemperate Patronage," 140.
128. Greg Kitsock, "All's Well That Ends with a Drink to Cogswell," *Washington City Paper*, March 6, 1992.
129. See Denis Hollier, *Against Architecture: The Writings of Georges Bataille*, trans. Betsy Wing (Cambridge, MA: MIT Press, 1989), 46–56.
130. "The Statoo Still with Us," *Brooklyn Eagle*, February 10, 1885, 2.
131. "Mark Twain, "The Famous Humorist Interviewed at Rochester," *Alta*, December 20, 1884, 9.
132. See, for example, "His Soul Was Sad," *Call*, March 2, 1893, 8; untitled article, *Brooklyn Eagle*, November 23, 1884, 6.
133. "Cogswell Fountain," *Alta*, June 15, 1979, 1. The WPC's platform included the prohibition of the appropriation of water supply by private interests; see "The Workingmen: Third Day's Session Now in Progress," *Examiner*, June 5, 1879, box 1, HDCTCC.

134. George, quoted in Benjamin R. Tucker, "On Picket Duty," *Liberty* 4, no. 20 (1887): 1, referring to Polytechnical College.
135. George, "Why Work is Scarce, Wages Low, and Labor Restless (Metropolitan Temple, San Francisco, March 26, 1878)," in *Henry George: Collected Journalistic Writings*, ed. Kenneth C. Wenzer, 4 vols. (Armonk, New York: M. E. Sharpe, 2003), 1:131–32.
136. Cogswell, "[Initial] Communication to the Board of Supervisors [final draft]"; Solnit, *River of Shadows*, 167, 169; Hales, *Silver Cities*, 167–71.
137. George, "Kearney Agitation," 451.
138. George, *Progress and Poverty*, 436, 437, 484.

CHAPTER THREE

1. See, respectively, Time capsule collection, Ancient and Honorable Artillery Company of Massachusetts, Faneuil Hall, Boston; *Proceedings of the Grand Council of Royal and Select Masters of the State of South Carolina* (Charleston: Lucas and Richardson, 1895), 51; "To Be Opened a Century Hence," *NYT*, July 20, 1888, 1; and Lester Reingold, "Capsule History," *American Heritage* 50, no. 7 (1999): 91. One-hundred-year vessels were also sealed in Huntington, New York (1889), Lyndon, Vermont (1891), and Ayers, Massachusetts (1898).
2. Edward Bellamy, *Looking Backward: 2000–1877* (Boston: Ticknor, 1888), 297, 351, 298; Bellamy, *Equality* (New York: D. Appleton, 1897), 70, 121.
3. Alvarado M. Fuller, *A.D. 2000* (Chicago: Laird and Lee, 1890), 83, 316, 411.
4. Fuller, 109, 342, 387–88, 390–91, 405, 411.
5. Mark Twain, "The Secret History of Eddypus, the World-Empire," in *The Science Fiction of Mark Twain*, ed. David Ketterer (Hamden, Conn.: Archon, 1984), 179, 181, 188, 196.
6. Critical neglect of "Secret History" dates back to Albert Bigelow Paine's dismissal of it as "not publishable matter . . . just one of the things which Mark Twain wrote to relieve mental pressure," in Albert Bigelow Paine, *Mark Twain: A Biography*, 4 vols. (New York: Harper, 1912), 3:1188.
7. "To 'Tribune' Men of 2001," *Chicago Tribune*, January 2, 1901, 10; all others cited below.
8. "Special Occasions," *Annual Report of the Peabody Historical Society* (1909–10), 78.
9. Harriet E. Smith, introduction to Mark Twain, *Autobiography of Mark Twain: The Complete and Authoritative Edition*, ed. Benjamin Griffin and Harriet E. Smith, 3 vols. (Berkeley: University of California Press, 2010–15), 1:1, 19.
10. Margaret Driscoll, signed note, in folder 97, CSCCC.
11. "Mark Twain's Bequest," interview in the *Times* (London), May 23, 1899, in Gary Scharnhorst, ed., *Mark Twain: The Complete Interviews* (Tuscaloosa: University of Alabama Press, 2006), 333.

12. Allen O. Myers, "Who Invented the Steamboat?," *Ohio Magazine* 2, no. 1 (1907): 5.
13. See Ellen F. Fitzpatrick, *History's Memory: Writing America's Past, 1880–1980* (Cambridge, MA: Harvard University Press, 2004), 29–43.
14. John Higham, *History: Professional Scholarship in America* (1965; Baltimore: Johns Hopkins University Press, 1989), 96, 113, 148–55, 158–59; Peter Novick, *That Noble Dream: The "Objectivity Question" and the American Historical Profession* (New York: Cambridge University Press, 1988), 47–60; Bonnie G. Smith, *The Gender of History: Men, Women, and Historical Practice* (Cambridge, MA: Harvard University Press, 2000), 9, 122, 130–33, 137–38, 148–49, 185–212; and Fitzpatrick, *History's Memory*, 31–33.
15. S. B. McCracken, ed., *Detroit in Nineteen Hundred: A Chronological Record of Events Both Local and State during the Closing Year of the Century* (Detroit: Evening News Association, 1901), 126–27.
16. "Of Its Aftermath: Sidelights on the Century Ball," *Kansas City Journal*, January 2, 1901, and "A Crowd of 15,000 Revelers in Old Convention Hall Placed Souvenirs in a Chest [. . .]," *Kansas City Star*, December 28, 1947, newspaper clippings related to the century box, KCCBC.
17. "Of Its Aftermath"; Austin Latchaw, untitled article on the Century Ball, *Kansas City Star*, December 29, 1940, newspaper clippings related to the century box, KCCBC.
18. Hillel Schwartz, *Century's End: An Orientation Manual toward the Year 2000* (New York: Currency Doubleday, 1996), 15, 191–93; *Yesterday's Future: The Twentieth Century Begins*, ed. Michael E. Stevens (Madison: State Historical Society of Wisconsin, 1999), 1–3; "The New Century," *Christian Science Journal* 18 (January 1901): 651–52.
19. J. E. S. [James E. Scripps], "The Twentieth Century: Some Reflections upon Its Incoming," newspaper clipping from *Detroit News Tribune*, December 30, 1900, item 132, DCBC.
20. Louis Shouse, untitled article, *Kansas City Star*, January 5, 1950, newspaper clippings related to the century box, KCCBC.
21. Schwartz, *Century's End*, 193.
22. "A Century's Progress," *Facts Illustrated*, folder 157, CSCCC; Charles E. Russell, "The Story of the Nineteenth Century," *Munsey's* 24 (January 1901): 551–59, folder 52, KCCBC.
23. Henry James, quoted in Schwartz, *Century's End*, 157.
24. Sarah Grand, "The New Aspect of the Woman Question," *North American Review* 158, no. 448 (1894): 270–76; Booker T. Washington, *A New Negro for a New Century* (Chicago: American, 1900); Mrs. Booker T. Washington [Margaret Murray Washington], "The New Negro Woman," *Lend-a-Hand* 15 (October 1895): 254–60; Henry Grady, "The New South," *Critic* 7, no. 10 (1887): 9–11; William DeWitt Hyde, *The New Ethics* (New York: Crowell, 1903); William Jethro Brown, *The New Democracy: A Political Study* (London: Macmillan, 1899).

25. Sir Walter Besant, "The Burden of the Twentieth Century," *North American Review* 173 (July 1901): 9–21; Elizabeth Bisland, "The Time-Spirit of the Twentieth Century," *Atlantic Monthly* 87 (January 1901): 15–22.
26. "The Nineteenth Century: A Review, an Interpretation, and a Forecast," *Outlook* 66 (December 29, 1900): 1019–33 (emphasis added).
27. Schwartz, *Century's End*, 12–13, 161.
28. Walter Laqueur, *Fin de Siècle and Other Essays on America and Europe* (New Brunswick, NJ: Transaction, 1997), 7.
29. John Jacob Astor IV, *A Journey into Other Worlds: A Romance of the Future* (New York: Appleton, 1894); Arthur Bird, *Looking Forward: A Dream of the United States of the Americas in 1999* (Utica, NY: I. C. Childs, 1899); George B. Lloyd's "Perchance to Dream: A Romance in the Year 2000," *Chicago Tribune*, beginning December 30, 1900, 40; Paul Devinne, *Day of Prosperity: A Vision of the Century to Come* (New York: Dillingham, 1902).
30. See, for example, H. G. Wells, "Area of Future Cities," *NYT*, May 26, 1901, 25; and "A Pictorial Forecast for the Coming Generations: New York in 1999; A Look 100 Years Ahead," special section of *New York World*, December 29, 1900.
31. J. F. Whiting, ed., *Twentieth Century*, mock newspaper, dated Saturn 43, 2000, folder 68, KCCBC.
32. Sackett Cornell, letter, June 3, 1879, box 1, folder 1, HDCTCC.
33. McCracken, *Detroit in Nineteen Hundred*, 128.
34. Adams Sherman Hill, diary, 4, 8, 10, box 59, HUC.
35. Patrick Parrinder, *Shadows of the Future: H.G. Wells, Science Fiction, and Prophecy* (Syracuse, NY: Syracuse University Press, 1995), 3.
36. Parrinder, 3; see also Howard P. Segal, *Technological Utopianism in American Culture* (Syracuse, NY: Syracuse University Press, 1985), 19–44. Note also the forerunners discussed in Paul K. Alkon, *Origins of Futuristic Fiction* (Athens: University of Georgia Press, 2010).
37. Wells, paraphrased in Parrinder, *Shadows*, 27–28.
38. See Susan Schulten, *The Geographical Imagination in America, 1880–1950* (Chicago: University of Chicago Press, 2001), 49, 72–81.
39. Theodore M. Porter, *The Rise of Statistical Thinking, 1820–1900* (Princeton, NJ: Princeton University Press, 1986); Richard Lewinsohn, *Science, Prophecy, and Prediction* (New York: Harper, 1961).
40. Elizabeth Grosz, "Thinking the New: Of Futures Yet Unthought," in *Becomings: Explorations in Time, Memory, and Futures*, ed. Elizabeth Grosz (Ithaca, NY: Cornell University Press, 1999), 16.
41. H. G. Wells, *The Discovery of the Future* (New York: Huebsch, 1913), 23.
42. Carlos E. Warner, "The Chamber of Commerce," 2, item 158, DCBC.
43. McCracken, *Detroit in Nineteen Hundred*, 128; Hugo Eyssell, calling card, album 56, KCCBC.
44. See especially Ernst Bloch, *The Principle of Utopia* (1954; Cambridge, MA: MIT Press, 1986).

45. Louis R. Ehrich, "To my descendants who may be living in the year 2001," 1, folder 49, CSCCC (emphasis added).
46. Eugène Minkowski, *Lived Time: Phenomenological and Psychopathological Studies*, trans. Nancy Metzel (1933; Evanston, IL: Northwestern University Press, 1970), 92, 95, 100.
47. See, for example, T. J. Jackson Lears, *No Place of Grace: Antimodernism and the Transformation of American Culture, 1880–1920* (1981; Chicago: University of Chicago Press, 1994).
48. "Messages to Posterity," *Kansas City Star*, January 3, 1901, newspaper clippings related to the century box, KCCBC.
49. See François Weil, *Family Trees: A History of Genealogy in America* (Cambridge, MA: Harvard University Press, 2013), 41–77, 143–79.
50. William F. Slocum, "To the President of Colorado College in the year 1999 [*sic*]," folder 14, and Mary Tenney Hatch, "To the President of the Young Women's Christian Association of Colorado Springs in the Year 2001," folder 12, CSCCC. John Hayes, "To the Chief of Police Kansas City, Mo.—A.D. 2001," folder 56; William Potts George, "To My Successor in the Pastorate of Westminster Congregational Church," folder 152; Charles W. Purvis, "To the President of the New Boys' Union One Hundred Years (2001) Hence," folder 81, KCCBC. Francis A. Blades, "Our Financial City Government," item 21; William C. Maybury, "To . . . the Mayor of Detroit in 2001," item 104, DCBC.
51. Ben Griffith, "To the male students of Colorado College in 2001 A. D.," folder 9; Mary Gilman Ahlers, "To the wife of a Professor in Colorado College in the year 2001," folder 15; Margaret T. Adams, "To any Native Americans . . . ," folder 92, CSCCC. E. B. Hubbard, "Greeting to the Christian Scientists of the 21st Century," folder 54, KCCBC. [Robert Fulton?], Secretary of the St. Andrew's Society, "Greetings to the Descendants of Early Scotchmen in Detroit," item 146; David W. Simons, "The Jewish People in Mercantile, Social and Professional Life in Detroit," item 142; David A. Boyd, "The Labor Movement in Detroit up to Date," item 34; Theodore A. McGraw, "Progress of Surgery in the 19th Century," item 102, DCBC.
52. Discussed below (p. 152).
53. On Christian Scientists' contributions to these vessels, see below (165). On Bellamy's spiritualism, see Catherine Tumber, *American Feminism and the Birth of New Age Spirituality: Searching for the Higher Self, 1875–1915* (Lanham, MD: Rowman and Littlefield, 2002), 69–108.
54. Mark Twain, "Mental Telegraphy: A Manuscript with a History," *Harper's New Monthly* 84 (December 1891): 95–104. Reference to daughter from Howard Kerr, *Mediums, and Spirit-Rappers, and Roaring Radicals: Spiritualism in American Literature, 1850–1900* (Urbana: University of Illinois Press, 1972), 181.
55. Margaret E. Ball, "To the Class of Two Thousand," *Mount Holyoke* 10, Commencement Number (June 1900): 15, MHC.

56. Samuel Le Nord Caldwell, "Colorado Springs Today," folder 52, CSCCC.
57. Frank I. Cobb, "The Newspapers of Detroit," 2, item 27, DCBC; Robert M. Brocket, calling card, album 61, KCCBC.
58. Twain, *Autobiography*, 1:221.
59. Twain, "Secret History," 197.
60. "History of the Ball: Will Be Preserved for Benefit of Next Generation," untitled newspaper clippings related to the century box, KCCBC.
61. James A. Reed, "To the Honorable Mayor of Kansas City, Mo. [in 2001]," 1, folder 25, KCCBC; R. L. Marshman of Kansas City, KS, concludes his letter with a rhetorical "shake [of] hands" ("The Hon. Mayor, Kansas City, U.S.A., January 1, 2001," 4, folder 101, KCCBC).
62. Hubbard, "Greeting to the Christian Scientists" (emphasis in original).
63. Twain, "Secret History," 197.
64. See, respectively, Caldwell, "Colorado Springs Today," 1, 40, and Myra Alline Levi, "With a loving kiss to my descendant," folder 141, CSCCC.
65. Ball, "To the Class of Two Thousand," 14.
66. On the sealing of messages, see p. 122 below.
67. George D. Gallaway, "To the Young Lady Who Stood Highest in the Class Graduated from Colorado College in the Year 2000," 1, 2, folder 31, CSCCC.
68. Calling card, album 40, KCCBC; Anna Belle Eisenhower, diary, box 59, HUC.
69. Archibald C. Coolidge, diary, box 58, HUC. Hill complains about the slow progress of his research in Hill, diary, 13.
70. William Garrott Brown, diary, 4, 6, 8–9, box 58, HUC.
71. Brown, 2–3, 10–11.
72. Brown, 12–13, 14.
73. Wendell H. Stephenson, "William Garrott Brown: Literary Historian and Essayist," *Journal of Southern History* 12, no. 3 (1946): 313–44.
74. Hill, diary, 9.
75. Barrett Wendell, diary, box 61, HUC.
76. See John C. Brereton, ed., *The Origins of Composition Studies in the American College, 1875–1925: A Documentary History* (Pittsburgh: University of Pittsburgh Press, 1995), 4, 8–13.
77. Wendell, diary.
78. Coolidge, diary.
79. Paul Henry Hanus, diary, 25–26, box 59, HUC.
80. Wendell, diary.
81. Hugo Münsterberg, diary, 10–11, box 60, HUC.
82. Albert C. Pearson, "Musical Matters," 1, folder 7, CSCCC.
83. Leah Ehrich, "Music in Colorado Springs," 1, folder 74, CSCCC.
84. Leah Ehrich, letter, 1, folder 49, CSCCC.
85. Dora Fletcher Noxon, "My Children," 1, folder 37, CSCCC.
86. Clarence and Louise Calhoun Arnold, "To [our] descendants," 1, folder 137, CSCCC.

87. Ehrich, "Music in Colorado Springs," 1, 2.
88. Walter Lawson Wilder, "The Century Chest," handwritten draft of editorial published in *Colorado Springs Gazette*, August 5, 1901, 1–2, folder 42; Bertha Francis Emery, "My dears," 1, folder 134, CSCCC.
89. Charles P. Bennett, "Colorado Springs Real Estate," 1, folder 78, CSCCC (emphasis added).
90. Ball, "To the Class of Two Thousand," 15.
91. Louis Wyborn Cunningham, "To [our] descendants," 2, folder 146, CSCCC.
92. J. R. Robinson, "To the Citizens of Colorado Springs of the Twenty-First Century," 17, folder 3, CSCCC.
93. Charles H. Williams, calling card, album 60, KCCBC.
94. Ehrich, "Music in Colorado Springs," 2.
95. Bellamy, *Equality*, 121–22.
96. Sigmund Freud, *Reflections on War and Death*, trans. A. A. Brill and Alfred B. Kuttner (1915; New York: Moffat, Yard, 1918), 41.
97. Rev. Charles L. Arnold, "The Arnold Home for the Aged and Hospital for Incurables," 3, item 29, and Francis L. York, "History and Prophecy of Music in Detroit," 10, item 153, DCBC.
98. Chester A. Snider, poem inscribed on note, folder 124, KCCBC.
99. Manly D. Ormes, "To the ministers residing in Colorado Springs in the year 2001," 8, folder 41, CSCCC; Ehrich, "Music in Colorado Springs," 2–3.
100. Warner, "Chamber of Commerce," 1.
101. Harriet Peck Farnsworth, "Dear Great Grandchildren," 1–2, folder 119, CSCCC.
102. Philippe Ariès, *Western Attitudes toward Death from the Middle Ages to the Present*, trans. Patricia M. Ranum (Baltimore: Johns Hopkins University Press, 1974), 85–108.
103. Franklin E. Brooks, letter, 1, folder 20, CSCCC.
104. Farnsworth, "Dear Great Grandchildren," 1; Mary Noyes Shaw, "To the woman in the great unknown who shall first receive this . . . ," 1, RCB.
105. W. Cunniff to William Coolidge Lane, March 1, 1900, 1, in "Information about the box," box 62, HUC.
106. Ehrich, "Music in Colorado Springs," 1.
107. William Coolidge Lane, diary, 38, box 59, HUC.
108. Student theme, box 56, HUC.
109. Louis R. Ehrich, "A Colorado National Park," *Magazine of Western History* 10, no. 4 (1889): 430, 432, 434.
110. Louis R. Ehrich, "'Posteritism': An Address Delivered at the Dedication Exercises of the Century Chest, on August 4th, 1901" (1901), 5–6, 7, ms. 1349, Ehrich Papers, Yale University Library.
111. Ehrich, "'Posteritism,'" 8, 9; see also Ehrich, "Colorado National Park," 435.

112. C. A. M., "Caring for the Future," *Out West* 16, no. 1 (1902): 67–68.
113. Manley D. Ormes, "To the Pastor and members of the Second Congregational Church in . . . 2001," 12, folder 41, CSCCC.
114. Michael W. O'Brien, "Banking in Prospective in Detroit," 6, item 122, DCBC.
115. Wilder, "Century Chest," 4.
116. "Messages to Posterity," *Kansas City Star*, January 3, 1901, newspaper clippings related to the century box, KCCBC.
117. Twain, "Secret History," 190, 195.
118. Inventories in folder 2, KCCBC; item 105, DCBC.
119. Ehrich, "'Posteritism,'" 3.
120. John Higham, *History: Professional Scholarship in America* (1965; Baltimore: Johns Hopkins University Press, 1989), 148, 151; see also Bonnie G. Smith, *The Gender of History: Men, Women, and Historical Practice* (Cambridge, MA: Harvard University Press, 2000), 130–31, 148–49.
121. Carol Kammen, *On Doing Local History* (Walnut Creek, CA: AltaMira, 2003), 21–23, 28–33.
122. Contemporary history only gained legitimacy after the Second World War. Smith, *Gender of History*, 155; Higham, *History*, 109–10; Arthur M. Schlesinger Jr., "On the Writing of Contemporary History," *Atlantic Monthly* 219 (March 1967): 69–74; although see Henry Rousso, *The Latest Catastrophe: History, the Present, the Contemporary* (Chicago: University of Chicago Press, 2016), 51–53.
123. William Coolidge Lane, circular announcing the century box, February 22, 1900, 1, 2, "Information about the box," box 62, HUC; Ehrich's goal was similarly a "complete picture of the City at the present time; Ehrich, "'Posteritism,'" 3.
124. E. L. Pearson, theme, box 55, HUC.
125. Twain, *Autobiography*, 3:329.
126. Ball, account book, MHC (emphasis added).
127. Items in folders 15, 49, 81, 53, CSCCC; main box, folder 156, 73, KCCBC, respectively.
128. Teresa Barnett, *Sacred Relics: Pieces of the Past in Nineteenth-Century America* (Chicago: University of Chicago Press, 2013), 173–96.
129. Smith, *Gender of History*, 122, 133, 137–38. Amateur historians were beginning to explore social and cultural history in the late nineteenth century, but they had little influence on the increasingly professional-dominated American Historical Association; see Fitzpatrick, *History's Memory*, 29–43.
130. Lucy Maynard Salmon, *History and the Texture of Modern Life: Selected Essays*, ed. Nicholas Adams and Bonnie G. Smith (Philadelphia: University of Pennsylvania Press, 2001).
131. Michael Kammen, *Mystic Chords of Memory: The Transformation of Tradition in American Culture* (New York: Vintage Books, 1991), 343.

132. Gabrielle M. Spiegel, "In the Mirror's Eye: The Writing of Medieval History in America," in *Imagined Histories: American Historians Interpret the Past*, ed. Anthony Molho and Gordon S. Wood (Princeton, NJ: Princeton University Press, 1998), 241.
133. Charles H. Haskins, "The Life of Medieval Students as Illustrated by Their Letters," *American Historical Review* 3, no. 2 (1898): 203, 204–5, 208, 216–17, 226, 228.
134. Maybury, "To . . . the Mayor of Detroit in 2001," 1.
135. See, for example, Irene Williams Chittenden, "The American Mother of the Nineteenth Century," 1, item 25, DCBC.
136. Lane, circular, 1, 3.
137. Hanus, diary, 1.
138. "The Keeping of Journals," *Washington Times*, March 12, 1900, clipping preserved by Lane, in "Information about the box," box 62, HUC.
139. Wendell, quoted in Brereton, *Origins of Composition Studies*, 130; on themes, see ibid., 11.
140. Walter Prichard Eaton, "The Daily Theme Eye," *Atlantic Monthly* 99 (March 1907): 427–28.
141. Charles William Eliot, *Addresses at the Inauguration of Charles William Eliot* (Cambridge, MA: Sever and Francis, 1869), 34.
142. Eaton, "Daily Theme Eye," 427–28.
143. See themes in boxes 55 and 56, HUC.
144. Lane, circular, 1, 2.
145. Ehrich, "'Posteritism,'" 3.
146. Lane, circular, 1; Eliot Spalding to Lane, April 23, 1900, in "Information about the box," box 62, HUC; and Lane's inscription on Hill's class list in box 56, HUC: "Diary of Percy W. Long withdrawn at his urgent request, Aug. 1913."
147. Raymond H. Geselbracht, "The Origins of Restrictions on Access to Personal Papers at the Library of Congress and the National Archives," *American Archivist* 49, no. 2 (1986): 142–62.
148. McCracken, *Detroit in Nineteen Hundred*, 127; see also Reed's (folder 25, KCCBC) and Marshman's (folder 101, KCCBC) orations in Kansas City; Ehrich's in Colorado Springs ("'Posteritism'"); and Margaret Ball's at Mount Holyoke (Ball, "To the Class of Two Thousand").
149. John Orne to Lane, May 18, 1900, 1, in "Information about the box," box 62, HUC; Coolidge, diary.
150. Lane, circular, 1; Lane, diary, 3–4.
151. "Of Its Aftermath: Sidelights on the Century Ball," *Kansas City Journal*, January 2, 1901, newspaper clippings related to the century box, KCCBC.
152. See, for example, Williamina Fleming's addendum to her diary, 22, box 59, HUC.
153. On the patriotism and boosterism of those other commemorative endeavors, see Kammen, *On Doing Local History*, 13–20, 22–26; and David

Glassberg, *American Historical Pageantry: The Uses of Tradition in the Early Twentieth Century* (Chapel Hill: University of North Carolina Press, 1990), esp. 1–67.

154. "Mark Twain's Bequest," in Scharnhorst, *Mark Twain: The Complete Interviews*, 333.
155. Twain, *Autobiography*, 1:221; influences cited in Smith's introduction, 1:5, 6, 15.
156. Smith, introduction to Twain, *Autobiography*, 1:16; on Howells's skepticism, see ibid., 21.
157. "Mark Twain's Bequest," in Scharnhorst, *Mark Twain: The Complete Interviews*, 334.
158. Mark Twain, "The Privilege of the Grave" (1905), *New Yorker*, December 22, 2008, 50.
159. Twain, "Secret History," 195.
160. Lane, circular, 2.
161. Smith, introduction to Twain, *Autobiography*, 1:19–20. Twain's claim that he submitted *Tom Sawyer* (1876) as a typed manuscript has been debunked.
162. Twain, "Secret History," 197, 196. On Twain and the Paige typesetter, see Paine, *Mark Twain*, 3:903–14.
163. Twain, *Autobiography*, 1:221–22. Publisher's suggestion in Smith, introduction to ibid., 19. On Twain's views about copyright, see Susan Gillman, *Dark Twins: Imposture and Identity in Mark Twain's America* (Chicago: University of Chicago Press, 1989), 183–88.
164. Twain, "Secret History," 196.
165. JoAnne Yates, *Control through Communication: The Rise of System in American Management* (Baltimore: Johns Hopkins University Press, 1993), 131.
166. Ehrich, "Music in Colorado Springs," 2.
167. Bellamy, *Equality*, 122, 123.
168. Martin Heidegger, quoted in Friedrich A. Kittler, *Gramophone, Film, Typewriter*, trans. Geoffrey Winthrop-Young and Michael Wutz (Stanford, CA: Stanford University Press, 1999), 199.
169. "What the Americans Have Done: No. 1.—The Remington Typewriter," *Review of Reviews* 24 (1901): 106.
170. Anthony Enns, "Spiritualist Writing Machines: Telegraphy, Typtology, Typewriting," *communication +1* 4 (September 2015): 14–19.
171. Alexander Graham Bell, notes, January 22, 1888, Bell family papers, 1834–1974, Manuscript Division, LC. See also Jonathan Sterne, *The Audible Past: Cultural Origins of Sound Reproduction* (Durham, NC: Duke University Press, 2003), 307–11; and Kittler, *Gramophone, Film, Typewriter*, 51–55.
172. Phonograph recordings, folder 159, CSCCC.
173. "Voices of the Dead," *Phonoscope* 1, no. 1 (1896): 5.
174. On Vienna's Phonogrammarchiv (1899), which inspired a Berlin Phonogram Archive in 1900, see "Wonders of the Graphophone: Proposition

to Preserve Accurate Records of Nineteenth Century Life," *Phonoscope* 4, no. 4 (1900): 6; on Alfred Kroeber's phonographic collections, also begun in 1900, see Richard Keeling, *A Guide to Early Field Recordings (1900–1949) at the Lowie Museum of Anthropology* (Berkeley: University of California Press, 1991), xiv–xvii. See also Paula Amad, *Counter-Archive: Film, the Everyday, and Albert Kahn's Archives de la Planète* (New York: Columbia University Press, 2010), 153.

175. Photograph of the class of 1900 (MHC); photographs of church officials (RCB); photographs of fire chief and five-year-old girl in folder 90 and folder 5, respectively, KCCBC.
176. Thirteen unidentified portraits of "Citizens of Colorado Springs," folder 151, CSCCC.
177. See "John the Orangeman, Harvard's 'Mascot,'" *Cambridge Chronicle*, March 12, 1898, 33.
178. Lane, diary, 52. Lane forgot to include the photographs in the chest when it was sealed, subsequently placing them next to it (Lane, note, in finding aid, HUC).
179. Sarah Anne Carter, "Picturing Rooms: Interior Photography 1870–1900," *History of Photography* 34, no. 3 (2010): 252; the contributor was Julian Burroughs, discussed on 258–59. Burroughs's photographs are in envelopes 8A, 8B, 8C, HUC.
180. See Nick Yablon, *Untimely Ruins: An Archaeology of American Urban Modernity, 1819–1919* (Chicago: University of Chicago Press, 2009), 212–27.
181. Lane, circular, 3. H. W. Eliot Jr. announced the competition in "Club Memoranda," *Photographic Times* 32, no. 7 (1900): 327. A circular was also issued: "Camera Club Photographic Contest," in "Information about the box," box 62, HUC.
182. Elizabeth Edwards, *The Camera as Historian: Amateur Photographers and Historical Imagination, 1885–1918* (Durham, NC: Duke University Press, 2012), xi, 2–3, 9–10, 14, 34, 46, 164–66; *Times* quoted on 207; eschewal of pictorialism, 79–121; archival labeling, 109–19.
183. On the camera's perceived threat to the archive, see Amad, *Counter-Archive*. On archive style, see Robin Kelsey, *Archive Style: Photographs and Illustrations for U.S. Surveys, 1850–1890* (Berkeley: University of California Press, 2007).
184. "Old Boston Photograph Collection" (rediscovered 2007), Print Department, Boston Public Library.
185. George E. Francis, "Photography as an Aid to Local History," *Proceedings of the American Antiquarian Society*, 2nd ser., 5 (1888): 274–75, 278–79. Francis's vision came to fruition in the 1930s with the launch of projects such as the Historic American Buildings Survey. On historians' neglect of photographs until the 1920s, see Thomas J. Schlereth, "Mirrors of the Past: Historical Photography and American History," in *Artifacts and the American Past*, ed. Thomas J. Schlereth (Nashville, TN: American Association for State and Local History, 1980), 14.

186. See Sterne, *Audible Past*, 332; *Kansas City Star* quoted on 302.
187. See, for example, Wales C. Martindale, "Public Schools in Detroit," item 96, DCBC.
188. Edwards, *Camera as Historian*, 104.
189. Alvarado M. Fuller, *A.D. 2000* (Chicago: Laird and Lee, 1890), 337.
190. Bellamy, *Equality*, 123.
191. Ehrich, "'Posteritism,'" 3. Twenty-one letters in the Detroit vessel were on linen paper. Harvard instructions in "Camera Club Photographic Contest," in "Information about the box," box 62, HUC.
192. Bellamy, *Equality*, 92–94.
193. On Colorado Springs' chest, see Ehrich, "'Posteritism,'" 4; on Detroit's, see McCracken, *Detroit in Nineteen Hundred*, 129; on Kansas City's, see "Of its Aftermath: Sidelights on the Century Ball," *Kansas City Journal*, January 2, 1901, and "Messages to Posterity," *Kansas City Star*, January 3, 1901, newspaper clippings related to the century box, KCCBC; on the Mount Holyoke box, see Mike Plaisance, "Mystery, History Unveiled," *Springfield* [MA] *Union News*, April 1, 2000, newspaper clipping, MHC; on Rockland's, see Paul E. Kandarian, "Glimpses of Bygone Day Come to Light," *Boston Globe*, January 10, 2002, GS2; on Harvard's, see Lane, circular, 1.
194. Ehrich mentions the chest's cost in "'Posteritism,'" 4.
195. Wilder, "Century Chest," 2.
196. McCracken, *Detroit in Nineteen Hundred*, 129; Lane, circular, 1; http://www.firstchurchrockland.org/history.htm; minutes of the reunion meeting of the class of 1900, June 8, 1946, 2, MHC.
197. *Ceremony at the Sealing of the Century Box by the Ancient and Honorable Artillery Company in Faneuil Hall, Boston* [. . .] (Boston: Alfred Mudge, 1882), 11.
198. See, respectively, folders 18, 39, 43, 57, 19, 68, 76, CSCCC; items 89, 148, 140, 144, 25, DCBC.
199. See below, p. 166–172.
200. See, respectively, items 108, 45, 34, 82, DCBC; and folder 29, CSCCC.
201. Twain, "Secret History," 194. "Mark Twain's Bequest," in Scharnhorst, *Mark Twain: The Complete Interviews*, 334.
202. Arthur N. Taft, "Life in Colorado Springs," folder 108, CSCCC.
203. "Pepys to Be Made by Order: Harvard to Collect Records," *Washington Post*, March 11, 1900, newspaper clipping in "Information about the box," box 62, HUC. On original proposal, see Lane, diary, 1.
204. Lane, diary, 19; Lane, circular, 2. Lucy Sprague, diary, box 61, HUC.
205. Lane, diary, 25; Frederick Wilkey, "An account of the organization and methods in use in Randall Hall," and W. H. Cutler, "Organization of the Janitors' Department," box 62, HUC.
206. Alpheus Hyatt to Lane, 1, in "Information about the box," box 62, HUC; Lane, circular, 2.
207. Theodore Case, *History of Kansas City* (Syracuse, NY: D. Mason: 1888), 529–31.

208. Clarence M. Burton, ed., *Compendium of History and Biography of the City of Detroit and Wayne County, Michigan* (Chicago: Henry Taylor, 1909), 503, 504; Caryn Hannan et al., *Michigan Biographical Dictionary* (St. Clair Shores, MI: Somerset, 2008), 113.
209. Walter B. Briggs, "William Coolidge Lane, 1859–1931," *Publications of the Cambridge Historical Society* 21 (1936): 72–75.
210. Louis R. Ehrich, "To my descendants who may be living in the year 2001," 9, folder 49, CSCCC; he sold the collection after the 1893 panic. The only nonelite centurial vessel architect was Shaw, a retired dressmaker in Rockland (Mary Noyes Shaw, "To the woman in the great unknown who shall first receive this [. . .]," 1, RCB).
211. "Taking Out the Greens," *Kansas City Journal*, January 4, 1901, newspaper clippings related to the century box, KCCBC; folders 49, 74, CSCCC.
212. References to this committee in J. R. Robinson, "To the Citizens of Colorado Springs of the Twenty-First Century," 1, folder 3, CSCCC.
213. Jay Shatz to Ehrich, folder 153, CSCCC.
214. Calling cards in folder 104, CSCCC. "A Heritage to Posterity," *Colorado Springs Gazette*, August 5, 1901, 8, listed eighty contributions, but the finding aid contains as many as 160 folders (one for each contributor or family of contributors).
215. "Heritage to Posterity," 8.
216. "Articles for Century Box," *Kansas City Times*, January 6, 1901; "Taking Out the Greens."
217. "Of Its Aftermath: Sidelights on the Century Ball," *Kansas City Journal*, January 2, 1901, newspaper clippings related to the century box, KCCBC; finally sealed on January 16, 1901.
218. Williamina Fleming, diary, 22, box 59, HUC.
219. According to the *Gazette*, Ehrich's project elicited "favorable comment in all parts of the country" ("Heritage to Posterity," 7). For a favorable response to the Harvard project, see Malcolm Storer to Lane, February 26, 1900; someone who felt "complimented at being asked" was John S. Laurence, letter to Lane, n.d., both in "Information about the box," box 62, HUC.
220. See, for example, George Herbert Palmer, diary, 29, box 60, HUC.
221. Edward Channing to Lane, February 24, 1900, and Albert Bushnell Hart to Lane, February 24, 1900, "Information about the box," box 62, HUC.
222. See Wallace C. Sabine to Lane, February 27, 1900, "Information about the box," box 62, HUC; for the latter, see Brown, diary, 7.
223. Lane, diary, 65–67, 94.
224. John Winthrop Platner to Lane, March 1, 1900, in "Information about the box," box 62, HUC.
225. L.B.R.B. [LeBaron Russell Briggs] to Lane, March 2, 1900, in "Information about the box," box 62, HUC.
226. William J. Ashley to Lane, February 26, 1900, 1, HUC.

227. H. W. Bymer, weekly theme, box 55, HUC.
228. Paine, *Mark Twain*, 4:1489.
229. Lucy Allen Paton, diary, box 60, HUC (emphasis added).
230. "The Keeping of Journals," *Washington Times*, March 12, 1900, clipping preserved by Lane, in "Information about the box," box 62, HUC.
231. Hugo Münsterberg, diary, 5, box 60, HUC.
232. Horace Lunt, "A.D. 1901 to A.D. 2001," 1, folder 22, CSCCC.
233. Lunt, 2; Henry C. Hall, letter, 5, folder 20, CSCCC.
234. Arthur Searle to Lane, February 24, 1900, in "Information about the box," box 62, HUC.
235. Adams Sherman Hill, diary, 10, 17, box 59, HUC.
236. Barrett Wendell, diary, box 61, HUC.
237. Lane, diary, 8.
238. Smith, *Gender of History*, 119, 136, 146.
239. Higham, *History*, 17.
240. William James to Lane, March 21, 1908, in "Information about the box," box 62, HUC.
241. Shatz to Mr. Ehrich, 1, folder 153, CSCCC; Lunt, 1 (emphasis added).
242. "Pepys to Be Made by Order: Harvard to Collect Records," *Washington Post*, March 11, 1900, newspaper clipping in "Information about the box," box 62, HUC.
243. Lane, diary, 3. See "Information about the box," box 62, HUC.
244. Twain, "Secret History," 194–96 (emphasis in original).
245. Twain, 195–96.
246. Fuller, *A.D. 2000*, 109–11, 115.
247. Twain, "Secret History," 177, 180–81, 183, 184, 189, 190–91, 195–97.
248. Twain, 179, 190, 191.
249. Charles H. Haskins, "The Life of Medieval Students as Illustrated by Their Letters," *American Historical Review* 3, no. 2 (1898): 205n2.

CHAPTER FOUR

1. All cited below.
2. Ernst Bloch, *The Principle of Hope*, trans. Neville Plaice, Stephen Plaice, and Paul Knight, 3 vols. (1959; Cambridge, MA: MIT Press, 1995), 2:623, 1:145–46.
3. Elizabeth Grosz, *The Nick of Time: Politics, Evolution, and the Untimely* (Durham, NC: Duke University Press, 2004), 253; see also 113–19.
4. Bloch, *Principle of Hope*, 1:3–18.
5. John M. Donaldson, "Rise of Architecture in Detroit," 1, item 86, and William Carson, letter on "the progress of trade," 1, item 14, HUC.
6. James A. Reed, "To the Honorable Mayor of Kansas City, Mo. [in 2001]," 1–2, folder 25, KCCBC.

7. *Members of Kansas City Live Stock Exchange, June 1, 1900* (directory), folder 28; C. C. Christie, "To the Mayor of Kansas City, Mo., in the Year 2000," 1, folder 122; "The California Limited 1900–1901 Santa Fe Route [via Kansas City]," folder 21; and commercial ephemera in folders 32, 58, 59, 104, KCCBC.
8. Harry Haskell, *Boss-Busters and Sin Hounds: Kansas City and Its Star* (Columbia: University of Missouri Press, 2007), 83.
9. Irving Howbert, "Reminiscences Connected with Early Life in Colorado," 10, folder 75, and J. R. Robinson, "To the Citizens of Colorado Springs of the Twenty-First Century," 2–3, folder 3, CSCCC. For population statistics, see Nancy Capace, ed., *Encyclopedia of Colorado* (St. Clair Shores, MI: Somerset, 1999), 175.
10. Charles P. Bennett, "Colorado Springs Real Estate," 15, folder 78, and John G. Shields, "To the Business Men of Colorado Springs in the Year Two Thousand and One," 2, folder 96, CSCCC.
11. William J. Palmer, untitled letter, folder 2, CSCCC.
12. Jeremiah Dwyer, "Growth of Detroit's Manufactories during the 19th Century," 5, item 56, DCBC.
13. "Hale Harness and Supply Co." (catalog, n.d., Kansas City, MO), folder 89, and "Yearbook of the Manufacturers Association of Kansas City" (1900), folder 61, KCCBC. It was the nation's leading city for agricultural implements by 1890; James Shortridge, *Kansas City and How it Grew, 1822–2011* (Lawrence: University Press of Kansas, 2012), 61.
14. W. W. Hassell, "To the Citizens of 2001," 2, 4, folder 40, CSCCC.
15. Samuel Le Nord Caldwell, "Colorado Springs Today," 7, folder 52, CSCCC; Shields, "To the Business Men," 2–3; see also the mining certificate discussed below, p. 188. See Kathleen A. Brosnan, *Uniting Mountain and Plain: Cities, Law, and Environmental Change along the Front Range* (Albuquerque: University of New Mexico Press, 2002), 92, 110. On Colorado City, see Capace, *Encyclopedia of Colorado*, 174.
16. Armond H. Griffith, "What Detroit Has Accomplished in Art," item 37; Francis L. York, "History and Prophecy of Music in Detroit," 2, item 153; Bertram C. Whitney, "The Stage in Detroit," 9, item 173, DCBC. For similar claims for Colorado Springs, see Mary A. Bartow, "Art in Colorado Springs," folder 50, CSCCC, and for Kansas City, see "Western Gallery of Art" (catalog, Kansas City Art Association, 1897), folder 75, KCCBC.
17. See flyer advertising the Century Ball, folder 109, KCCBC.
18. Isaac N. Stevens, "The Press of Colorado Springs," 1, 2, folder 28, CSCCC. James Schermerhorn, "Newspapers Now and Then," 2, 3, item 127, DCBC.
19. Robert Barrie, "Our Legislative City Government," 2, item 12, DCBC; Bennett, "Colorado Springs Real Estate," 13, 20.
20. Michael W. O'Brien, "Banking in Prospective in Detroit," 4, item 122, DCBC.

21. Dwyer, "Growth of Detroit's Manufactories," 5, 6.
22. Orrin Baldwin, letter on "Detroit's industrial and commercial possibilities," 1, item 2, DCBC.
23. Charles L. Freer, letter on "Possibilities of Detroit as a manufacturing center," 2, item 80, DCBC.
24. Steven Watts, *The People's Tycoon: Henry Ford and the American Century* (New York: Vintage Books, 2005), 51–54.
25. See R. Richard Wohl and A. Theodore Brown, "The Usable Past: A Study of Historical Traditions in Kansas City," *Huntington Library Quarterly* 23 (May 1960): 237–59.
26. I. B. Marlatt, "To My Descendants of 2001 AD" (January 4, 1901), 2–3, folder 30, KCCBC.
27. Marlatt, 2; H. H. King, "To the Honorable Mayor of Kansas City, U.S.A," postdated January 1, A.D. 2001, 3, folder 88, KCCBC.
28. Estimates in Desire B. Willemin, "Foundation of Detroit by the French," 3, item 151, and J. E. S. [James E. Scripps], "The Twentieth Century: Some Reflections upon Its Incoming," *Detroit News Tribune*, Sunday, December 30, 1900, item 132, DCBC; quotation from latter.
29. Scripps, "Twentieth Century."
30. Edwin Abbott, "The Suburbs of Detroit," 7, item 98, DCBC.
31. James E. Scripps, "A Prophecy for Detroit as a Metropolis," 1, item 130, DCBC.
32. Reed, "To the Honorable Mayor," 2–3. On aspirations to become the metropolis of the west, see Wohl and Brown, "Usable Past," 15.
33. Moss cited in "Messages to Posterity," *Kansas City Star*, January 3, 1901, newspaper clippings related to the century box, KCCBC.
34. "World Capital," *Twentieth Century*, ed. J. F. Whiting, mock newspaper, dated Saturn 43, 2000, folder 68, KCCBC.
35. Shortridge, *Kansas City*, 7–8.
36. R. L. Marshman, "The Hon. Mayor, Kansas City, U.S.A., January 1, 2001," 1, folder 101, KCCBC.
37. 1900 telephone directory, folder 70, KCCBC.
38. Arthur N. Taft, "Observations of Life in Colorado Springs," 1–2, folder 108, CSCCC.
39. Thomas MacLaren, "An Account of the Architecture of the City," 8, folder 17; Caldwell, "Colorado Springs Today," 7; Walter Lawson Wilder, "The Century Chest," draft editorial for *Colorado Springs Gazette*, 4, folder 42, CSCCC.
40. Marlatt, "To My Descendants," 1; Shortridge, *Kansas City*, 29, 30.
41. York, "Music in Detroit," 8–9.
42. Frank I. Cobb, "The Newspapers of Detroit," 1–2, item 27, DCBC.
43. Caldwell, "Colorado Springs Today," 29.
44. MacLaren, "Architecture of the City," 3, 4; Taft, "Life in Colorado Springs," 5.

45. "Mrs. Nation Fired in Police Court," *Kansas City World*, April 15, 1901, 2C; Mrs. Henry L. McCune [Helen McCune], calling card, album 74, KCCBC.
46. Joseph Rowntree and Arthur Sherwell, *Temperance Problem and Social Reform* (London: Hodder and Stoughton, 1900), 289.
47. Caldwell, "Colorado Springs Today," 22, 23–24; Franklin Noxon, "To my own posterity," 7, folder 37, CSCCC.
48. Manly D. Ormes, "Ministers in Colorado Springs and Congregational Church," 7, folder 41, CSCCC.
49. Noxon, "To my own posterity," 7.
50. See Lawrence H. Larsen and Nancy J. Hulston, *Pendergast!* (Columbia: University of Missouri Press, 2013), 12–43.
51. Carlos E. Warner, "The Chamber of Commerce," 6, item 158, DCBC.
52. Louis Ehrich, "'Posteritism': An Address Delivered at the Dedication Exercises of the Century Chest, on August 4th, 1901" (1901), 9, ms. 1349, Ehrich Papers, Yale University Library.
53. Francis A. Blades, "Our Financial City Government," 1–2, item 21, DCBC.
54. Frederick F. Ingram, "Public Lighting of Detroit," 2, item 50, and Edward W. Pendleton, "The Water Supply of Detroit" (letter listed in the inventory, but not found in the vessel), DCBC. On Maybury's endorsement of Pingree's policies, see "William C. Maybury," in *Biographical Dictionary of American Mayors, 1820–1980*, ed. Melvin G. Holli and Peter Jones (Westport, CT: Greenwood Press, 1981), 250.
55. Ehrich, "'Posteritism,'" 10–11.
56. Edgar T. Ensign, "Public Parks and Parkways in Colorado Springs, Colo.," 5, folder 44, CSCCC. On the wide streets in Palmer's original plat as evidence of "building for the future," see Taft, "Life in Colorado Springs," 10–11. On Mayor Robinson's park plan, see "To the Citizens," 12–17.
57. Myrtle P. Hurlbut, "The Parks and Boulevards of Detroit," 2, item 89, DCBC.
58. John M. Donaldson, "Rise of Architecture in Detroit," 6, item 86, DCBC.
59. Scripps, "Twentieth Century."
60. William Potts George, "To my Successor in the Pastorate of Westminster Congregational Church," 1, folder 152, KCCBC; King, "To the Honorable Mayor," 4.
61. Robert J. Mason, inscription on reverse of calling card, album 58, and J. Logan Jones, letter, folder G, KCCBC.
62. Mr. and Mrs. John H. Powell, inscription on reverse of calling card, album 52, KCCBC.
63. Bennett, "Colorado Springs Real Estate," 2–3, and Leah Ehrich, "Music in Colorado Springs," 3, folder 74, CSCCC.
64. François Hartog, *Regimes of Historicity: Presentism and Experiences of Time* (New York: Columbia University Press, 2015), 203.
65. Taft, "Life in Colorado Springs," 9, 13; Bertha Francis Emery, "My dears," 10–11, folder 134, CSCCC.

66. See Mark Twain, *The Devil's Race-Track: Mark Twain's Great Dark Writings*, ed. John S. Tuckey (Berkeley: University of California Press, 1980), 373–83.
67. See *Mark Twain's Weapons of Satire: Anti-Imperialist Writings on the Philippine-American War*, ed. Jim Zwick (Syracuse, NY: Syracuse University Press, 1992).
68. Louis Ehrich et al., "To the President of the United States," March 24, 1898, folder 49, CSCCC.
69. See Jim Zwick, "The Anti-Imperialist Movement, 1898–1921," in *Whose America? The War of 1898 and the Battles to Define the Nation*, ed. Virginia M. Bouvier (Westport, CT: Praeger, 2001), 171–92.
70. Louis Ehrich, "To my Fellow-Townspeople of Colorado Springs in the Year 2001," 3, folder 49, CSCCC.
71. Theodore Roosevelt, letter, folder 1, CSCCC.
72. Ehrich, "To my Fellow-Townspeople," 5, 6, 8. McKinley proclamation in *Encyclopedia of the Spanish-American and Philippine-American Wars*, ed. Spencer Tucker, 2 vols. (Santa Barbara, CA: ABC-CLIO, 2009), 1:924.
73. A. R. Kieffer, "To the Rector of Grace Church, Colorado Springs, Colo., in the Year 2001," 2, folder 60, CSCCC. For a proimperialist message, see William Alexander Platt, "A Message on National Politics," 4, 6, folder 25, CSCCC.
74. See, for example, William Garrott Brown, diary, 5, box 58, HUC.
75. Paul Revere Frothingham, diary, 8–9, box 59, and Joseph Henry Thayer, diary, 4 (verso), box 61, HUC.
76. Daniel Kittredge, theme, and Augustine Derby, theme, box 55, HUC.
77. Derby, theme; Thayer, diary, 2.
78. Louis Blitz, "Jewish People in Detroit in the 19th Century," 2, item 76, DCBC.
79. Roosevelt quoted in Stuart Creighton Miller, *Benevolent Assimilation: The American Conquest of the Philippines, 1899–1903* (New Haven, CT: Yale University Press, 1982), 238.
80. On the discourse of experts, see Carolyn Marvin, *When Old Technologies Were New: Thinking about Electric Communication in the Late Nineteenth Century* (New York: Oxford University Press, 1988); on corporate networks, see Richard John, *Network Nation: Inventing American Telecommunications* (Cambridge, MA: Harvard University Press, 2010); on cultural representations, see Cecelia Tichi, *Shifting Gears: Technology, Literature, Culture in Modernist America* (Chapel Hill: University of Carolina Press, 1987).
81. See, for example, Claude Fischer, *America Calling: A Social History of the Telephone to 1940* (Berkeley: University of California Press, 1992).
82. See Howard P. Segal, *Technological Utopianism in American Culture* (Syracuse, NY: Syracuse University Press, 1985); and Carolyn de la Peña, *The Body Electric: How Strange Machines Built the Modern American* (New York: New York University Press, 2005).

83. Baldwin, letter on "Detroit's industrial and commercial possibilities," 2; Warner, "Chamber of Commerce," 6–7.
84. Miriam Storrs Washburn, "Colorado Springs from a Woman's Standpoint," 6, folder 66, CSCCC.
85. Marshman, "The Hon. Mayor," 2; Police Commissioners, "The Police Department of Detroit," 1, item 160, and William C. Maybury, "To His Honor, The Mayor of Detroit in 2001," 2, item 104, DCBC.
86. Elizabeth Cass Goddard, "Social Life," 2, folder 13, CSCCC.
87. Telephone directories in folder 70, KCCBC; folder 154, CSCCC. On the elite's growing preference to unlist, see Marvin, *When Old Technologies Were New*, 104.
88. Scripps, "Twentieth Century"; Maybury, "To His Honor, The Mayor," 3; Moss's message, cited in "Messages to Posterity," *Kansas City Star*, January 3, 1901, newspaper clippings related to the century box, KCCBC.
89. Margaret E. Ball, "To the Class of Two Thousand," *Mount Holyoke* 10, Commencement Number (June 1900): 15, MHC.
90. Dora Noxon, "Metaphysics—True and False," 3, folder 37, CSCCC.
91. Radiographs in folder 118, CSCCC.
92. "World Capital," *Twentieth Century*, ed. J. F. Whiting, mock newspaper, dated Saturn 43, 2000, folder 68, KCCBC.
93. George Rex Buckman, "Notes on the Lighting and Heating of Colorado Springs," 1, 2, 4, folder 23, CSCCC; Caldwell, "Colorado Springs Today," 8.
94. See Jessie Adella Aiken, "Day and Night Nursery Association," 4, folder 68, and Fred Stevens photographs, nos. 83–86, folder 160, CSCCC.
95. Marshman, "The Hon. Mayor," 2–3; Scripps, "Twentieth Century."
96. "An Exciting Elopement," *Twentieth Century*, ed. J. F. Whiting, mock newspaper issue, dated Saturn 43, 2000, folder 68, KCCBC.
97. See, for example, Williamina Fleming, diary, 22, box 59, HUC; Robinson, "To the Citizens," 11–12.
98. Caldwell, "Colorado Springs Today," 5.
99. Theodore A. McGraw, "Progress of Surgery in the 19th Century," 5, item 102, and Morse Stewart, "Medical Reminiscences of the Nineteenth Century," 3, item 137, DCBC.
100. Robinson, "To the Citizens," 11–12; Kittredge, theme. Predictions of longevity in William Coolidge Lane, diary, 64, box 59, HUC; and "Local Mention," *Twentieth Century*, ed. J. F. Whiting, mock newspaper, dated Saturn 43, 2000, folder 68, KCCBC.
101. Bennett, "Colorado Springs Real Estate," 2–3.
102. Jeremiah Dwyer, "Growth of Detroit's Manufactories during the 19th Century," item 56, DCBC.
103. Caldwell, "Colorado Springs Today," 17.
104. Washburn, "Colorado Springs from a Woman's Standpoint," 7.
105. Elizabeth Solly, "Housekeeping," 1, folder 39, CSCCC.

106. John Potter, "Early Settlers and Their Hopes," 6, folder 48, CSCCC.
107. "Local Mention" and untitled news item, *Twentieth Century*, ed. J. F. Whiting, mock newspaper, dated Saturn 43, 2000, folder 68, KCCBC.
108. "A Curious Machine" and untitled news item, *Twentieth Century*, ed. J. F. Whiting, mock newspaper, dated Saturn 43, 2000, folder 68, KCCBC.
109. Bennett, "Colorado Springs Real Estate," 5.
110. See Elizabeth Edwards, *The Camera as Historian: Amateur Photographers and Historical Imagination, 1885–1918* (Durham, NC: Duke University Press, 2012), 7.
111. Fredric Jameson, *Postmodernism, or The Cultural Logic of Late Capitalism* (Durham, NC: Duke University Press, 1991), 279.
112. D. P. Alderson, inscription on reverse of business card, folder 3, KCCBC.
113. Dwyer, "Growth of Detroit's Manufactories," 6; he hoped they could be converted to make different products.
114. Caldwell, "Colorado Springs Today," 18; number of automobiles in John Ludlow Pendery, "Biography of J. L. Pendery," 13, folder 148, CSCCC.
115. Buckman, "Lighting and Heating of Colorado Springs," 6.
116. Baldwin, "Detroit's industrial and commercial possibilities," 2; Kittredge, theme; Scripps, "Twentieth Century"; Frederick F. Ingram, "Public Lighting of Detroit," 3, item 50, DCBC.
117. George Herbert Palmer, diary, 4–5, box 60, HUC.
118. Oscar LeSeur, "Growth of Homeopathic Practice in Detroit," 1, 2, item 73, DCBC. On homeopathy's rise and fall, see Kevin Starr, *The Social Transformation of American Medicine* (New York: Basic Books, 1982), 96–109.
119. Samuel S. Stephenson, "The Future of Biopathy," 1–2, item 133, DCBC. De la Peña's omission of biopathy from *Body Electric* indicates its obscurity.
120. Rennie B. Schoepflin, *Christian Science on Trial: Religious Healing in America* (Baltimore: Johns Hopkins University Press, 2003), 23–24.
121. Dora Noxon, "My Children," 3, folder 37, CSCCC.
122. Dora Noxon, "Metaphysics," 11, folder 37; Edson M. Cole, "Ten Reasons Why Christian Science Will Be the Religion of Your Day," 6, folder 45, CSCCC.
123. Annie M. Knott, "Growth of Christian Science in Detroit," 1, item 71, DCBC.
124. Patty Stuart Jewett, "To Church of Christ Scientist of Colorado Springs in the Year 2001," 7, folder 19, CSCCC; E. B. Hubbard, "Greeting to the Christian Scientists of the 21st Century," 1, folder 54, KCCBC; Cole, "Ten Reasons," 2. Percentage calculated on the basis of figures from Raymond Cunningham, "The Impact of Christian Science on the American Churches, 1880–1910," *American Historical Review* 72 (1967): 890, 893.
125. Knott, "Christian Science in Detroit," 4.
126. Desire B. Willemin, "Foundation of Detroit by the French," 3, item 151, and [Robert Fulton?], Secretary of the St. Andrew's Society, "Greetings to the Descendants of Early Scotchmen in Detroit," 1, item 146, DCBC.

127. Frank I. Cobb, "The Newspapers of Detroit," 1–2, item 27, DCBC.
128. David W. Simons, "The Jewish People in Mercantile, Social and Professional Life in Detroit," 1–3, item 142, DCBC.
129. Louis Blitz, "Jewish People in Detroit in the 19th Century," 2, item 76, DCBC.
130. On this division with Detroit's Jewish community, see Sidney M. Bolkosky, *Harmony and Dissonance: Voices of Jewish Identity in Detroit, 1914–1967* (Detroit: Wayne State University Press, 1991), 19–67; on Simons's standing in the community, see 37.
131. Olivier Zunz, *The Changing Face of Inequality: Urbanization, Industrial Development, and Immigrants in Detroit, 1880–1920* (Chicago: University of Chicago Press, 1982), 104.
132. Zunz, esp. 178–95.
133. See folders 4, 10, 14, 15, 31, 41, 54, 60, 65, 69, 71, 76, 80, 82, 86, 100, 108, 121–24, 133, 135, 159; genealogical records included in, among others, Charles Fox Gardiner, "To My Professional Brothers of the Year 2001," folder 62; DeWitt C. and Sarah Maria Smith Jencks, message to their descendants, folder 131, CSCCC. See also François Weil, *Family Trees: A History of Genealogy in America* (Cambridge, MA: Harvard University Press, 2013), 31–34, 112–42.
134. Harriet Peck Farnsworth, "Dear Great Grandchildren," 8, folder 119, CSCCC.
135. Adams Sherman Hill, *Principles of Rhetoric* (New York: Harper, 1896), 2; Hill, *Our English* (New York: Harper, 1897), 102–40.
136. Hill, *Our English*, 72–101.
137. Amy Zenger, "Race, Composition, and 'Our English': Performing the Mother Tongue in a Daily Theme Assignment at Harvard, 1886–87," *Rhetoric Review* 23, no. 4 (2004): 334–35.
138. Francis Galton, *Inquiries in Human Faculty and Its Development* (London: Macmillan, 1883), 8–19, 339–63; the term *eugenic* is introduced on 24. On Galton's composite photography, see, among others, Anne Maxwell, *Picture Imperfect: Photography and Eugenics, 1870–1940* (Brighton: Sussex Academic, 2010), 79–107; and Shawn Michelle Smith, *American Archives: Gender, Race, and Class in Visual Culture* (Princeton, NJ: Princeton University Press, 1999), 86–93.
139. Henry P. Bowditch, "Are Composite Photographs Typical Pictures?" *McClure's* 3, no. 4 (1894): 331–42. His letter declining Lane's invitation is in "Information about the box" folder, box 62, HUC.
140. Periodization from Maxwell, *Picture Imperfect*, 7–8, 109–21.
141. Maxwell, 7.
142. Galton, *Inquiries*, 14.
143. See, for example, Robert DeC. Ward, "The Immigration Problem: Its Present Status and Its Relation to the American Race of the Future," *Charities* 12, no. 6 (1904): 138–51, quotation on 149.

144. John Ames Mitchell, *The Last American: A Fragment from the Journal of Khan-li* (New York: Frederick A. Stokes and Brother, 1889), 33.
145. Edward Bellamy, *Equality* (New York: D. Appleton, 1897), 365; Arthur Bird, *Looking Forward: A Dream of the United States of the Americas in 1999* (Utica, NY: I. C. Childs, 1899), 229–31.
146. Theodore Roosevelt, letter, 1–2, folder 1, CSCCC; Roosevelt, letter to Marie Von Horst, October 18, 1902, in *Works of Theodore Roosevelt*, 14 vols. (New York: Collier, 1910), 14:508–9.
147. Hamlin Garland, "The Mystery of the Mountains," manuscript, folder 79, CSCCC.
148. Ehrich, "'Posteritism,'" 5, 8, 11; Ehrich, "Posteritism," *Conservative* 4, no. 17 (1901): 4–5.
149. See Laura Otis, *Organic Memory: History and the Body in the Late Nineteenth and Early Twentieth Centuries* (Lincoln: University of Nebraska Press, 1994).
150. Ehrich, "Posteritism," 10.
151. Margaret T. Adams, "To Any Native Americans [. . .] living in the year 2000," 5, 6, folder 92, CSCCC.
152. (Harry) Kellogg Durland, diary, box 58, HUC. See Werner Sollors, Caldwell Titcomb, and Thomas A. Underwood, eds., *Blacks at Harvard: A Documentary History of African-American Experience at Harvard and Radcliffe* (New York: New York University Press, 1993).
153. D. Augustus Straker, "The Past, Present, and Future of the Colored Race in Detroit," 2, 3, item 156, DCBC. Ben DeBaptiste's brother George was the better-known activist. On Straker, see Susan D. Carle, *Defining the Struggle: National Racial Justice Organizing, 1880–1915* (New York: Oxford University Press, 2013), 65–69; and David M. Katzman, *Before the Ghetto: Black Detroit in the Nineteenth Century* (Urbana: University of Illinois Press, 1973), 189–94; statistic on 62.
154. Katzman, *Before the Ghetto*, 95–97.
155. Straker, "Past, Present, and Future," 4.
156. Glenn O. Phillips, "The Response of a West Indian Activist: D. A. Straker, 1842–1908," *Journal of Negro History* 66, no. 2 (1981): 136.
157. Charles W. Chestnutt, "The Future American," in *Charles W. Chesnutt: Essays and Speeches*, ed. Joseph McElrath Jr., Robert Leitz III, and Jesse Crisler (Stanford, CA: Stanford University Press, 1999), 121, 122–23, 125, 132–33, 135.
158. Adams, "To Any Native Americans," 5–6 (emphasis added).
159. Michelle Elizabeth Tusan, "Inventing the New Woman: Print Culture and Identity Politics during the Fin-de-Siecle," *Victorian Periodicals Review* 31, no. 2 (1998): 170, 175.
160. G. Stanley Hall, *Adolescence: Its Psychology and Its Relations to Physiology, Anthropology, Sociology, Sex, Crime, Religion and Education*, vol. 2 (New York: D. Appleton, 1905), 634; Roosevelt, letter to Von Horst, 508–9. See also

Carroll Smith-Rosenberg, *Disorderly Conduct: Visions of Gender in Victorian America* (New York: Oxford University Press, 1985), 245–96.

161. On the myth of "frontier egalitarianism," see Rebecca J. Mead, *How the Vote Was Won: Woman Suffrage in the Western United States, 1868–1914* (New York: New York University Press, 2004), 17–18.
162. Caldwell, "Colorado Springs Today," 23–24, 37–38.
163. Bertha Francis Emery, "My dears," 9, folder 134, CSCCC.
164. Washburn, "Colorado Springs from a Woman's Standpoint," 12. On the meanings of the shirtwaist, see Lois W. Banner, *American Beauty* (New York: Knopf, 1983), 148–50.
165. Irene Williams Chittenden, "The American Mother of the Nineteenth Century," 1, 3, item 25, DCBC.
166. Mrs. John Vallé Moran [Emma Moran], "The American Mother in Detroit," 1–2, 8–9, item 183, DCBC.
167. On feminist reappropriations in Britain, see Tusan, "Inventing the New Woman," 169–82.
168. University of Michigan admitted women in 1870 and Cornell in 1872; Helen Lefkowitz Horowitz, *Campus Life: Undergraduate Cultures from the End of the Eighteenth Century to the Present* (Chicago: University of Chicago Press, 1988), 193. Statistics from Theodore Caplow, Louis Hicks, and Ben J. Wattenberg, *The First Measured Century: An Illustrated Guide to Trends in America, 1900–2000* (Washington, DC: AEI Press, 2001), 54.
169. Lynn D. Gordon, "Education and the Professions," in *A Companion to American Women's History*, ed. Nancy A. Hewitt (Oxford: Blackwell, 2002), 239–40.
170. Horowitz, *Campus Life*, 195, 197–98; generational model from Gordon, "Education and the Professions," 239.
171. Horowitz, *Campus Life*, 195, 201–2. John R. Thelin, *A History of American Higher Education* (Baltimore: Johns Hopkins University Press, 2004), 97–98, 143–45.
172. See especially Archibald C. Coolidge, diary, box 58, HUC. See Laurel Ulrich, ed., *Yards and Gates: Gender in Harvard and Radcliffe History* (New York: Palgrave Macmillan, 2004), 87–158.
173. Barrett Wendell, diary, box 61, HUC. On the 1999 merger, see Adam A. Sofen, "Radcliffe Enters Historic Merger with Harvard," *Harvard Crimson*, April 21, 1999, 7.
174. Lucy Allen Paton, diary, box 60, HUC.
175. Ben Griffith, "To the male students of Colorado College in 2001 A. D.," 7, folder 9, and Ella Graber, "To the girls of Colorado College in the Year Two Thousand," 2, folder 76, CSCCC. For an example of the characterization of greater liberality in the west, see Andrea G. Radke-Moss, *Bright Epoch: Women and Coeducation in the American West* (Lincoln: University of Nebraska Press, 2008).

176. Finding aid for Lucy Allen Paton Papers, 1920–1943, Radcliffe College Archives, Schlesinger Library, Radcliffe Institute, Harvard University. On gender discrimination, see Thelin, *History of American Higher Education*, 143–45; Margaret W. Rossiter, *Women Scientists in America: Struggles and Strategies to 1940* (Baltimore: Johns Hopkins University Press, 1984), esp. 56. Harvard did not appoint a female professor until 1919.
177. Rossiter, *Women Scientists in America*, 53; the story may be apocryphal.
178. Williamina Fleming, diary, 9–10, box 59, HUC. On Fleming's promotion to curator, see "Harvard honors women," *Boston Herald*, January 16, 1899, HUF 165.87.37, Harvard University Archives.
179. See Ruth Bordin, *Alice Freeman Palmer: The Evolution of a New Woman* (Ann Arbor: University of Michigan Press, 1993), 140.
180. Mandolin club in Mount Holyoke College year book, *Llamarada*, 1899, and photograph of "dramatics" in anonymous, untitled blue book, MHC.
181. "Mount Holyoke College Catalogue, 1899–1900," 14, MHC.
182. Graber, "To the Girls," 3, 4, 8.
183. A. B. Wolfe, themes, box 55, HUC.
184. Margaret E. Ball, "To the Class of Two Thousand," *Mount Holyoke* 10, Commencement Number (June 1900): 15, MHC. Ball did become professor at Mount Holyoke.
185. M. L. E., "The Educated Unemployed," in *Mount Holyoke* 9, no. 5 (1900): 166–67, MHC.
186. Graber, "To the Girls," 13–14.
187. Fleming, diary, 1, 18–19. Julie Des Jardins, *The Madame Curie Complex: The Hidden History of Women in Science* (New York: Feminist Press, 2010), 99–100.
188. One exception is Anne Ruggles Gere, *Intimate Practices: Literacy and Cultural Work in U.S. Women's Clubs, 1880–1920* (Urbana: University of Illinois Press, 1997).
189. Elizabeth Cass Goddard, "Social Life," 4, folder 13, CSCCC.
190. Edward Evans-Carrington, "Office of Associated Charities," folder 5, and Stella A. Kyle et al, letters regarding the Women's Relief Corps, folder 47, CSCCC. James A. Post, "The Charity Organization Movement," item 11; Florine Smith Stoddard, "Y.W.C.A.," item 148; Mrs. Jacob F. Teichner [Fanny Teichner], "Hebrew Homes and Charities in Detroit," 140, DCBC.
191. Pamela A. Patterson, "The Growth of the Woman's Club Movement," 1, item 117, DCBC.
192. Fonetta Flansburg, "To the 'Club Women' of 2001," 2, folder 116, CSCCC.
193. Patterson, "Woman's Club Movement," 3–6.
194. Virginia Donaghe McClurg, "The Mesa Verde Cliff Dwellings and the Women's Park," folder 43, CSCCC. McClurg is mentioned only in passing in Kathleen Fiero, *Dirt, Water, Stone: A Century of Preserving Mesa Verde*

(Durango, CO: Durango Herald Small Press, 2006), 29. Unfortunately, internal divisions led to a national rather than a state park in 1906.

195. Patterson, "Woman's Club Movement," 4.
196. Bertha Francis Emery, "My dears," 9–10, folder 134, CSCCC. On derision toward club women and estimate of two million, see Martha H. Patterson, ed., *The American New Woman Revisited: A Reader, 1894–1930* (New Brunswick: Rutgers University Press, 2008), 10.
197. Flansburg, "To the 'Club Women,'" 1.
198. William Potts George, "Men as Domestic Servants: How Is This as a Solution of the Housekeeping Problem?," *NYT*, October 18, 1905, 10; Elizabeth Solly, "Housekeeping," 3, folder 39, CSCCC.
199. Washburn, "Colorado Springs from a Woman's Standpoint," 4.
200. See, for example, Anne Firor Scott and Andrew MacKay Scott, *One Half the People: The Fight for Woman Suffrage* (Urbana: University of Illinois Press, 1982), 22, 24, and Ellen Carol DuBois, *Woman Suffrage and Women's Rights* (New York: New York University Press, 1998), 160–75.
201. See Carolyn Stefanco, "Networking on the Frontier: The Colorado Women's Suffrage Movement, 1876–1893," in Elizabeth Jameson and Susan H. Armitage, *Writing the Range: Race, Class, and Culture in the Women's West* (Norman: University of Oklahoma Press, 1997), 265–76; and Mead, *How the Vote Was Won*, 53–72. On the connection between the New Woman and the winning of suffrage, see Holly McCammon, Karen Campbell, Ellen Granberg, and Christine Mowery, "How Movements Win: Gendered Opportunity Structures and U.S. Women's Suffrage Movements, 1866 to 1919," *American Sociological Review* 66, no. 1 (2001): 49–70.
202. Dora Noxon, "My Children," 5–6, folder 37, CSCCC; Washburn, "Colorado Springs from a Woman's Standpoint," 8.
203. Noxon, "My Children," 6.
204. Washburn, "Colorado Springs from a Woman's Standpoint," 8.
205. Photograph in folder 110, CSCCC.
206. DuBois, *Woman Suffrage*, 160–75.
207. Sara M. P. Skinner, "Women's Suffrage—Retrospect and Prophecy," 1, 13–14, item 144, DCBC.
208. Ball, "To the Class of Two Thousand," 15.
209. Sherry Schirmer, *A City Divided: The Racial Landscape of Kansas City, 1900–1960* (Columbia: University of Missouri Press, 2002), 40; James Shortridge, *Kansas City and How it Grew, 1822–2011* (Lawrence: University Press of Kansas, 2012), 78–79.
210. Nancy Capace, ed., *Encyclopedia of Colorado* (St. Clair Shores, MI: Somerset, 1999), 175.
211. Olivier Zunz, *The Changing Face of Inequality: Urbanization, Industrial Development, and Immigrants in Detroit, 1880–1920* (Chicago: University of Chicago Press, 1982), 6, 32–33, 95, 97, 101, 200, 204, 209.

212. "Century Ball Statement," *Kansas City World*, November 27, 1900, newspaper clippings related to the century box, KCCBC.
213. Mrs. Sidney Trowbridge Miller [Katherine Miller], "Detroit's Social Life," esp. 7–8, item 154, DCBC; Goddard, "Social Life of Colorado Springs," 1–11.
214. Ben Griffith, "To the male students of Colorado College in 2001 A. D.," 10–11, folder 9.
215. Jeremiah Dwyer, "Growth of Detroit's Manufactories during the 19th Century," 5, item 56, DCBC.
216. James E. Scripps, "A Prophecy for Detroit as a Metropolis," 1, item 130, DCBC.
217. See the letters about charities cited above in note 190; quotation from Post, "Charity Organization Movement," 1. The motto of Detroit's United Jewish Charities (whose founding president contributed a letter) was "help the poor to help themselves"; Sidney M. Bolkosky, *Harmony and Dissonance: Voices of Jewish Identity in Detroit, 1914–1967* (Detroit: Wayne State University Press, 1991), 37.
218. Harriet Peck Farnsworth, "Dear Great Grandchildren," 8, folder 119, CSCCC.
219. Solly, "Housekeeping," 1–2, 5; W. W. Hassell, "To the Citizens of 2001," 4, folder 40, CSCCC. See also Kathleen A. Brosnan, *Uniting Mountain and Plain: Cities, Law, and Environmental Change along the Front Range* (Albuquerque: University of New Mexico Press, 2002), 97–98.
220. John Potter, "Early Settlers and Their Hopes," 5, folder 48, CSCCC.
221. Wales C. Martindale, "Public Schools in Detroit," 2, item 96, DCBC.
222. Paul Revere Frothingham, diary, 12, box 59, HUC.
223. W. H. Cutler, "Organization of the Janitors' Department," 1–3, box 62, HUC.
224. See, among others, Seymour Martin Lipset and Gary Marks, *It Didn't Happen Here: Why Socialism Failed in the United States* (New York: W. W. Norton, 2000), and Robin Archer, *Why Is There No Labor Party in the United States?* (Princeton, NJ: Princeton University Press, 2010).
225. Elizabeth Jameson, *All that Glitters: Class, Conflict, and Community in Cripple Creek* (Urbana: University of Illinois Press, 1998), 49–61.
226. Jameson, 62–86, 179. On the 1901 strike, see Guy E. Miller, "The Telluride Strike," in *The Cripple Creek Strike: A History of Industrial Wars in Colorado, 1903–4–5*, ed. Emma Florence Langdon (Denver, CO: Great Western, 1904–1905), 200–12.
227. Thomas G. Andrews, "'Made by Toile'? Tourism, Labor, and the Construction of the Colorado Landscape, 1858–1917," *Journal of American History* 92, no. 3 (2005): 837–63.
228. Taft, "Life in Colorado Springs," 6–7.
229. William Alexander Platt, "A Message on National Politics," 7–9, folder 25, CSCCC.

230. Caldwell, "Colorado Springs Today," 9.
231. Zunz, *Changing Face of Inequality*, 219, 224, 227; statistics on 224.
232. William Carson, letter on "the progress of trade," 6, item 14, HUC.
233. Charles W. Purvis, "To the President of the New Boys' Union One Hundred Years (2001) Hence," and union materials in folder 81, KCCBC.
234. Malcolm McLeod, "Condition of Labor Men in the City of Detroit," 1, 2, item 108, DCBC.
235. David A. Boyd, "The Labor Movement in Detroit Up to Date," 2, item 34, and Boyd, "The Council of Trades and Labor Unions and the Labor Movement of Detroit," *Labor Day Review*, n.d., 27, 29, item 36, DCBC.
236. Charles G. Collais, "Present Conditions of Labor Organizations," 2–4, 6–9, folder 29, CSCCC.
237. See, for example, David Brundage, "Gilded Age," in *Encyclopedia of U.S. Labor and Working-class History*, ed. Eric Arnesen, 2 vols. (New York: Routledge, 2007), 1:526.
238. Collais, "Present Conditions," 11; Boyd, "Labor Movement in Detroit," 1; McLeod, "Condition of Labor Men," 2.
239. Albert Boesel, letter, 1, folder 83, CSCCC; Boyd, "Labor Movement in Detroit," 1.
240. McLeod, "Condition of Labor Men," 1.
241. Collais, "Present Conditions," 10.
242. Boyd, "Council of Trades," 25.
243. [Louis F. Post?], untitled editorial, *The Public*, no. 187 (November 2, 1901): 471.
244. Henry W. Farnam, "Labor Legislation and Economic Progress," in *Proceedings of the Annual Meeting, American Association for Association for Labor Legislation* (Princeton, NJ: Princeton University Press, 1909), 36–37, 38, 42. Prominent labor historian John R. Commons called it a "remarkable address" in "The American Association for Labor Legislation," *Charity Organization Review* 21, no. 15 (1909): 665–66. Frank Tracy Carlton deployed the idea in his *History and Problems of Organized Labor* (Boston: D. C. Heath, 1911), 347. See also the glowing review in "Notes," *Nation*, May 8, 1913, 481. D. H. MacGregor, "Some Ethical Aspects of Industrialism," *International Journal of Ethics* 19, no. 3 (1909): 290–91.
245. Mary Noyes Shaw, "To the woman in the great unknown," 1, RCB.
246. Dora Noxon, "My children," 6; Franklin Noxon, "To my own posterity," 9, folder 37, CSCCC.
247. Solly, "Housekeeping," 1–5.
248. Daniel Kittredge, theme, box 55, HUC.
249. See, for example, Archer, *Why Is There No Labor Party*, esp. 182–84. On religious historians' neglect of the socialist strain in the Social Gospel movement, see Gary Scott Smith, *The Search for Social Salvation: Social Christianity and America, 1880–1925* (Lanham, MD: Lexington, 2000), 2–3, 6.

250. A. R. Kieffer, "To the Rector of Grace Church, Colorado Springs, Colo., in the Year 2001," 1–4, folder 60, CSCCC.
251. Benjamin Brewster, "A Minister's Life in Colorado Springs," 8–9, folder 69, CSCCC.
252. W. H. W. Boyle, "Outlook and Uplook: Twentieth Century Citizenship—Its Opportunity and Its Responsibility," *Facts*, n.d., 8, folder 125, CSCCC.
253. William Coolidge Lane, diary, 87, box 59, HUC. On Frothingham's political views, see Howard Chandler Robbins, *The Life of Paul Revere Frothingham* (Boston: Houghton Mifflin, 1935), 54, 57, 59–77.
254. Folder 18, CSCCC.
255. Washburn, "Colorado Springs from a Woman's Standpoint," 14. See Edward K. Spann, *Brotherly Tomorrows: Movements for a Cooperative Society in America, 1820–1920* (New York: Columbia University Press, 1989).
256. See especially Ernst Bloch, *Traces*, trans. Anthony A. Nassar (Stanford, CA: Stanford University Press, 2006).
257. Bellamy, *Equality*, 94, 95, 101–102. Bellamy avoided the term *socialism*.
258. See "News," *Public* 3, no. 121 (1900): 250. Ehrich deposited his writings on "sound money" (folder 49, CSCCC).
259. Louis Ehrich, "Stock-Sharing as a Preventive of Labor Troubles," *Forum* 18 (December 1894): 435–36, 438.

CHAPTER FIVE

1. "Record of Our Time to Be Imperishable," *NYT*, December 10, 1911, 17.
2. "New England and Other Matters," *Youth's Companion* 86, no. 41 (1912): ii; "Tough Questions on Time Capsule," *American Journal* (Westbrook), July 24, 2014, http://news.keepmecurrent.com/tough-questions-on-time-capsule/; "Two 'Century Boxes' to Be Opened in 2015," *NYT*, July 9, 1916, X10; OCCCC; 1914 time capsule collection, NYHS.
3. Virginia Sohlberg, journal, 1913, OCCCC.
4. Invitation in MHRA, Miscellaneous Personal Name Files, box 73, miscellaneous file, Manuscripts and Archives Division, NYPL.
5. "An American Pantheon Suggested," *New York Observer and Chronicle*, September 7, 1911, 313.
6. See, for example, Jonathan Sterne, *The Audible Past: Cultural Origins of Sound Reproduction* (Durham, NC: Duke University Press, 2003), esp. 298, 324–25.
7. Harold A. Innis, *The Bias of Communication* (1951; Toronto: University of Toronto Press, 2008), esp. 33–60.
8. "Record of Our Time," 17.
9. "Alexander Konta, Banker, Is Dead," *NYT*, April 29, 1933, 13.
10. Untitled pamphlet detailing MHRA's aims (New York: MHRA, 1912), 2, Miscellaneous Personal Name Files, box 73, miscellaneous file, Manuscripts and Archives Division, NYPL.

11. "American Pantheon," 313; "Alexander Konta, Banker," 13.
12. Frederic Harrison, "A Pompeii for the Twenty-Ninth Century," *Nineteenth Century* 28, no. 163 (1890): 381, 383, 390. It was reprinted in Harrison, *Realities and Ideals: Social, Political, Literary and Artistic* (London: Macmillan, 1908), 467–83.
13. W. T. Larned, letter to *The Dial*, September 16, 1912, 186; "Wants Photographs to Last Forever," *NYT*, November 30, 1912, 22.
14. See, for example, Randall Mason, *The Once and Future New York: Historic Preservation and the Modern City* (Minneapolis: University of Minnesota Press, 2009), xxiv, 7, 19, 204.
15. "Celebrities Write on Parchment for Future Ages," *NYT*, June 2, 1912, SM9; "Records for Posterity," *NYT*, July 13, 1911, 8 (emphasis added).
16. Edward Bellamy, *Looking Backward, 2000–1887* (Boston: Ticknor, 1888), 203.
17. See chapters 2 and 3, above, and Nick Yablon, *Untimely Ruins: An Archaeology of American Urban Modernity, 1819–1919* (Chicago: University of Chicago Press, 2009), 183–89.
18. H. G. Wells, *The Time Machine* (New York: Henry Holt, 1895), 160.
19. H. G. Wells, *The War in the Air* (New York: Macmillan, 1908), 351.
20. H. G. Wells, *A World Set Free: A Story of Mankind* (New York: E. P. Dutton, 1914), 222–23.
21. Richard Jefferies, *After London, or Wild England* (London: Cassell, 1885), 82–86.
22. [Ignatius Donnelly], *Caesar's Column: A Story of the Twentieth Century* (Chicago: Schulte, 1890), 285, 363.
23. M. P. Shiel, *The Purple Cloud* (London: Chatto and Windus, 1901), 113.
24. Van Tassel Sutphen, *The Doomsman* (New York: Harper, 1906), 70, 71.
25. Jack London, *The Scarlet Plague* (1912; New York: Macmillan, 1915), 175–76. On his earlier faith, see Yablon, *Untimely Ruins*, 163–67.
26. John Ames Mitchell, *The Last American: A Fragment from the Journal of Khan-li* (New York: Frederick A. Stokes and Brother, 1889), 18, 33, 51, 54, 55.
27. Mitchell, 51.
28. Jefferies, *After London*, 85.
29. Harrison, "A Pompeii," 390–91.
30. George Allan England, *Darkness and Dawn* (Boston: Small, Maynard, 1914), 34, 48, 66–67.
31. England, 445–47.
32. C. E. Vail, "Our Duty to the Future," *Scientific Monthly* 3 (1916): 585, 589.
33. Stephen Jay Gould, *Time's Arrow, Time's Cycle: Myth and Metaphor in the Discovery of Geological Time* (Cambridge, MA: Harvard University Press, 1987), 1–20.
34. Harrison, "A Pompeii," 383.
35. Vail, "Our Duty," 585, 594.

36. "American Pantheon," 313; "Record of Our Time," 17.
37. "MHRA," *World Almanac and Encyclopedia* (New York: Press, 1914), 529.
38. J. E. Chamberlin, *The Ifs of History* (Philadelphia: Altemus, 1907); J. N. Larned, *Seventy Centuries of the Life of Mankind* (Springfield, MA: C. A. Nichols, 1905).
39. Jelliffe's collection is now at Hartford Hospital's Institute of Living; Plimpton's is at Columbia University.
40. See Mason, *Once and Future New York*, xiv, 11–12, 14.
41. The classic articulation of this idea (and these terms) is T. R. Schellenberg, "The Appraisal of Modern Public Records," in *A Modern Archives Reader*, ed. Maygene F. Daniels and Timothy Walch (Washington: NARA, 1984), 57–70. See also Aleida Assmann, *Cultural Memory and Western Civilization: Functions, Media, Archives* (New York: Cambridge University Press, 2011), 328–30; on broader challenges to the traditional archive, see Paula Amad, *Counter-Archive: Film, the Everyday, and Albert Kahn's Archives de la Planète* (New York: Columbia University Press, 2010).
42. Untitled MHRA pamphlet, 7.
43. Ibid., 11. See Donald R. McCoy, "The Struggle to Establish a National Archives in the United States," in *Guardian of Heritage: Essays on the History of the National Archives*, ed. Timothy Walch (Washington, DC: NARA, 1985), 7–10.
44. Untitled MHRA pamphlet, 6.
45. "Celebrities Write on Parchment," SM9.
46. MHRA, "Message of the Records" (ca. 1913), MssCol 2025, Manuscripts and Archives Division, NYPL.
47. The word *systematic* appears in numerous articles about the MHRA, for example, "A Plan for the Preservation of Historic Records," *Catholic Fortnightly Review* 18, no. 18 (1911): 533.
48. Harrison, "A Pompeii," 383.
49. Untitled MHRA pamphlet, 3.
50. "American Pantheon," 313.
51. Fernand Braudel, *On History*, trans. Sarah Matthews (Chicago: University of Chicago Press, 1980), 3.
52. "Celebrities Write on Parchment," SM9.
53. Ibid.
54. Frederic Harrison, "Culture: A Dialogue," *Fortnightly Review* 11 (November 1867): 603–14.
55. Harrison, "A Pompeii," 383, 386.
56. "American Pantheon," 313.
57. Ibid. (emphasis added); Konta to [Robert] Underwood Johnson, July 15, 1911, 3, Century Company records, MssCol 504, Manuscripts and Archives Division, NYPL.
58. "Celebrities Write on Parchment," SM9.
59. Epigraph on front cover of untitled MHRA pamphlet.

60. Untitled MHRA pamphlet, 6; "What to Preserve or Destroy," *NYT*, December 17, 1911, 14.
61. "Chairman Konta's Views," *Talking Machine World* 8, no. 3 (1912): 23.
62. "Servia's Claims Absurd, Says Konta," *NYT*, January 8, 1913, 8.
63. George F. Kunz, "The Projected Museum of the Peaceful Arts," *Proceedings of the American Association of Museums* 6 (1912): 30–37, quotation on 30. Alexander Konta, "Modern Records Museums," *NYT*, November 19, 1913, 8.
64. Alex Wright, *Cataloging the World: Paul Otlet and the Birth of the Information Age* (New York: Oxford University Press, 2014).
65. Amad, *Counter-Archive*, 102, 148–49; on Otlet's "Universal Decimal Classification," see Wright, *Cataloging the World*, 81–83, 106–7.
66. Untitled MHRA pamphlet, 12.
67. "Messages for Posterity a Century Hence," *Sun*, March 9, 1913, 3.
68. See Lyle Dick, "Robert Peary's North Polar Narratives and the Making of an American Icon," *American Studies* 45, no. 2 (2004): 5–34.
69. "Wants Photographs to Last Forever," *NYT*, November 30, 1912, 22.
70. See Leo Braudy, *The Frenzy of Renown: Fame and Its History* (New York: Vintage Books, 1997), 226ff., 239–50, 362–64, respectively.
71. See Eveline G. Bouwers, *Public Pantheons in Revolutionary Europe: Comparing Cultures of Remembrance, c. 1790–1940* (New York: Palgrave Macmillan, 2012), 45–90, 91–131, 161–212, respectively.
72. John Reps, *The Making of Urban America: A History of City Planning in the United States* (Princeton, NJ: Princeton University Press, 1965), 252.
73. Complaints about Statuary Hall cited in Charles E. Fairman, *Art and Artists of the Capitol of the United States of America* (Washington, DC: GPO, 1927), 493.
74. George Kunz, *The Hall of Fame* (New York: New York University, 1908), 3–5, 8.
75. *Report of the Senate Committee on the District of Columbia on the Improvement of the Park System of the District of Columbia* (Washington, DC: GPO, 1902), 48; Fairman, *Art and Artists*, 493–96.
76. See Kunz, *Hall of Fame*; "American Pantheon," 313.
77. Konta to Johnson, 4.
78. Untitled MHRA pamphlet, 12.
79. Larned, cited in "The Appraisal of Contemporary Greatness," *Dial*, August 16, 1912, 90.
80. William N. Thompson and Ernita Joaquin, "The Hall of Fame for Great Americans: Organizational Comatosis or Hibernation," *History News Network* (2010); Larned, "Immortals of Modern Times," *Morning Oregonian*, July 25, 1912, 8; finding aid, 1, Pirie MacDonald Portrait Photograph Collection [1885]–1942, NYHS.
81. W. T. Larned, letter to *The Dial*, September 16, 1912, 186.
82. "American Pantheon," 313.

83. "Wants Photographs to Last Forever," 22.
84. On the kinetophone, see Rick Altman, *Silent Film Sound* (New York: Columbia University Press, 2004), 175–78; on Edison's earlier (1895) kinetophone, see 80.
85. "Mayor in Talking Pictures," *NYT*, April 18, 1913, 1.
86. "Films for Posterity," *NYT*, May 1, 1913, 6.
87. "Posterity Gets First Kinetophone Records," *Oregon Daily Journal* (Portland), March 2, 1913, 46; the New-York Historical Society was also specified as recipient. The MHRA did receive those "first dozen" kinetophone recordings; "Will Preserve Records of War," *Pittsburgh Press*, November 29, 1914, 17.
88. Konta to Johnson, 2.
89. Anthony Slide, *Nitrate Won't Wait: A History of Film Preservation in the United States* (Jefferson, NC: McFarland, 1992), 9–24; Janna Jones, *The Past Is a Moving Picture: Preserving the Twentieth Century on Film* (Gainesville: University Press of Florida, 2014), 25–51; Caroline Frick, *Saving Cinema: The Politics of Preservation* (New York: Oxford University Press, 2011), 29–52. On earlier efforts in France, see Amad, *Counter-Archive*, esp. 152.
90. Frick, *Saving Cinema*, 41.
91. Slide, *Nitrate Won't Wait*, 4; Jones, *Past Is a Moving Picture*, 25.
92. "Big Men in Talking Pictures," *New York Clipper*, May 10, 1913, 16; "Andrew Carnegie before the Camera," *Moving Picture World*, February 7, 1914, 654.
93. Thomas Edison, "The Phonograph and Its Future," *North American Review* 126, no. 262 (1878): 534.
94. See, for example, Harrison, "A Pompeii," 382.
95. "An Interesting Ceremony," *Teachers Magazine* 36, no. 5 (1914): 154. Gaston Leroux, *Le Fantôme de l'Opéra* (Paris: Lafitte, 1910), 11.
96. E. W. Scripture, "The German Emperor's Voice," *Century* 73, no. 1 (1906): 135–39.
97. Harrison, "A Pompeii," 386.
98. Untitled MHRA pamphlet, 3.
99. "Modern Historic Records," *Moving Picture News*, January 20, 1912, 13.
100. "Will Preserve Records of War," 17.
101. Untitled MHRA pamphlet, 13; Alexander Konta, "Records of Indian Life," *Moving Picture News*, March 9, 1912, 10.
102. "Wonders of the Graphophone: Proposition to Preserve Accurate Records of Nineteenth Century Life," *Phonoscope* 4, no. 4 (1900): 6.
103. "Messages for Posterity a Century Hence," *Sun*, March 9, 1913, 3.
104. Konta to Johnson, 1.
105. Konta, "Records of Indian Life," 10.
106. "Immortality?" *Edison Phonograph Monthly* 10, no. 2 (1912): 9.
107. "MHRA," *World Almanac and Encyclopedia* (New York: Press, 1914), 529.
108. "To Record History for Eye and Ear," *Indianapolis Star*, November 4, 1911, 8.

109. "Preserving Records," *Tennessean* [Nashville], October 30, 1911, 4; "Moving Pictures to Revolutionize History," *St. Louis Post-Dispatch*, May 26, 1912, SM6. On political uses of film by 1912, see Jeff Menne and Christian B. Long, ed., *Film and the American Presidency* (New York: Routledge, 2015).
110. "Wants Photographs to Last Forever," 22; the "indestructible" records were presumably Edison Diamond Disc Records, introduced 1912.
111. Untitled MHRA pamphlet, 3; "Records for Posterity," 8.
112. "Will Preserve Records of War," 17. On that invention, see Frederick A. Talbot, "Moving Pictures for the Amateur," *Technical World Magazine* 18, no. 1 (1912): 70–73.
113. "What to Preserve or Destroy," 14.
114. "American Pantheon," 313; *New York Tribune*, quoted in untitled MHRA pamphlet, 15.
115. Frederic Luther, *Microfilm: A History, 1839–1900* (Annapolis: National Microfilm Association, 1959), 32–33, 47–82, quotation on 25.
116. Reported in George Kunz, "The Imperishable Records of the Ancients, Compared with Methods in Use up to the Present Time," in *Documents of the Assembly of the State of New York* 32, no. 59 (1912): 384–85.
117. Untitled MHRA pamphlet, 4.
118. "In Honor of an Old Coffee House," *Simmons' Spice Mill* 37, no. 6 (1914): 572.
119. C. E. Vail, "Our Duty to the Future," *Scientific Monthly* 3 (1916): 593.
120. William M. Reedy, quoted in "A Plan for the Preservation of Historic Records," *Catholic Fortnightly Review* 18, no. 18 (1911): 533.
121. "To Record History for Eye and Ear," 8.
122. "The Mayor In the 'Movies,'" *NYT*, May 6, 1913, 20.
123. See, for example, "Observations By Our Man About Town," *Moving Picture World*, May 3, 1913, 476.
124. Altman, *Silent Film Sound*, 175, 177.
125. "Immortality?," *Edison Phonograph Monthly* 10, no. 2 (1912): 9; "Modern Historic Records," *Moving Picture News*, January 20, 1912, 13.
126. See profile of Saunders in *Moving Picture News*, June 8, 1912, 12–13.
127. Gaynor, letter to the board of aldermen, December 27, 1912, in *Some of Mayor Gaynor's Letters and Speeches* (New York: Greaves, 1913), 130, 132, 133.
128. Edward S. Curtis, "Outline of the North American Indian Project" (1906), http://xroads.virginia.edu/~ma02/daniels/curtis/proposal2.html.
129. Edward S. Curtis, *The North American Indian*, 20 vols. and 20 portfolios (Cambridge, MA: University Press, 1907–1930), 4:24.
130. Edward S. Curtis, inscription in margin beneath *The Oath*, in MHRA, "Message of the Records" (ca. 1913), MssCol 2025, Manuscripts and Archives Division, NYPL.
131. "Chairman Konta's Views," *Talking Machine World* 8, no. 3 (1912): 23.

132. Konta, "Records of Indian Life," 10. On Curtis's use of those other media, see Mick Gidley, *Edward S. Curtis and the North American Indian, Incorporated* (New York: Cambridge University Press, 1998), 199–230.
133. Kunz, "Imperishable Records," 384–85. See Kim Walters and Michael Khanchalian, "Keeping Archaeology Alive: The Race against Time to Save the Lummis Sound Recordings," *Convergence* (Winter 2009): 14–21. On the use of the phonograph to record Indian languages since 1890, see Brian Hochman, *Savage Preservation: The Ethnographic Origins of Modern Media Technology* (Minneapolis: University of Minnesota Press, 2014), 73–114.
134. Konta, "Records of Indian Life," 10; Curtis, inscription beneath photograph, in MHRA, "Message of the Records" (ca. 1913), MssCol 2025, Manuscripts and Archives Division, NYPL.
135. See, for example, Brian W. Dippie, *The Vanishing American: White Attitudes and U.S. Indian Policy* (1970; Lawrence: University Press of Kansas, 1991); on Curtis's embrace of salvage ethnography, see Gidley, *Edward S. Curtis*, 22, 75, 102.
136. Curtis, *North American Indian*, 1:xv.
137. Curtis, 4:40–41.
138. The most famous instance is Curtis's erasure of the clock from the published version of *In a Piegan Lodge* (1910).
139. Gidley, *Edward S. Curtis*, 81–108.
140. Both Grant and Brainard fought in the Bannock Indian War of 1878. Kenn Harper, *Give Me My Father's Body: The Life of Minik, the New York Eskimo* (1986; New York: Washington Square Press, 2001), 94.
141. Hochman, *Savage Preservation*, 1–33, quotation on 14.
142. Kunz, "Imperishable Records," 384.
143. Alice Palmer Henderson, *The Rainbow's End: Alaska* (New York: Herbert S. Stone, 1898), 143.
144. Richard Irving Dodge, *Our Wild Indians: Thirty-Three Years' Personal Experience among the Red Men of the Great West* (Hartford, CT: Worthington, 1882), 248, 398.
145. Curtis, *North American Indian*, 4:42; Florence Curtis Graybill and Victor Boesen, *Visions of a Vanishing Race* (1976; Boston: Houghton Mifflin, 1986), 13.
146. See Alan Trachtenberg, *Shades of Hiawatha: Staging Indians, Making Americans, 1880–1930* (New York: Hill and Wang, 2004), 207.
147. H. G. Wells, handwritten statement for MHRA (ca. 1911), Special Collections, University of Illinois at Urbana Champaign.
148. Sherry L. Smith, *Reimagining Indians: Native Americans through Anglo Eyes, 1880–1940* (New York: Oxford University Press, 2000), 121, 127, 134–38. Smith acknowledges that Lummis sometimes "demonized" or "infantilized" Indians (120, 129).
149. Shamoon Zamir, *The Gift of the Face: Portraiture and Time in Edward S. Curtis's* The North American Indian (Chapel Hill: University of North

Carolina Press, 2014), 133, 135; see also Martha A. Sandweiss, "Picturing Indians: Curtis in Context," in *The Plains Indian Photographs of Edward S. Curtis* (Lincoln: University of Nebraska Press, 2001), 13–37.

150. Two Leggings's autobiography (narrated 1919–23) was later published as *Two Leggings: The Making of a Crow Warrior*, ed. Peter Nabokov (New York: Crowell, 1967); Peggy Albright, *Crow Indian Photographer: The Work of Richard Throssel* (Albuquerque: University of New Mexico Press, 1997).
151. Trachtenberg, *Shades of Hiawatha*, 204. Fatimah Tobing Rony, *The Third Eye: Race, Cinema, and Ethnographic Spectacle* (Durham, NC: Duke University Press, 1996), 70, 214.
152. Gidley refers to it as the project's "keynote" (*Edward S. Curtis*, 23, 74, 280) and Sandweiss as its "signature image" (*Print the Legend: Photography and the American West* [New Haven, CT: Yale University Press, 2004], 219). Zamir, by contrast, juxtaposes it with a photograph that appears near the end of the first portfolio, "Out of the Darkness," which depicts Navajo riders coming out into sunlight (*Gift of the Face*, 273).
153. See Christopher M. Lyman, *The Vanishing Race and Other Illusions: Photographs of Indians by Edward S. Curtis* (New York: Pantheon Books, 1982), 82–85; but also Gidley, *Edward S. Curtis*, 69–70, 273–74; Trachtenberg, *Shades of Hiawatha*, 176.
154. Jonathan Sterne, *The Audible Past: Cultural Origins of Sound Reproduction* (Durham, NC: Duke University Press, 2003), 27, 288–89.
155. See, for example, "May Bar Moving Pictures," *NYT*, November 17, 1909, 1.
156. Harold Innis, *The Bias of Communication* (1951; Toronto: University of Toronto Press, 2008), 172–73. Estimate in "MHRA," *World Almanac and Encyclopedia* (New York: Press, 1914), 529; quotation in untitled MHRA pamphlet, 1. A landmark article about the problem was "Will Future Generations Lose Historical Records of To-Day?" *NYT*, July 24, 1910, SM15.
157. Untitled MHRA pamphlet, 5. See David W. Dunlap, "The Record of the Newspaper of Record," *NYT*, May 3, 2017, A2.
158. Untitled MHRA pamphlet, 4.
159. "To Record History by Moving Pictures," *Brooklyn Daily Eagle*, March 31, 1912, 13; "Films for Posterity," *NYT*, May 1, 1913, 6.
160. England, *Darkness and Dawn*, 477.
161. England, 480–84. On the invention of an "indestructible" metal disc, see Harry A. Knauss, "Metal Phonograph Records," *Science and Invention* 9, no. 6 (1921): 520, 572.
162. England, *Darkness and Dawn*, 212–13.
163. Allan Benson, "The Wonderful New World Ahead of Us," *Cosmopolitan* 50, no. 3 (1911): 299.
164. England, *Darkness and Dawn*, 289.
165. England, 392–99.
166. Kunz, "Imperishable Records," 380, 374, 376.

167. "To Teach History with 65 Plaques," *Brooklyn Daily Eagle*, July 18, 1915, 60. On plaques as tools for Americanization, see Mason, *Once and Future New York*, 47–49.
168. Kunz, "Imperishable Records," 383–84.
169. "Wants Photographs to Last Forever," *NYT*, November 30, 1912, 22.
170. "Will Preserve Records of War," *Pittsburgh Press*, November 29, 1914, 17; Harrison, "A Pompeii," 385.
171. Kunz, "Imperishable Records," 385.
172. Letter to *Science*, April 13 1917, 363.
173. C. E. Vail, "Our Duty to the Future," *Scientific Monthly* 3 (1916): 595, 589, 592.
174. Harrison, "A Pompeii," 386.
175. Kunz, "Imperishable Records," 384.
176. Konta quoted in "How Will the Future Know Us," *Sunday Times-Tribune* (Waterloo, Iowa), June 8, 1913, sec. 2, p. 1.
177. "Messages for Posterity a Century Hence," *Sun*, March 9, 1913, 3.
178. Ibid., 3.
179. Kunz, "Imperishable Records," 368.
180. Harrison, "A Pompeii," 383.
181. R. Lamb, reading a paper prepared by Arthur Dillon, in "Record of Our Time to Be Imperishable," *NYT*, December 10, 1911, 17; Boyden Sparkes, "Civilization, He Fears, Is Slipping," *New York Tribune*, November 13, 1921, 7.
182. "Record of Our Time," 17.
183. Harrison, "A Pompeii," 384.
184. England, *Darkness and Dawn*, 445, 446, 458.
185. Adrian Forty, *Concrete and Culture: A Material History* (London: Reaktion, 2012), 169–72; quotation on 169. There were one-off experiments and demonstrations, such as the William E. Ward House (1871).
186. "Record of Our Time," 17.
187. Forty, *Concrete and Culture*, 23–28; Arthur Dillon, "The Proposed Tilden Trust Library," *Architectural Review* 1, no. 8 (1892): 70; "Record of Our Time," 17.
188. Walter Benjamin, *The Arcades Project*, trans. Howard Eiland and Kevin McLaughlin (Cambridge, MA: Belknap Press of Harvard University Press, 1999), 150–70.
189. Harrison, "A Pompeii," 384; Vail, "Our Duty," 590; statement by Ladies Aid Society about the century chest (1913), OCCCC.
190. England, *Darkness and Dawn*, 60, 77–78, 164, 175, 463, 469.
191. Auguste Choisy, "The Art of Building among the Romans" (trans. Arthur Dillon), *Brickbuilder* 6, no. 7 (1897): 144–48 (continued in subsequent numbers); "Durability of Concrete," *Cement Age* 8, no. 6 (1909): 397–98.
192. Dillon, read by Lamb, cited in "Record of Our Time," 17; Konta advocated fireproofing in "Celebrities Write on Parchment for Future Ages," *NYT*,

June 2, 1912, SM9, and copying in "Alexander Konta Plans to Preserve History by Camera and Phonograph," *St. Louis Post-Dispatch*, September 10, 1911, 3.

193. Untitled MHRA pamphlet, 6.
194. "Record of Our Time," 17.
195. Harrison, "A Pompeii," 384–85.
196. Sparkes, "Civilization, He Fears, Is Slipping," 7. On these terms, see p. 48 above.
197. Kunz, "Imperishable Records," 383.
198. Card in MHRA collection, Manuscripts and Archives Division, NYPL (see also facsimiles there).
199. "Celebrities Write on Parchment," SM9; see also "To Bury Records in Egypt," *NYT*, January 26, 1913, 5.
200. England, *Darkness and Dawn*, 441, 445.
201. On Curtis's waning publicity by 1914, see Trachtenberg, *Shades of Hiawatha*, 175.
202. "Will Preserve Records of War," *Pittsburgh Press*, November 29, 1914, 17.
203. Sparkes, "Civilization, He Fears, Is Slipping," 7.
204. Receipt was confirmed in "Gifts," *Bulletin of New York Public Library* 17 (1913): 118.
205. "Mrs. Konta Sues Husband," *NYT*, January 3, 1914, 18.
206. "Hungarians Visit Wilson to Tell of Loyalty to U.S.," *St. Louis Post-Dispatch*, February 23, 1916, 2; George Creel, *How We Advertised America* (New York: Harper and Brothers, 1920), 187.
207. "Konta Tells of His Attempts to Assist Germans," *St. Louis Post-Dispatch*, December 5, 1918, 2; Creel, *How We Advertised America*, 187.
208. Rick Altman, *Silent Film Sound* (New York: Columbia University Press, 2004), 177–78.
209. Eastman Kodak's Recordak camera was introduced in 1928, and Verneur Pratt's Optigraph reader patented in 1923.
210. *Tribune*, cited in "Modern Historic Records," *Moving Picture News*, January 20, 1912, 13.
211. This conception of the archivist is mostly associated with Hilary Jenkinson, *A Manual of Archive Administration* (Oxford: Clarendon, 1922), but was circulating a decade earlier through the writings of Franklin Jameson and others; see Francis X. Blouin Jr. and William G. Rosenberg, *Processing the Past: Contesting Authorities in History and the Archives* (New York: Oxford University Press, 2011), 38–39.
212. E. H. Carr, *What Is History?* (New York: Random House, 1961), 3–35; Arthur C. Danto, *Analytic Philosophy of History* (New York: Cambridge University Press, 1965), 149.
213. Lewis Mumford, *Technics and Civilization* (1934; Chicago: University of Chicago Press, 2010), 244; Mumford, *The City in History: Its Origins, Its Transformations, and Its Prospects* (New York: Harcourt, 1961), 98.

214. "Saving for the Year 2000," *Boston Evening Transcript*, August 5, 1913, 11.
215. "Appraisal of Contemporary Greatness," *Dial*, August 16, 1912, 90; W. T. Larned, letter to *The Dial*, September 16, 1912, 186; editor's response, 186.
216. "Wants Photographs to Last Forever," *NYT*, November 30, 1912, 22.
217. "Saving for the Year 2000," 11.

CHAPTER SIX

1. "Awed Fair Visitors Peer into Future at Shrine of the Time Capsule," New York World's Fair 1939–1940 records, MssCol 2233, Manuscripts and Archives Division, NYPL.
2. David S. Youngholm, "The Time Capsule," *Science* 92, no. 2388 (1940): 301; Stanley Hyman and St. Clair McKelway, "The Time Capsule," *New Yorker*, December 5, 1953, 194–95; and see above, p. 317, n. 1.
3. There were occasional "century box" deposits in the 1920s, for example, in Turner, Maine (1923), and Cincinnati, Ohio (1928).
4. G. Edward Pendray, "For 8113 A.D.," *Literary Digest*, October 31, 1936, 19–20. Thornwell Jacobs, *Step Down, Dr. Jacobs: The Autobiography of an Autocrat* (Atlanta: Westminster, 1945), 527. Paul Stephen Hudson provides an overview in "The 'Archaeological Duty' of Thornwell Jacobs: The Oglethorpe Atlanta Crypt of Civilization Time Capsule," *Georgia Historical Quarterly* 75, no. 1 (1991): 121–38. No scholars have mentioned Ward's project until now.
5. Duren J. H. Ward, *Our Stage of Civilization: Data for Its History with Letters to Future Ages* (Denver, CO: privately printed, [1955?]), 6, SHSI; Jacobs, *Step Down*, 908. The crypt's archivist T. K. Peters acknowledged just two precursors—Deihm and Harvey—and criticized both for failing to go beyond cornerstones in what they preserved, thus giving Jacobs full "credit for the conception of the greatest historical project in all history" (Jacobs, 881–82); see also Jacobs's acknowledgment of a Japanese effort to memorialize the victims of the 1923 earthquake (Jacobs, 888).
6. Jacobs, "Today—Tomorrow," 260. Westinghouse Electric Corporation, *The Book of Record of the Time Capsule of Cupaloy* (New York: Westinghouse, 1938), 5–6.
7. Duren J. H. Ward, *Helping the Future*, pamphlet [1927?], and "A Page from Our Suggestions to the Future Generations," Shambaugh Family Papers, 1880s–1953, series 1, box 17, UISC University Archives (hereafter Shambaugh Papers). After a first opening in 2131, subsequent openings would be at midcentury. Later target dates in Ward, "Letters to Future Ages: Introductory" (prospectus; Denver: privately printed, 1926), 6. Ward cites Petrarch in *Letters to Future Ages* ([Denver?]: privately printed, 1955), 1.
8. W. H. Harvey, *The Pyramid Booklet* (Monte Ne, AR: Pyramid Association, [1928?]), 12–13.

9. Duren J. H. Ward, *Two Letters about a Long-Range Purpose and the Invention of a Social Thermostat* (Denver: privately printed, 1933), 9, 11, 39.
10. Jacobs conceived of the crypt while researching for *The New Science and the Old Religion* (Atlanta: Oglethorpe University Press, 1927) and began formal plans in 1935; see Jacobs, *Step Down*, 880. Harvey announced the Pyramid in *Common Sense, or The Clot on the Brain of the Body Politic* (Monte Ne, AR: Harvey, [1920]), 2.
11. [G. Edward Pendray], *The Story of the Westinghouse Time Capsule* (East Pittsburgh, PA: Westinghouse, [1939]).
12. A. W. Robertson, cited in "A Message to the Future," *Amazing Stories* 13, no. 1 (1939): 21.
13. Youngholm, "Time Capsule," 302. On the manager's warnings, see Hyman and McKelway, "Time Capsule," 199.
14. "'Magic House' Will Exhibit Time Capsule," *Washington Post*, May 5, 1940, R10. Hyman and McKelway, "Time Capsule," 212–15.
15. Hyman and McKelway, "Time Capsule," 194.
16. Roland Marchand, *Corporate Soul: The Rise of Public Relations and Corporate Imagery in American Big Business* (Berkeley: University of California Press, 1998), 203, 245, 248.
17. Marchand, 3–4.
18. Hyman and McKelway, "Time Capsule," 194.
19. Marchand, *Corporate Soul*, 245.
20. Hyman and McKelway, "Time Capsule," 211–12.
21. Westinghouse, *Book of Record*, 15.
22. "$75 in Cornerstone Bank to Grow to $1,843,200," *NYT*, April 20, 1930, 2. I am grateful to Louie Lazar for bringing this vessel to my attention.
23. Acquisitions ledger, box 10, TJC. Numbers from Peters, "The Corridors of Time: A New Museum Depicting the Evolution of Civilization" (typed prospectus), 4, Crypt of Civilization filing cabinet, OU; "Crypt-Sealing Ceremony Set for Saturday," undated newspaper clipping, box 1, TJC.
24. Westinghouse, *Book of Record*, 15–16.
25. Arthur P. Molella, Karen Fiss, Morris Low, and Robert Hugh Kargon, *World's Fairs on the Eve of War: Science, Technology, and Modernity, 1937–1942* (Pittsburgh, PA: University of Pittsburgh Press, 2015), 61, 66. Quotes from A. Joan Saab, *For the Millions: American Art and Culture between the Wars* (Philadelphia: University of Pennsylvania Press, 2004), 142.
26. Marchand, *Corporate Soul*, 249–52, 267.
27. Howard P. Segal, *Technological Utopianism in American Culture* (Syracuse, NY: Syracuse University Press, 1985), 125.
28. Westinghouse, *Book of Record*, 5.
29. A. W. Robertson, quoted in "Record of Today Buried for 6939," *NYT*, September 24, 1938, 19.
30. Robertson, "Message to the Future," 21.
31. Westinghouse, *Book of Record*, 18.

32. Robertson, "Message to the Future," 19. Robert Moore Williams, "The Warning from the Past," *Thrilling Wonder Stories* 14, no. 1 (1939): 34–49.
33. See, for example, Gawain Edwards [G. Edward Pendray], "A Rescue from Jupiter," *Science Wonder Stories* 1, no. 9, 10 (February, March 1930): 786, 913–17, 920, 933, discussed below.
34. Westinghouse, *Book of Record*, 11–12, 18.
35. Hyman and McKelway, "Time Capsule," 201 (emphasis added).
36. Anonymous, "Journey into Time," *Archaeology* 1, no. 3 (1948): 166.
37. Ward to Shambaughs, September 15, 1937, Shambaugh Papers; Jacobs, *Step Down*, 497n2; Lois Snelling, *Coin Harvey, Prophet of Monte Ne* (Point Lookout, MS: S of O Press, 1973), 25–26, 35–36; Duren J. H. Ward, *Ward-Munger, Varnum-Martin Genealogy* (Denver: Up the Divide, 1926).
38. Snelling, *Coin Harvey*, 36–40. Sales figures from Richard Hofstadter, *The Paranoid Style in American Politics, and Other Essays* (New York: Knopf, 1965), 242.
39. Ward, *Ward-Munger, Varnum-Martin Genealogy*; Ward to Shambaughs, March 5, 1934, 3, Shambaugh Papers.
40. Jacobs, *Step Down*, 910, 1011.
41. Jacobs, 910.
42. Jacobs, 933.
43. Paul K. Saint-Amour, *Tense Future: Modernism, Total War, Encyclopedic Form* (New York: Oxford University Press, 2015), 1–43.
44. Jacobs, *Step Down*, 907; "Crypt-Sealing Ceremony."
45. Harvey, *Common Sense*, 13.
46. Harvey, 27 (emphasis in original).
47. Harvey, *Pyramid Booklet*, 17; Harvey, *Common Sense*, 2.
48. Both cited in "A Banquet of Progressives" (program for a symposium, May 31, 1935), Shambaugh Papers.
49. Ward to Shambaughs, March 27, 1934; November 1934, 3, Shambaugh Papers.
50. Duren J. H. Ward, "The Far Reaching Foundation toward Safer Civilization," circular, n.d., Shambaugh Papers.
51. Ward, *Letters to Future Ages*, 136.
52. Ward, *Our Stage of Civilization*, 8; Harvey, *Pyramid Booklet*, 13; Jacobs, "Today—Tomorrow," 261.
53. Harvey, *Pyramid Booklet*, 12–15.
54. Harvey, 14.
55. Ward, *The Far-Reaching Foundation toward Safer Civilization: Records for Future Ages*, pamphlet, [1931?], 3, author's possession. *Fairmount Mausoleum*, prospectus; n.d.; *A Wonderful Tomb: Tower of Memories*, prospectus, Denver, Crown Hill Memorial Park, n.d.; *The Tower of Memories*, pamphlet, n.d., Shambaugh Papers. Jessica Mitford, *The American Way of Death Revisited* (New York: Knopf Doubleday, 2011), 85–86.
56. Jacobs, *Step Down*, 486, 898–900.

57. Williams, "Warning from the Past," 40.
58. Jacobs, 897, 920–21; T. K. Peters, "The Preservation of History in the Crypt of Civilization," *Journal of the Society of Motion Picture Engineers* 34 (February 1940): 211.
59. Anthony Paredes, "Oglethorpe University Class of 1961: A Fiftieth Anniversary Historical Remembrance," http://www.alumni.oglethorpe.edu/paredes.
60. Ward to Shambaughs, April 5, 1934, 3, Shambaugh Papers; Ward, *Our Stage of Civilization*, 6; *Fairmount Mausoleum*.
61. Ward, *Far-Reaching Foundation*, 13–14; Ward to Shambaughs, February 19, 1934, 1, 2, Shambaugh Papers.
62. Ward to Shambaughs, March 27, 1934, 2; Ward, *Far-Reaching Foundation*, 6, 11.
63. Ward to Shambaughs, October 20, 1934, 2, Shambaugh Papers.
64. Ward to Shambaughs, April 19, 1934, 4–5; April 27, 1934; Shambaugh Papers.
65. Albert Ingalls, memorandum, n.d., in box 10, TJC.
66. Jacobs, *Step Down*, 895, 898, 903.
67. Keith Bresnahan, "'An Unused Esperanto': Internationalism and Pictographic Design, 1930–70," *Design and Culture* 3, no. 1 (2011): 5–24.
68. Those artists were George Carlson, John Joffre Brock, and Fred Chance.
69. Harvey, *Pyramid Booklet*, 14.
70. Jacobs, *Step Down*, 898; Peters, "Preservation of History," 211; originally 1,500 words, the list expanded to 3,000. On Ogden, see Susan Shaw Sailer, "Universalizing Languages: *Finnegans Wake* Meets Basic English," *James Joyce Quarterly* 36, no. 4 (1999): 853–68. H. G. Wells, *The Shape of Things to Come* (New York: Macmillan: 1933), 415–20. Westinghouse's *Book of Record* deployed pictograms and "neo-phonetic spelling" to teach a "vocabulary of high-frequency English" (19–37).
71. Jacobs, *Step Down*, 1088.
72. Peters, "The Preservation of This Dynamic Epoch," in Jacobs, *Step Down*, 889.
73. David Sarnoff, "Six Thousand Years," in *The World's Great Speeches*, ed. Lewis Copeland (Garden City, NY: Garden City Publishing, 1942), 431.
74. Jacobs, *Step Down*, 918–19.
75. Jacobs, 907, 910.
76. Peters, "Preservation of History," 210.
77. Arthur G. Stangland, "The 35th Millennium," *Wonder Stories* 3, no. 3 (1931): 331, 335, 350, 351.
78. [G. Edward Pendray], *Story of the Westinghouse Time Capsule* (East Pittsburgh, PA: Westinghouse Electric and Manufacturing Company, [1939]).
79. Duren J. H. Ward, *Two Letters about a Long-Range Purpose and the Invention of a Social Thermostat* (Denver: privately printed, 1933), 9–11, 38.
80. Ward to Shambaughs, April 19, 1934, 2.

81. Ward to Shambaughs, April 5, 1934, 2, 4.
82. Ward, *Letters to Future Ages*, 201; Ward, *Tower of Memories* (capitals in originals).
83. Ward to Shambaughs, November 1934, 2–3, Shambaugh Papers; Harvey, *Pyramid Booklet*, 15; W. H. Harvey, *The Book* (Rogers, AR: Mundus Publishing, 1930), 133, 201.
84. Ward, *Far-Reaching Foundation*, 15.
85. Ward, *Tower of Memories.*
86. Ward, *Helping the Future*, pamphlet [1927?], Shambaugh Papers.
87. Ward, *Two Letters*, 17, 21–22.
88. Duren J. H. Ward, *Temple to Civilization*, pamphlet, n.d., 15, SHSI. Ward cites Franklin in "Page from Our Suggestions."
89. Harvey, *Pyramid Booklet*, 14.
90. "$75 in Cornerstone Bank to Grow to $1,843,200," *NYT*, April 20, 1930, 2.
91. Ward, "To 20th Century Unitarians," conference paper, n.d., Shambaugh Papers.
92. Ward, *Letters to Future Ages*, 105, 180.
93. Jacobs, *Step Down*, 288–89. The General Assembly of the Presbyterian Church had by then declined to accredit Oglethorpe.
94. See Jacobs's textbook, *The New Science and the Old Religion* (Atlanta: Oglethorpe University Press, 1927).
95. Ward, "Far Reaching Foundation"; Jacobs, *New Science*, 453.
96. One exception was the rag paper on which Westinghouse printed some copies of its *Book of Record.*
97. Jeffrey Meikle, *American Plastic: A Cultural History* (New Brunswick, NJ: Rutgers University Press, 1995), 78–82; Anthony Slide, *Nitrate Won't Wait: A History of Film Preservation in the United States* (Jefferson, NC: McFarland, 1992), 1, 5.
98. Peters, "Corridors of Time," 34, 36; Jacobs, *Step Down*, 903.
99. Peters, "Corridors of Time," 15.
100. Jacobs, *Step Down*, 893–95; Peters, "Preservation of History," 207–8; total numbers from Paul Stephen Hudson, "The 'Archaeological Duty' of Thornwell Jacobs: The Oglethorpe Atlanta Crypt of Civilization Time Capsule," *Georgia Historical Quarterly* 75, no. 1 (1991): 125–26. Peters to F. D. McHugh, September 6, 1939, box 10, TJC.
101. [Pendray], *Story of the Westinghouse Time Capsule.*
102. Lewis Mumford, *Technics and Civilization* (1934; Chicago: University of Chicago Press, 2010), 244.
103. Peters, "Preservation of History," 208–10; Jacobs, *Step Down*, 893; reference to secrecy in "Vault to Be Opened in 8113 to Contain Records of Our Time," *Times* (Ashville, NC), September 17, 1937, box 1, folder 2, TJC.
104. Jacobs, *Step Down*, 893, 899; Peters, "Corridors of Time," 35; [Pendray], *Story of the Westinghouse Time Capsule.*
105. Harvey, *Pyramid Booklet*, 13–14.

106. T. K. Peters, "The Preservation of This Dynamic Epoch," in Jacobs, *Step Down*, 891; *The Stream of Knowledge*, directed by T. K. Peters (1938), film; Mumford, *Technics and Civilization*, 239–45.
107. [Pendray], *Story of the Westinghouse Time Capsule*; Westinghouse, *Book of Record*, 50–51.
108. Correspondence with *Scientific American* in box 10, TJC; Harvey, *Pyramid Booklet*, 13.
109. Westinghouse, *Book of Record*, 51; Jacobs, *Step Down*, 580.
110. Ward, "Page from Our Suggestions"; Ward to Shambaughs, February 19, 1934, 1; March 5, 1934, 1.
111. Ward to Shambaughs, February 19, 1934, 1; September 15, 1937, 1. AAAS's official magazine, *Science*, also promoted the capsule.
112. For the assumption that the technocracy movement had faded by the mid-1930s, see Howard P. Segal, *Technological Utopianism in American Culture* (Syracuse, NY: Syracuse University Press, 1985), 120–24, and William Akin, *Technocracy and the American Dream: The Technocrat Movement, 1900–1941* (Berkeley: University of California Press, 1977), xi.
113. Jacobs, *Step Down*, 918; George R. Leighton to Peters, July 28, 1939, in Crypt of Civilization filing cabinet, OU.
114. Albert Blakeslee, "Science Five Thousand Years Hence," *American Biology Teacher* 3, no. 4 (1941): 119.
115. Segal, *Technological Utopianism*, 125–26.
116. Ward, *Letters to Future Ages*, 85.
117. Program for symposium on "Civilization, Science, Education," June 23, 1934, Shambaugh Papers.
118. Ward to Shambaughs, n.d., Shambaugh Papers.
119. Segal, *Technological Utopianism*, 121–22.
120. Westinghouse, *Book of Record*, 49.
121. D. J. H. W. [Duren Ward], "My Most Fundamental Social Hope," *Up the Divide* 1, no. 9 (1910): 368–69.
122. Ward, *Two Letters*, 26.
123. Ward, *Letters to Future Ages*, 2; Ward, *Our Stage of Civilization*, 4.
124. Segal, *Technological Utopianism*, 123; Akin, *Technocracy*, 112.
125. Duren J. H. Ward, *Will Civilization Collapse?*, pamphlet, n.d., Shambaugh Papers; "Civilization, Science, Education."
126. Jacobs, *Step Down*, 923–40. "An 'Upper Class,' Eh?" *Miami Herald*, March 16, 1940, 6, and "Educator Debunks Democracy," undated newspaper clipping, box 1, folder 5, TJC.
127. Ward, *Our Stage of Civilization*, 6–8.
128. F. D. McHugh, memorandum to Orson Munn, box 1, folder 10, TJC.
129. D. S. Youngholm, "A Cross-Section of Our Time," *Science* 88, no. 2277 (1938): 168.
130. "Record of Today Buried for 6939," *NYT*, September 24, 1938, 19.

131. On Harvey's association, see *Pyramid Booklet*, 19. On Ward's, see "Volunteer Committee," circular, [1928?], Shambaugh Papers; Ward, *Two Letters*, 8.
132. Ward "[owned a] printing auxiliary and shop of some sort [for] 50 years—12 presses. Put forth thousands of leaflets," from *A Four-Score-Year Book*, pamphlet, 7, SHSI. Harvey first founded the Coin Publishing Company and later the Mundus Publishing Company. Jacobs founded Oglethorpe University Press, which published his *New Science*.
133. See, for example, Duren J. H. Ward, *Letters to Future Ages and Library of Data for Historical Comparison*, undated pamphlet, Shambaugh Papers; Harvey, *Book*, 11; Harvey, *Common Sense*, 2.
134. Ward to Shambaughs, March 5, 1934, 3; April 5, 1934, 3. Harvey, *Pyramid Booklet*, 15.
135. "Dedication and Prospectus of the Far-Reaching Foundation [. . .]," invitation, October 1, 1935, Shambaugh Papers; Hudson, "Archaeological Duty," 125–26; Albert Blakeslee's address was broadcast live on NBC radio network (Blakeslee, "Science Five Thousand Years Hence," 117).
136. Roland Marchand, *Corporate Soul: The Rise of Public Relations and Corporate Imagery in American Big Business* (Berkeley: University of California Press, 1998), 283.
137. See "Dedication and Prospectus."
138. Ward, "Page from Our Suggestions"; Ward, *Letters to Future Ages*, 165–66.
139. W. H. Harvey, *The Remedy* (Chicago: Mundus Publishing, 1915), 119, 121–22.
140. A. W. Robertson, quoted in "Message to the Future," 20–21; Blakeslee was also a leading eugenicist.
141. Christina Cogdell, *Eugenic Design: Streamlining America in the 1930s* (Philadelphia: University of Pennsylvania Press, 2004).
142. Ward, *Letters to Future Ages*, 115 (emphasis added). Jacobs, *Step Down*, 918.
143. Daniel Kevles, *In the Name of Eugenics: Genetics and the Uses of Human Heredity* (Berkeley: University of California Press, 1985) is typical in characterizing the movement as led by scientists and social scientists, many of whom rejected "church orthodoxies" (12), and as opposed by religious figures (118–19).
144. Ward described the book project in *The Far-Reaching Foundation toward Safer Civilization: Records for Future Ages*, pamphlet, [1931?], 11, author's possession. E. A. Ross, *The Old World in the New* (New York: Century, 1914). Ross had famously been fired from Stanford University for his eugenicist views.
145. Jacobs, *Step Down*, 537.
146. William Swinton, *Swinton's Outlines of the World's History* (Halifax: Mackinlay, 1883), 2.
147. Gawain Edwards [Edward Pendray], *Earth-Tube* (1929; New York, Arno Press, 1975).

148. Harvey, *A Tale of Two Nations* (Chicago: Coin, 1894); Richard Hofstadter, *The Paranoid Style in American Politics, and Other Essays* (New York: Knopf, 1965), 300–301; Harvey, *Common Sense*, 18.
149. Ward, *Letters to Future Ages*, 121–22; see also 92, 151.
150. Jacobs, *Step Down*, 374–76, 535–36; Jacobs, diary entry, December 7, 1938, box 1, folder 3, TJC.
151. Jacobs, *Step Down*, 1084.
152. Thornwell Jacobs, *Red Lanterns on St. Michael's* (New York: E. P. Dutton, 1940); Jacobs, letter to Atlanta *Constitution*, in *Step Down*, 1010; Jacobs, editorial in *Georgian*, in *Step Down*, 1019.
153. Jacobs, diary entry in *Step Down*, 1087.
154. Jacobs, *Step Down*, 896–97.
155. Gawain Edwards [Edward Pendray], "A Rescue from Jupiter," *Science Wonder Stories* 1, no. 9 (February 1930): 786; no. 10 (March 1930): 913–17, 920, 933.
156. Ward, *Our Stage of Civilization*, 1–2.
157. Ward, *Helping the Future*, pamphlet [1927?], Shambaugh Papers.
158. The crypt and time capsule did include addresses by President Franklin Roosevelt and other world leaders (Jacobs, *Step Down*, 886; [Pendray], *Story of the Westinghouse Time Capsule*).
159. Ward, *Helping the Future*; Peters, in Jacobs, *Step Down*, 881; Jacobs, "Today—Tomorrow," 260.
160. Peters paraphrasing Jacobs in *Step Down*, 881; Ward, *Our Stage of Civilization*, 2.
161. Ward, *Two Letters*, 24–25; Ward, *Tower of Memories*, pamphlet, n.d., Shambaugh Papers; Ward, *Far-Reaching Foundation*, 4–5.
162. Ward, *Two Letters*, 34–35.
163. Jacobs, quoted in Marguerite Steedman, "Vault to be Sealed until 8113," *Atlanta Journal*, June 27, 1937, box 1, folder 2, TJC; Jacobs, "Today—Tomorrow," 260; Westinghouse, *Book of Record*, 13; Youngholm, "Cross-Section of Our Time."
164. [Pendray], *Story of the Westinghouse Time Capsule*; Ward, *Far-Reaching Foundation*, 5. On Borges' encyclopedia, see Michel Foucault, *The Order of Things: An Archaeology of the Human Sciences* (1966; New York: Vintage Books, 1994), xvi.
165. [Pendray], *Story of the Westinghouse Time Capsule*; Harvey, *Pyramid Booklet*, 14; James Harvey Robinson, *Civilization* (London: Encyclopaedia Britannica, 1929), 10.
166. Paul K. Saint-Amour, *Tense Future: Modernism, Total War, Encyclopedic Form* (New York: Oxford University Press, 2015), 190–203, quotation on 182.
167. Saint-Amour detects this impulse in 1930s modernist literary epics (Saint-Amour, 185–86).
168. H. G. Wells, *The Shape of Things to Come* (New York: Macmillan: 1933), 420.

169. Isaac Asimov, "Foundation" (completed September 1941), *Astounding Science-Fiction* 29, no. 3 (1942): 38–52.
170. Philip Wylie and Edwin Balmer, *When Worlds Collide* (1933; Lincoln: University of Nebraska Press, 1999), 143. Jorge Luis Borges, *The Library of Babel*, trans. Andrew Hurley (Jaffrey, NH: David R. Godine, 2000).
171. Harvey, *Pyramid Booklet*, 14.
172. [Pendray], *Story of the Westinghouse Time Capsule*.
173. Inventory of Crypt of Civilization, box 10, TJC; Peters, "Preservation of History," 210.
174. Jacobs, *Step Down*, 880, 893–94, 896.
175. Jacobs, 536, 882, 895–97, 903, 1053; "6,000-Year Test for Music," *Atlanta Constitution*, April 11, 1939, box 1, folder 4, TJC. On "America's discovery of popular culture" in the interwar period, see Reuel Denney, *The Astonished Muse* (1957; New Brunswick, NJ: Transaction, 1989), xxiv.
176. [Pendray], *Story of the Westinghouse Time Capsule*.
177. [Pendray]; Jacobs, *Step Down*, 778.
178. Jacobs, *Step Down*, 897; [Pendray], *Story of the Westinghouse Time Capsule*.
179. Jacobs, quoted in Steedman, "Vault to be Sealed."
180. John Gilkesen, *Anthropologists and the Rediscovery of America, 1886–1965* (New York: Cambridge University Press: 2010), 12–13, 35–39.
181. See Wissler's correspondence with Pendray, box 5, folders 9, 11, 12, 15, 18, Clark Wissler Collection, Ball State University Special Collections. Stanley Hyman and St. Clair McKelway, "The Time Capsule," *New Yorker*, December 5, 1953, 209–10.
182. Wissler, *Man and Culture* (New York: Crowell, 1923), 55–61, 73–74, 254. Several of Pendray's fifteen categories (language, art and entertainment, information, religion) correspond to Wissler's; see [Pendray], *Story of the Westinghouse Time Capsule*.
183. Helen Taylor to Peters, January 9, 1940, Crypt of Civilization filing cabinet, OU.
184. See David W. Noble, *Death of a Nation: American Culture and the End of Exceptionalism* (Minneapolis: University of Minnesota Press, 2002), 79–105; Warren I. Susman, *Culture as History: The Transformation of American Society in the Twentieth Century* (New York: Pantheon, 1984), 157–58; Sarah E. Igo, *The Averaged American: Surveys, Citizens, and the Making of a Mass Public* (Cambridge, MA: Harvard University Press, 2009), 103–49; Laura L. Lovett, *Conceiving the Future: Pronatalism, Reproduction, and the Family in the United States, 1890–1938* (Chapel Hill: University of North Carolina Press, 2009), 164.
185. Lovett, *Conceiving the Future*, 164. Luke E. Lassiter, *The Other Side of Middletown: Exploring Muncie's African American Community* (Oxford: Altamira, 2004).
186. Inventory of Crypt of Civilization; Gunther Schuller, *The Swing Era: The Development of Jazz, 1930–1945* (New York: Oxford University Press, 1991), 201.

187. Jane Norman Smith to Jacobs, December 12, 1937, procurement correspondence, Crypt of Civilization filing cabinet, OU.
188. Rose Arnold Powell, telegram to Pendray, box 18, WC.
189. Will Durant, *The Story of Civilization*, vol. 1, *Our Oriental Heritage* (1935: New York: Simon and Schuster, 1942), 1–4. See Susman, *Culture as History*, 66–74, 105–21, 156–57; Gilkesen, *Anthropologists and the Rediscovery*, 14–16, 200–49; John Raeburn, *A Staggering Revolution: A Cultural History of Thirties Photography* (Urbana: University of Illinois Press, 2006), 131–34; Raymond Williams, *Keywords: A Vocabulary of Culture and Society* (Oxford: Oxford University Press, 1985), 57–60.
190. Robinson, *Civilization*, 13–15, 34, 43–44.
191. See, for example, Edward Sapir, "Culture, Genuine and Spurious," *American Journal of Sociology* 29, no. 4 (1924): 401–29.
192. Jacobs, *Step Down*, 905. On the emerging study of folk art and crafts, see Michael Kammen, *Mystic Chords of Memory: The Transformation of Tradition in American Culture* (New York: Vintage Books, 1991), 407–43.
193. Westinghouse, *Book of Record*, 16–27.
194. "Archaeologist Reverses Job: Buried Relics of Today," *Science News Letter*, October 5, 1940, 22; Clark Wissler, "Man and His Baggage," *Natural History* 55, no. 7 (1946): 325, 327–28, 330.
195. "Time Capsule Sunk till 6939 at Fair," *NYT*, September 24, 1940, 33. Jay R. Tunney, *The Prizefighter and the Playwright: Gene Tunney and Bernard Shaw* (Richmond Hill, ON: Firefly Books, 2010).
196. Karen Lucic, *Charles Sheeler and the Cult of the Machine* (Cambridge, MA: Harvard University Press, 1991), 103–4.
197. Westinghouse, *Book of Record*, 47.
198. C. Douglas Smith, letter to *NYT*, September 25, 1938, 163.
199. Ralph Farley [Roger Sherman Hoar], "The Time Capsule," *Astonishing Stories* 2, no. 4 (1941): 30, 32, 33–34.
200. "Diaries Kept for Posterity," *NYT*, April 17, 1938, 158; Victoria Crosby, "How to Be History," *Argosy*, March 2, 1940, 100.
201. *Izvestia*, December 12, 1938, paraphrased in "Palace of the Soviet [*sic*] Adopts N.Y. 'Time Capsule' Idea," *Archaeology* 1, no. 4 (1948): 224.
202. E. L. Doctorow, *World's Fair* (1985; New York: Random House, 2007), 277–78, 283–84, 287–88.

CHAPTER SEVEN

1. C. F. Deihm, ed., *President James A. Garfield's Memorial Journal* (New York: C. F. Deihm, 1882), 198; "Messages to Posterity," *Kansas City Star*, January 3, 1901, newspaper clippings related to the century box, KCCBC; Thornwell Jacobs, "Today—Tomorrow," *Scientific American* 155, no. 5 (1936): 260.
2. Alvarado M. Fuller, *A.D. 2000* (Chicago: Laird and Lee: 1890), 114–17.
3. George Allan England, *Darkness and Dawn* (Boston: Small, Maynard, 1914), 464–65.

4. J. C. Cowdrick, "The Bronze Box," *Munsey's* 11, no. 1 (1894): 66–69.
5. Sundry Civil Act, March 3, 1879, *The Statues at Large of the United States of America, from October 1877 to March 1879* (Washington: GPO, 1879), 391. Deihm, *Garfield's Memorial Journal*, 196–97.
6. "1876 Century Safe Lost in Oblivion," *Washington Post and Times Herald*, September 12, 1954, Centennial Safe—Correspondence folder, Art and Reference Files (hereafter A&RF), AOC; Fuller, *A.D. 2000*, 115.
7. "14 Years to Glory," newspaper clipping, [1962?], Folder—Centennial Safe—Correspondence folder, A&RF, AOC.
8. Margi Hofer, curator at NYHS, personal communication with author, September 2013.
9. "Gifts," *Bulletin of New York Public Library* 17 (1913): 118.
10. http://crypt.oglethorpe.edu/international-time-capsule-society/most-wanted-time-capsules.
11. Paul Hudson, quoted in "Crazy for Time Capsules," American Public Media, January 3, 2009, http://weekendamerica.publicradio.org/display/web/2009/01/03/time_capsule.html.
12. Judy L. Thomas, "History Revealed: Present Is Sealed," *Kansas City Star*, January 2, 2001.
13. "Old Photographs Arouse Memories," *Chicago Tribune*, August 12, 1908, 5.
14. *Chicago City Manual* (Chicago: Bureau of Statistics and Municipal Library, 1915), 119.
15. Alan Riding, "From a Vault in Paris, Sounds of Opera 1907," *NYT*, February 16, 2009, C3.
16. "A Look at Boston of Years Gone By," *Boston Globe*, September 16, 1980, 27.
17. "Ben Franklin's Downfall," newspaper clipping, March 6, 1879, CHS. *Historic Celebration: 25th Anniversary of the Telegraph Hill Dwellers and Opening of the Dr. H.D. Cogswell* [. . .] *100 Year Time Capsule, April 21 & 22, 1979* (program), CHS.
18. Mark Twain, *Autobiography of Mark Twain: The Complete and Authoritative Edition*, ed. Benjamin Griffin and Harriet E. Smith, 3 vols. (Berkeley: University of California Press, 2010–15).
19. Martine Costello, "College Finds Pieces of Past," *Holyoke Transcript-Telegram*, November 30, 1988, MHC.
20. "Spirit of Class of 1900 Felt during March 31 Time Capsule Opening," *College Street Journal* 13, no. 25 (2000); David Arnold, "Capt. Forbes Had a Noble Vision," *Boston Globe*, September 18, 1980, 1, 8.
21. "Chicago Has Relic Safe," *NYT*, July 14, 1907, 12.
22. "Some Genealogical Notes on the Charles F. Deihm Family" (1972), Centennial Safe—Correspondence folder, A&RF, AOC.
23. "Smithsonian Can't Find Keys to Centennial Safe," *NYT*, May 29, 1971, 20; "Congress Warming to a Long-Held Gift of 1876 Mementos," *NYT*, October 18, 1974, 87. H. Con. Res. 84 was passed by the Senate on October 16, 1974.

24. Jeff Jarvis, "The 100-Year War," *San Francisco Examiner*, March 2, 1979, 5, box 1, HDCTCC.
25. "100-Year-Old Letters Revealed in Detroit Time Capsule," CNN, January 1, 2001, http://edition.cnn.com/TRANSCRIPTS/0101/01/tod.13.html.
26. "S.F. Time Capsule Unveiled," undated newspaper clipping, ephemera collection, CHS.
27. "Capitol Safe Mystery Solved: It Can't Be Opened till 1976," *Times-Herald*, September 16, 1941, AOC.
28. Florian H. Thayn to Mario Campioli, Centennial Safe—Correspondence folder, A&RF, AOC.
29. "Ford Bids Nation 'Keep Reaching into the Unknown,'" *NYT*, July 2, 1976, 35.
30. "Mrs. Deihm's Centennial Safe," *Bicentennial Times*, November 1976, A&RF—Display, AOC.
31. "History in Capsule Form," A&RF—Display, AOC.
32. Marie Czach, "Time Capsule: 'C. D. Mosher's Bicentennial Gift to Chicago,' at the Chicago Historical Society," *Afterimage* 4, no. 5 (1976): 15–16.
33. Robert C. Post, ed., *1876: A Centennial Exhibition* (Washington, DC: Smithsonian Institution, 1976), 25. On nostalgia for 1876 more generally, see David Lowenthal, "The Bicentennial Landscape: A Mirror Held Up to the Past," *Geographical Review* 67, no. 3 (1977): 253–67.
34. Michael Kammen, *Mystic Chords of Memory: The Transformation of Tradition in American Culture* (New York: Vintage Books, 1991), 572, 588; Lowenthal, "Bicentennial Landscape," 265; John Bodnar, *Remaking America: Public Memory, Commemoration, and Patriotism in the Twentieth Century* (Princeton, NJ: Princeton University Press, 1993), 229–32, 238, 243. On the decline of world's fairs as representations of the nation, see Lynn Spillman, *Nation and Commemoration: Creating National Identities in the United States and Australia* (New York: Cambridge University Press, 1997), 195n8.
35. The Architect of the Capitol received several suggestions about how to restock the safe for 2076; see, for example, Marian Fitzhenry, letter to George White, July 1, 1976, Centennial Safe—Correspondence folder, A&RF, AOC.
36. US Bureau of the Census statistics, https://www.census.gov/population/www/documentation/twps0027/tab20.txt; for prediction, see above, pp. 151–52.
37. Andrea Cecil, "100-year-old time capsule opened in Detroit," *Kalamazoo Gazette*, January 2, 2001, D6.
38. Jason Roe, "Ringing in the New," KC History, Missouri Valley Special Collections, Kansas City Public Library, http://www.kclibrary.org/?q=blog/week-kansas-city-history/ringing-new.
39. Statistics from the Center for American Women and Politics, Rutgers University, http://www.cawp.rutgers.edu/women-elective-office-2018.

40. *Report of the Independent Monitor for the Detroit Police Department* (Detroit: Kroll, 2004), 1.
41. "S.F. Time Capsule Unveiled."
42. See Brian Durrans, "Posterity and Paradox: Some Uses of Time Capsules," in *Contemporary Futures: Perspectives from Social Anthropology*, ed. Sandra Wallman (New York: Routledge, 1992), 51.
43. Deihm's safe and its contents were briefly displayed in a Capitol corridor; the Kansas City box at Union Station; the Harvard chest at the University Archives; and the Mosher photographs at the Chicago History Museum (in 1926 and 1976).
44. Fitzhenry, letter. Facts regarding the fate of Deihm's safe and its contents from Pam McConnell, personal communication with author, June 2010.
45. Lesley Martin, research librarian at CHM, personal communication with author; Dana Gerber, research librarian at CHS, personal communication with author.
46. Pamela McConnell, archivist at AOC, discussion with author, June 2010; Tanya Hollis, archivist at CHS, discussion with author, May 2010.

EPILOGUE

1. David Lowenthal, "Posterity: Present Concerns with the Future," British Academy report (2007); Lowenthal, "Stewarding the Future," *Norwegian Journal of Geography* 60, no. 1 (2006): 20; Peter Van Wyck, *Signs of Danger: Waste, Trauma, and Nuclear Threat* (Minneapolis: University of Minnesota Press, 2004), 32.
2. "Archivists Seal 'Time Capsule' with Bicentennial Mementos," *Daytona Beach Morning Journal*, July 5, 1977, 5; "Time Capsule Is Set for the Tricentennial," *NYT*, September 12, 1976, 59; Monica Hesse, "Bicentennial Wagon Train Signatures Are Lost Pieces of American Past," *Washington Post*, July 3, 2011.
3. On its organizer's resignation amid allegations of mismanagement, see Paul Gray, "The Trouble with Columbus," *Time*, October 7, 1991.
4. Knute Berger, "The Future of the Future: Are Time Capsules Becoming a Thing of the Past?," *Seattle Weekly News*, October 9, 2006.
5. Ralph Ellsworth to S. E. Hyman, November 5, 1952, inserted in copy of Westinghouse Electric Corporation, *The Book of Record of the Time Capsule of Cupaloy* (New York: Westinghouse, 1938), UISC.
6. "Objects to be Included in Westinghouse Time Capsule II," Westinghouse press release, [1964?], 1–2, box 20, folder 2, WC; *The Official Record of Time Capsule Expo '70* (Kodama, Japan: Matushita Electric Industrial Company, 1980).
7. "Objects to be Included in Westinghouse Time Capsule II," 3. Committee members involved in national defense included James B. Conant, Vanne-

var Bush, Glenn Seaborg, and Hugh L. Dryden. On these themes within the larger fair, see Michael L. Smith, "Making Time: Representations of Technology at the 1964 World's Fair," in *The Power of Culture: Critical Essays in American History*, ed. T. J. Jackson Lears and Richard Wrightman Fox (Chicago: University of Chicago Press, 1993), 223–44.

8. Walter L. Miller, *A Canticle for Leibowitz* (1961; New York: HarperCollins, 2006).
9. "Objects to be Included in Westinghouse Time Capsule II," 4–5.
10. "Apollo 11 Goodwill Messages," NASA press release no. 69-83F, July 13 1969, NASA Technical Reports Server.
11. Jim Bell, *The Interstellar Age: The Story of the Men and Women Who Flew the Forty-Year Voyager Mission* (New York: Dutton, 2015), 71–102. Carl Sagan et al., *Murmurs of Earth: The Voyager Interstellar Record* (New York: Random House, 1978), 3.
12. Term used by William E. Jarvis, *Time Capsules: A Cultural History* (Jefferson, NC: McFarland, 2003), 138–74, and W. L Rathje, "America's Authentic Time Capsules," in *Encyclopedia of Consumption and Waste: The Social Science of Garbage*, vol. 1, ed. Carl A. Zimring and William L. Rathje (Los Angeles: Sage, 2012), 1041.
13. Ann Beardsley, C. Tony Garcia, and Joseph Sweeney, eds., *Historical Guide to NASA and the Space Program* (Lanham, MD: Rowman and Littlefield, 2016), 139.
14. Jon Lomberg, "A Portrait of Humanity" (2007), http://www.jonlomberg.com/articles/a_portrait_of_humanity.html.
15. For Apollo plaques, see http://airandspace.si.edu/multimedia-gallery/5515hjpg; for the LAGEOS plaque, see NASA press release no. 76-67, [1976?], 12–13.
16. Arthur C. Clarke's "History Lesson," in *Collected Stories of Arthur C. Clarke* (New York: Tom Doherty Associates, 2000), 92–98; Robert Nathan, *The Weans* (New York: Knopf, 1960); David Macaulay, *Motel of the Mysteries* (New York: Houghton Mifflin, 1979).
17. Thomas Sebeok, "Pandora's Box: How and Why to Communicate 10,000 Years into the Future," in *On Signs*, ed. Marshall Blonsky (Baltimore: Johns Hopkins University Press, 1985), 448–66; Abraham Weitzberg, *Building on Existing Institutions to Perpetuate Knowledge of Waste Repositories* (Rockville, MD: NUS, 1982); Martin Pasqualetti, "Landscape Permanence and Nuclear Warnings," *Geographical Review* 87, no. 1 (1997): 74, 78. Despite these warnings, the DOE adhered to the markers, retaining faith in written languages and pictograms and in durable materials (Van Wyck, *Signs of Danger*, 74–75).
18. Michel Foucault, *The Order of Things: An Archaeology of the Human Sciences* (1966; New York: Vintage Books, 1994).
19. Susan Sontag, *On Photography* (Harmondsworth: Penguin, 1979), 32.
20. Lomberg, "Portrait of Humanity."

21. "Bicentennial Opens Up New Interest in Time Capsules," *NYT,* June 21, 1976, 62.
22. Prices from MyLegacy.org.
23. See, for example, Abby Smith Rumsey, *When We Are No More: How Digital Memory Is Shaping Our Future* (New York: Bloomsbury, 2016); and Damon Krukowski, *The New Analog: Listening and Reconnecting in a Digital World* (Cambridge, MA: MIT Press, 2017).
24. http://www.xantheberkeley.com/creating-tangible-time-capsules (no longer posted).
25. Jennifer Rauch, *Slow Media: Why Slow Is Satisfying, Sustainable, and Smart* (New York: Oxford University Press, 2018).
26. David Lowenthal, "The Bicentennial Landscape: A Mirror Held Up to the Past," *Geographical Review* 67, no. 3 (1977): 263.
27. John W. Smith, "Saving Time: Andy Warhol's Time Capsules," *Art Documentation* 20, no. 1 (2001): 8–10.
28. Jonathan Harris, "Reflections on the Time Capsule" (2006), https://yodel.yahoo.com/blogs/general/reflections-time-capsule-188.html (accessed May 2017; no longer posted); Timothy Webmoor, "From Silicon Valley to the Valley of Teotihuacan: The 'Yahoo!s' of New Media and Digital Heritage," *Visual Anthropology Review* 24, no. 2 (2008): 183–200.
29. Saba Hamedy, "Paramount, Google Start on 'Interstellar'-Sparked 'Time Capsule' Film," *Los Angeles Times,* November 20, 2014.
30. Lunar Mission One, with its payload of "digital memory boxes," is scheduled for launch in 2024; see also the Time Capsule to Mars (http://www.timecapsuletomars.com) and the MoonArk (http://moonarts.org), which are intended to launch in 2019 and 2020, respectively.
31. Lomberg, "Portrait of Humanity."
32. http://www.earthtapestry.org; http://number27.org/tenbyten.
33. Lester A. Reingold, "Capsule History," *American Heritage* 50, no. 7 (1999): 93.
34. Richard Kemeny, "All of Human Knowledge Buried in a Salt Mine," *Atlantic Monthly,* January 9, 2017, https://www.theatlantic.com/technology/archive/2017/01/human-knowledge-salt-mine/512552/.
35. Pamela Licalzi O'Connell, "Time in a Bottle," *NYT,* April 22, 1999, G1.
36. Matt Sly and Jay Patrikios, *Dear Future Me: Hopes, Fears, Secrets, Resolutions* (Cincinnati: F+W, 2007), 4; on "internet time," see John Potts, *The New Time and Space* (New York: Palgrave Macmillan, 2015), 48.
37. David Lowenthal, "The Forfeit of the Future," *Futures* 27, no. 4 (1995): 385–95; Andrew Bennett, paraphrasing Zygmunt Bauman, in "On Posterity," *Yale Journal of Criticism* 12, no. 1 (1999): 141–42. On architectural short-termism, see Vincent Scully, "Tomorrow's Ruins Today," *NYT,* December 5, 1999, SM38.
38. Henry Steele Commager, "Commitment to Posterity: Where Did It Go?," *American Heritage* 27, no. 5 (1976): 5–6.

39. Christopher Lasch, "The Narcissist Society," *New York Review of Books*, September 30, 1976, 5–12, and Lasch, *The Culture of Narcissism: American Life in an Age of Diminishing Expectations* (1978; New York: W. W. Norton 1991), 51.
40. Carl Becker, "Everyman His Own Historian," *American Historical Review* 37, no. 2 (1932): 221–36.
41. Paula Lichtenberg, "Lover of Her Own Sex: The Life of Laura DeForce Gordon," *S.F.B.A.G.L.H.S. Newsletter*, 1985, 7.
42. https://www.madre.org/time-capsule-future.
43. https://lens.blogs.nytimes.com/2011/10/31/picturing-7-billion.
44. http://www.nytimes.com/interactive/2010/05/03/blogs/a-moment-in-time .html; quotation from archive.onedayonearth.org.
45. "Swedish Photo Project Aims to capture May 15," CBC, May 15, 2012, http://www.cbc.ca/news/world/swedish-photo-project-aims-to-capture -may-15-1.1186797; http://www.earthpyramid.org/faq.
46. Yasuko Kamiizumi, "In Deep Freeze, a Little Air and DNA for the Future," *NYT*, February 21, 1998, B9; Zin-ichi Karube, Atsushi Tanaka, Akinori Takeuchi, and Yasuyuki Shibata, "Three Decades of Environmental Specimen Banking at the National Institute for Environmental Studies, Japan," *Environmental Science and Pollution Research* 22 (2015): 1587–96.
47. Edward O. Wilson, *The Diversity of Life* (New York: W. W. Norton, 1999), 331.
48. Alan DeMarche, *Generative Ruin: A Monumentality of our Time* (Master of Architecture thesis, University of California Berkeley, 2015).
49. John Guillebaud, "History of the Environment Time Capsule Project," http://www.ecotimecapsule.com/capsule.shtml#HISTORY.
50. Stewart Brand, *The Clock of the Long Now: Time and Responsibility* (New York: Basic Books, 2000), 2–3, 51–52, 97–99.
51. See above, p. 321, n. 46.
52. Hans Jonas, "Ontological Grounding of a Political Ethics: On the Metaphysics of Commitment to the Future of Man," in *The Public Realm: Essays on Discursive Types in Political Philosophy*, ed. Reiner Schurmann (Buffalo, NY: SUNY Press, 1989), 155 (emphasis in original).
53. Suzaan Boettger, *Earthworks: Art and the Landscape of the Sixties* (Berkeley: University of California Press, 2004), 89–90, 144, 277n44.
54. Joseph Beuys, *7000 Oaks: Long-Term Installation* (New York: DIA Center for the Arts, 1982); John K. Grande, *Art Nature Dialogues: Interviews with Environmental Artists* (Binghamton, NY: SUNY Press, 2012), 1, 3.
55. Trevor Paglen, *The Last Pictures* (Berkeley: University of California Press, 2012), 4, 11, 12.
56. Katie Paterson and David Mitchell, *Future Library*, pamphlet, 2015.

Index

Page numbers in italics refer to illustrations.